£30.00
D0301701
FROM THE LIBRARY

Numerical Algorithms

MANCHESTER/LIVERPOOL SUMMER SCHOOLS IN NUMERICAL ANALYSIS

L. M. Delves and J. Walsh: *Numerical solution of integral equations*

G. Hall and J. M. Watt: *Modern numerical methods for ordinary differential equations*

I. Gladwell and R. Wait: *A survey of numerical methods for partial differential equations*

Christopher T. H. Baker and Chris Phillips: *The numerical solution of nonlinear problems*

J. L. Mohamed and J. E. Walsh: *Numerical algorithms*

Numerical Algorithms

Edited by

J. L. MOHAMED
Department of Statistics and Computational Mathematics,
University of Liverpool

and

J. E. WALSH
Department of Mathematics,
University of Manchester

CLARENDON PRESS · OXFORD
1986

Oxford University Press, Walton Street, Oxford OX2 6DP
Oxford New York Toronto
Delhi Bombay Calcutta Madras Karachi
Petaling Jaya Singapore Hong Kong Tokyo
Nairobi Dar es Salaam Cape Town
Melbourne Auckland
and associated companies in
Beirut Berlin Ibadan Nicosia

Oxford is a trade mark of Oxford University Press

Published in the United States
by Oxford University Press, New York

British Library Cataloguing in Publication Data
Numerical algorithms.
1. Numerical analysis
I. Mohamed, J.L. II. Walsh, J.E.
519.4 QA297
ISBN 0–19–853364–0

Printed in Great Britain by St Edmundsbury Press,
Bury St Edmunds, Suffolk

PREFACE

This book is based on the material presented at a joint Summer School in July 1984, organised by the Department of Mathematics, University of Manchester and the Department of Statistics and Computational Mathematics, University of Liverpool. The aim of the book is to provide, for a wide range of applied computational problems, descriptions of those algorithms which give cheap, reliable and stable solution procedures. Each chapter highlights currently available numerical software for the methods described, and detailed sources of the packages and program libraries referred to may be found in the Appendix.

The range of problems considered includes the solution of systems of dense and sparse linear algebraic equations in Chapter 1, and the solution of the algebraic eigenvalue problem which is covered in considerable depth in Chapter 2. Chapters 3 and 4 discuss the solution of initial-value and boundary-value problems in ordinary differential equations, while the use of extrapolation in numerical quadrature is surveyed in Chapter 5, together with automatic quadrature methods for the numerical approximation of integrals. The solution of integral equations of Fredholm and Volterra type is reviewed in Chapter 6, and Chapters 7, 8 and 9 describe methods for the solution of partial differential equations; Chapter 7 is devoted to an exposition of the highly popular and successful finite-element method and Chapter 8 gives other methods for elliptic equations. Parabolic problems are covered in Chapter 9. Recent advances in the design of algorithms for curve- and surface-fitting problems using splines are presented in Chapters 10 and 16. The theory and application of methods in the field of optimisation are treated in Chapters 11-14. Chapter 11 deals with L_1 and L_∞ curve-fitting and the use of linear programming. The minimisation of a nonlinear objective function is considered for unconstrained, constrained and least-squares problems in Chapters 12, 13 and 14. Finally, a self-contained description of problems in time-series analysis is provided in Chapter 15.

This book should serve as a useful reference text for applied mathematicians, scientists and engineers, in the practical solution of commonly occurring problems. It aims to provide both the specialist and non-specialist reader with a readable account of those theoretical and practical aspects which are important in the successful design and implementation of numerical algorithms. The book could also serve as a textbook for advanced undergraduates and postgraduate students attending courses in Computational Mathematics.

We are extremely grateful to all those who have made the production of this book possible; a full list of contributors follows this preface. Special thanks are due to the secretariat of the Department of Statistics and Computational Mathematics at Liverpool University, and to Stephen Smith and Michael McCrann, without whose patience and endurance the mounting of the typescript onto the departmental wordprocessor would not have been possible.

J.L. Mohamed
J. Walsh
August 1986

CHAPTER CONTENTS

CONTRIBUTORS

Dr. C.T.H. Baker (Chapters 5, 6)
Department of Mathematics, University of Manchester,
Oxford Road, Manchester M13 9PL, England.

Professor I. Barrodale (Chapter 11)
Department of Computer Science, University of Victoria,
P.O. Box 1700, Victoria, British Columbia, Canada V8W 2Y2.

Dr. J.A. Belward (Chapter 14)
Department of Mathematics, University of Queensland,
St. Lucia, Queensland 4067, Australia.

Professor L.M. Delves (Chapter 6)
Department of Statistics and Computational Mathematics,
University of Liverpool, Victoria Building,
Brownlow Hill, Liverpool L69 3BX, England.

Mr. J.J. Du Croz (Chapter 2)
NAG Central Office, Mayfield House, 256 Banbury Road,
Oxford OX2 7DE, England.

Dr. T.L. Freeman (Chapter 13)
Department of Mathematics, University of Manchester,
Oxford Road, Manchester M13 9PL, England.

Dr. G. Hall (Chapter 3)
Department of Mathematics, University of Manchester,
Oxford Road, Manchester M13 9PL, England.

Mr. S.J. Hammarling (Chapter 2)
NAG Central Office, Mayfield House, 256 Banbury Road,
Oxford OX2 7DE, England.

Dr. P.J. Harley (Chapter 12)
Department of Applied Mathematics, University of Sheffield,
Sheffield S10 2TN, England.

Mr. J.G. Hayes (Chapters 10, 16)
Division of Numerical Analysis, National Physical Laboratory,
Teddington, Middlesex TW11 0LW, England.

Dr. D. Kershaw (Chapter 8)
Department of Mathematics, University of Lancaster,
Bailrigg, Lancaster LA1 4YL, England.

Dr. J.N. Lyness (Chapter 5)
Argonne National Laboratory, Applied Mathematics Division,
Argonne, Illinois IL 60439, U.S.A.

Dr. J.L. Mohamed (Chapters 1, 4)
Department of Statistics and Computational Mathematics,
University of Liverpool, Victoria Building,
Brownlow Hill, Liverpool L69 3BX, England.

Dr. R.W. Thatcher (Chapter 7)
Department of Mathematics, UMIST, Sackville Street,
Manchester M60 1QD, England.

Dr. G. Tunnicliffe Wilson (Chapter 15)
Department of Mathematics, University of Lancaster,
Bailrigg, Lancaster LA1 4YL, England.

Dr. R. Wait (Chapter 1)
Department of Statistics and Computational Mathematics,
University of Liverpool, Victoria Building,
Brownlow Hill, Liverpool L69 3BX, England.

Professor J. Walsh (Chapters 8, 9)
Department of Mathematics, University of Manchester,
Oxford Road, Manchester M13 9PL, England.

Dr. J. Williams (Chapter 3)
Department of Mathematics, University of Manchester,
Oxford Road, Manchester M13 9PL, England.

Dr. C. Zala (Chapter 11)
Department of Computer Science, University of Victoria,
P.O. Box 1700, Victoria, British Columbia, Canada V8W 2Y2.

CHAPTER 1

Solution of Linear Equations

J.L. Mohamed and R. Wait

1. Introduction

We shall be concerned with the solution of the following systems of linear equations:

1. <u>The square system</u> $A\underline{x} = \underline{b}$ where A is a real matrix of order n. If A is nonsingular (i.e. it has rank n corresponding to n linearly independent columns) then we seek the unique solution $\underline{x} = A^{-1}\underline{b}$. If A is singular the system has no solution or an infinite number of solutions.

2. <u>The overdetermined system</u> $A\underline{x} = \underline{b}$ where A is a real $m \times n$ matrix with $m > n$, and $\underline{x}$, $\underline{b}$ are vectors of length n, m respectively. Here there are more equations than unknowns and the system will usually be inconsistent. There will be no exact solution and we seek a vector $\underline{x}$, a <u>least-squares solution</u> of $A\underline{x} = \underline{b}$, which minimises the L_2 norm of the residual vector

$$||\underline{r}||_2 = ||\underline{b} - A\underline{x}||_2 = \left[\sum_{i=1}^{m} r_i^2\right]^{1/2} .$$

(i) If rank(A) = n, there exists a unique least-squares solution of $A\underline{x} = \underline{b}$ given by $\underline{x} = (A^TA)^{-1}A^T\underline{b}$, i.e. $\underline{x}$ satisfies the <u>normal equations</u>

$$A^TA\underline{x} = A^T\underline{b} . \tag{1.1}$$

(ii) If rank(A) = $k < n$, there exist an infinite number of least-squares solutions. A unique solution, the <u>minimal least-squares solution</u> of $A\underline{x} = \underline{b}$, is defined if we seek the least-squares solution $\underline{x}$ which has the shortest length with respect to the L_2 norm. We may represent $\underline{x}$ as $A^+\underline{b}$ where A^+ is a unique real $n \times m$ matrix called the <u>pseudo-inverse</u> or the <u>Moore-Penrose generalised inverse</u> of A, which satisfies the following conditions:

$$AA^+A = A,\ A^+AA^+ = A^+,\ (AA^+)^T = AA^+,\ (A^+A)^T = A^+A. \qquad (1.2)$$

3. <u>The underdetermined system</u> $A\underline{x} = \underline{b}$ where A is a real $m \times n$ matrix with $m < n$, and $\underline{x}$, $\underline{b}$ are vectors of length n, m respectively. Here there are more unknowns than equations, and there is either no solution or an infinite number of solutions. Again we seek a least-squares solution of $A\underline{x} = \underline{b}$.

(i) If rank(A) = m, then a solution of $A\underline{x} = \underline{b}$ is given by $\underline{x} = A^T(AA^T)^{-1}\underline{b}$, corresponding to the minimal least-squares solution of $A\underline{x} = \underline{b}$.

(ii) If rank(A) = k < m, there exist an infinite number of least-squares solutions, and again we seek the minimal least-squares solution $\underline{x} = A^+\underline{b}$.

In discussing the solution of all three problems, we shall make use of an important decomposition of an $m \times n$ matrix A, namely the <u>singular value decomposition</u>, given by

$$A = U \begin{bmatrix} \Sigma \\ 0 \end{bmatrix} V^T \quad \text{or} \quad A = U\,[\,\Sigma \quad 0\,]\,V^T \qquad (1.3)$$

for $m \geq n$ or $m \leq n$ respectively, where $U = [\underline{u}_1\ \underline{u}_2\ \ldots\ \underline{u}_m]$ and $V = [\underline{v}_1\ \underline{v}_2\ \ldots\ \underline{v}_n]$ are orthogonal matrices of order m and n, and $\Sigma = \text{diag}\,[\sigma_1 \ldots \sigma_p]$ where $p = \min(m, n)$ with $\sigma_1 \geq \sigma_2 \geq \ldots \geq \sigma_p \geq 0$. The σ_i are the <u>singular values</u> of A and $\underline{u}_i$, $\underline{v}_i$ are referred to as the <u>left singular vectors</u> and <u>right singular vectors</u> of A. From (1.3), it follows that

$$A\underline{v}_i = \sigma_i\underline{u}_i \ , \quad A^T\underline{u}_i = \sigma_i\underline{v}_i \ , \quad i = 1,\ldots,p \ , \qquad (1.4)$$

and we have the result that rank(A) = k < p if and only if $\sigma_{k+1} = \sigma_{k+2} = \ldots = \sigma_p = 0$; thus the singular value decomposition of A can be used to determine rank(A). A measure of the sensitivity of the $m \times n$ linear system $A\underline{x} = \underline{b}$ to perturbations in A and/or $\underline{b}$ is given by the so-called condition number

$$K(A) = ||A||\ ||A^+||$$

with

$$K_2(A) = \frac{\sigma_1}{\sigma_k} \tag{1.5}$$

where $k = \text{rank}(A)$.

2. Dense equations

In each of the problems 1, 2 and 3 we shall assume initially that A has no special structure, and that it is <u>dense</u>, i.e. A contains few zero elements (typically, less than 10% of the elements of A might be zero). The solution of non-dense or sparse systems of equations will be discussed in section 3. We shall restrict the discussion to real coefficient matrices; the extension to include complex problems is straightforward.

Algorithms for the solution of problems 1, 2 and 3 should be carefully designed to be numerically stable, efficient with respect to operation counts and storage costs, and implemented in a portable form. Good algorithms for problem 1 will recognise singularity or near-singularity of the coefficient matrix, and will indicate the degree of accuracy achieved in the computed solution $\tilde{\underline{x}}$.

2.1 Gauss elimination

The method of <u>Gauss elimination with partial pivoting</u> has evolved as the standard method for the solution of problems of type 1. This direct method yields, in the absence of rounding and other errors, the exact solution in a finite number ($\frac{1}{3}n^3 + O(n^2)$) of arithmetic operations. The method is well understood (see Golub and Van Loan (1983), Ch. 4), using a finite sequence of elementary row operations to reduce, in a stable fashion, the original system to one whose coefficient matrix is upper triangular. The reduced system is then easily solved by backward substitution to yield, if A is nonsingular, the unique solution of $A\underline{x} = \underline{b}$. In practice however, because of the finite word-length of the computer, the method yields an approximation $\tilde{\underline{x}}$ to $\underline{x}$; the accuracy of this approximation is discussed in section 2.2.

The method of Gauss elimination with partial pivoting is mathematically equivalent to an LU factorisation of the matrix PA, where P is a permutation matrix which contains only one non-zero element, namely unity, in each row and column; L is a unit lower triangular matrix of Gauss multipliers and U is the reduced upper triangular matrix.

Before the method is applied to the solution of $A\underline{x} = \underline{b}$, it is usual to apply some form of scaling to the original problem, particularly if the elements of A are obtained from experiment or there is some reason to suspect that they are in error (e.g. if the elements have been rounded to t digits). If for example the rows and columns of A are scaled by factors $\alpha_1, \ldots, \alpha_n$ and $\gamma_1, \ldots, \gamma_n$ the resulting scaled matrix is given by $B = D_1AD_2$ where $D_1 = \mathrm{diag}\,[\alpha_i]$ and $D_2 = \mathrm{diag}\,[\gamma_i]$. Then the system $B\underline{y} = D_1\underline{b}$ is solved for $\underline{y} = D_2^{-1}\underline{x}$ and finally $\underline{x}$ is computed as $D_2\underline{y}$. Although the problems $A\underline{x} = \underline{b}$ and $B\underline{y} = D_1\underline{b}$ are mathematically equivalent the numerical performance of the Gauss algorithm on these problems can be very different. A particular form of scaling, known as equilibration (Wilkinson 1965, Ch. 4), chooses the α_i, γ_i so that the elements of B have approximately equal magnitude; the set of scaling factors, however, is not unique. Note that if the α_i, γ_i are chosen to be other than powers of β, the base of the floating-point numbers, then further unnecessary rounding errors are introduced.

2.2 How good is the computed solution $\tilde{\underline{x}}$?

Suppose $\tilde{\underline{x}}$ is computed using Gauss elimination with partial pivoting and t-digit, base β floating-point arithmetic. Then $\tilde{\underline{x}}$ can be shown to be the exact solution of a perturbed system

$$(A + \delta A)\,\tilde{\underline{x}} = \underline{b} \tag{1.6}$$

with

$$\frac{||\underline{x} - \tilde{\underline{x}}||}{||\underline{x}||} \leqslant \frac{||A^{-1}\delta A||}{1 - ||A^{-1}\delta A||}\,, \tag{1.7}$$

assuming $||A^{-1}\delta A|| < 1$. It follows that when $||A^{-1}||\ ||\delta A|| < 1$

$$\frac{||\underline{x} - \tilde{\underline{x}}||}{||\underline{x}||} \leqslant \frac{K(A)\,\dfrac{||\delta A||}{||A||}}{1 - K(A)\,\dfrac{||\delta A||}{||A||}} \tag{1.8}$$

where $K(A) = ||A|| \; ||A^{-1}||$ is the condition number of A and $(||\delta A||/||A||)$ is the relative perturbation of A. Thus a relatively small change in the elements of A will give rise to a corresponding large relative change in the solution if K(A) is large. If the elements of A are obtained by measurement then they will be subject to error; similarly, the elements of A may be numbers which do not have finite base β representations. It is thus important to know how small errors in A will affect the computed solution; if K(A) is large, corresponding to an ill-conditioned system, then the computed solution may bear no resemblance to the true solution.

Given an approximate solution $\tilde{\underline{x}}$, how can we determine its accuracy ? If we define $\underline{e} = \underline{x} - \tilde{\underline{x}}$ then we require an estimate of $\underline{e}$. The relation (1.8) implies that an estimate $\tilde{K}(A)$ of K(A) will be needed. If this is large, the system may be 'too ill-conditioned' for accurate solution to working precision. We can certainly form the residual vector $\underline{r} = \underline{b} - A\tilde{\underline{x}}$ from available information, with a view to discovering how well $\tilde{\underline{x}}$ satisfies the original system. Clearly, if the result is to be meaningful, $\underline{r}$ should be formed using higher precision; in practice $\underline{r}$ is formed using double precision arithmetic, i.e. 2t-digit arithmetic on a t-digit machine, with accumulation of inner products. Now $A\underline{e} = \underline{r}$ and it follows that

$$\frac{||\underline{x} - \tilde{\underline{x}}||}{||\underline{x}||} < \frac{||A^{-1}|| \;\; ||\underline{r}||}{||\underline{b}|| \; / \; ||A||} = K(A) \frac{||\underline{r}||}{||\underline{b}||} . \tag{1.9}$$

The relation (1.9) illustrates that, for an ill-conditioned system, the size of the residual vector does not give a good indication of the size of the error in the computed solution. We can use the result (Forsythe and Moler 1967)

$$||\underline{r}|| \simeq \beta^{-t} \; ||A|| \; ||\tilde{\underline{x}}|| \tag{1.10}$$

to show that

$$K(A) \simeq \frac{||\underline{e}|| \; \beta^{t}}{||\tilde{\underline{x}}||} . \tag{1.11}$$

Now given $\underline{r}$, it is possible to estimate $\underline{e} = \underline{x} - \underline{x}^{(0)}$, where $\underline{x}^{(0)} = \tilde{\underline{x}}$, by solving the linear system $A\underline{e} = \underline{r}$ using Gauss elimination with partial pivoting and t-digit, base β arithmetic. The L, U factors of a permuted form of A have already been calculated and we are thus able to calculate, in $O(n^2)$ arithmetic operations, an approximation to $\underline{e}$, given by $\underline{e}^{(0)} = \underline{x}^{(1)} - \underline{x}^{(0)}$ say (where $\underline{e}^{(0)}$ satisfies a perturbed system $(A+\delta A)\,\underline{e}^{(0)} = \underline{r}$). We can also estimate K(A) as $(||\underline{e}^{(0)}||/||\underline{x}^{(0)}||)\,\beta^{t} \simeq \beta^{p}$ and thus decide whether the working precision is sufficient to provide a sensible solution. If $p > t$, the system is 'too ill-conditioned' to be solved using this precision; if however $p < t$, we would expect $\tilde{\underline{x}}$ to agree with the first (t-p) digits of $\underline{x}$. The process can be repeated to provide a sequence of estimates $\{\underline{x}^{(i)}\}_{i=0,1...}$ to $\underline{x}$ given by $A\underline{e}^{(i)} = \underline{b} - A\underline{x}^{(i)}$, $\underline{x}^{(i+1)} = \underline{x}^{(i)} + \underline{e}^{(i)}$. This method, which can be used in conjunction with the method of Gauss elimination, and which simultaneously detects (by estimating the condition number of A) and improves the accuracy of an approximate solution of $A\underline{x} = \underline{b}$, is known as <u>iterative refinement</u>. After k iterations the computed solution $\underline{x}^{(k)}$ can be expected to agree with the first q digits of $\underline{x}$ where $q = \min\{t, k(t-p)\}$. The terminating condition in the algorithm will be satisfied either when $(||\underline{e}^{(i)}||_{\infty} / ||\underline{x}^{(i)}||_{\infty}) \leq \beta^{-t}$ or when a specified maximum number of iterations has been reached. The method may be implemented on a given machine provided that it has the facility for performing double precision arithmetic. There are clearly storage and portability problems associated with the method, because of the need to keep a copy of the original coefficient matrix A and right-hand side vector $\underline{b}$ so that the residual may be formed, and because of the need to calculate $\underline{r}$ using double precision arithmetic.

2.3 The LINPACK condition estimator

The LINPACK (Dongarra, Bunch, Moler and Stewart 1979) routines for the solution of linear systems and related problems, have been designed to be completely machine-independent and they therefore require an efficient alternative means of estimating the condition number of A. The condition estimator adopted in the LINPACK routines is based on the observation that if $\underline{z}$ satisfies the linear equations $A\underline{z} = \underline{y}$ then

$$||A^{-1}|| \geq \frac{||\underline{z}||}{||\underline{y}||}. \qquad (1.12)$$

Thus if $||A^{-1}||$ is estimated as $(||\underline{z}||/||\underline{y}||)$ the condition number of A may be estimated as $\tilde{K}(A) = (||A|| \; ||\underline{z}||/||\underline{y}||)$. Following Cline, Conn and Van Loan (1982), $\tilde{K}_p(A)$ is regarded as a reliable estimator of $K_p(A)$ if

$$c_1 K_p(A) \leqslant \tilde{K}_p(A) \leqslant c_2 K_p(A)$$

for 'reasonable' constants c_1 and c_2 independent of A. Thus for p = 1 (the L_1 norm can be quickly computed), we have

$$\frac{||\underline{z}||_1}{||A^{-1}||_1 \; ||\underline{y}||_1} K_1(A) = \tilde{K}_1(A) \leqslant K_1(A)$$

and satisfaction of the reliability criterion requires $(||\underline{z}||_1 \, / \, ||\underline{y}||_1)$ to be as close to $||A^{-1}||_1$ as possible. This requirement may be fulfilled (see the L_2 argument given below) by step (i) of the algorithm:

(i) Choose $\underline{d}$ such that the solution to $A^T\underline{y} = \underline{d}$ is large in the L_1 norm, relative to $\underline{d}$.

(ii) Solve $A\underline{z} = \underline{y}$.

(iii) Set $\tilde{K}_1(A) = ||A||_1 \, (||\underline{z}||_1 \, / \, ||\underline{y}||_1)$.

Consider for a moment the singular value decomposition of A given by

$$A = U \Sigma V^T$$

where $U = [\underline{u}_1 \; \ldots \; \underline{u}_n]$, $V = [\underline{v}_1 \; \ldots \; \underline{v}_n]$ are orthogonal matrices, and $\Sigma = \text{diag}\,[\sigma_i]$ is a n × n diagonal matrix with $\sigma_1 \geqslant \sigma_2 \geqslant \ldots \geqslant \sigma_n \geqslant 0$. Now if $\underline{d}$ has the orthogonal expansion

$$\underline{d} = \sum_{i=1}^{n} \alpha_i \underline{v}_i \tag{1.13}$$

where $\underline{v}_i = \frac{1}{\sigma_i} A^T\underline{u}_i$, $\underline{u}_i = \frac{1}{\sigma_i} A\underline{v}_i$ (see (1.4)) , it follows that

$$\underline{d} = A^T\underline{y} \quad , \quad \underline{y} = \sum_{i=1}^{n} (\alpha_i / \sigma_i)\, \underline{u}_i \tag{1.14}$$

$$\underline{y} = A\underline{z} \quad , \quad \underline{z} = \sum_{i=1}^{n} (\alpha_i / \sigma_i^2)\, \underline{v}_i \, . \tag{1.15}$$

Then it can be shown that

$$\frac{\|\underline{z}\|_2}{\|\underline{y}\|_2} \geq \|A^{-1}\|_2 \frac{|\alpha_n|}{\|\underline{d}\|_2} \, . \tag{1.16}$$

Thus if $(|\alpha_n|/\|\underline{d}\|_2)$ is close to unity (ie. if $\underline{d}$ has a significant component in the direction of $\underline{v}_n$) step (i) of the method tends to produce a vector $\underline{d}$ with the desired property. The method assumes that A has been factorised in the form A = LU and the system $A^T\underline{y} = \underline{d}$ is solved in the form of two triangular systems:

(a) $U^T\underline{w} = \underline{d}$

where each d_i is chosen as ± 1 so as to maximise each w_i (see Cline et al. (1982) for further details);

(b) $L^T\underline{y} = \underline{w}$.

Step (ii) solves the system $A\underline{z} = \underline{y}$ as two triangular systems:

(a) $L\underline{v} = \underline{y}$

(b) $U\underline{z} = \underline{v}$.

The LINPACK routine SGECO uses the method of Gauss elimination with partial pivoting to solve problem 1 but the method of iterative refinement is not used in this machine-independent routine. An estimate of the <u>reciprocal</u> condition number of A is computed (to avoid overflow) and if $1/\tilde{K}(A) \simeq \beta^{-p}$ (where β is the arithmetic base), the elements of the computed solution $\tilde{\underline{x}}$ can usually be expected to have p fewer significant figures of accuracy than the elements of A.

The condition estimators described here and in section 2.2 both require

$O(n^2)$ arithmetic operations.

2.4 Direct factorisation techniques

Gauss elimination without pivoting is equivalent to an LU factorisation of A, and this factorisation, if it exists, is unique if we take $\ell_{ii} = 1$. The elements of L and U can easily be determined using the matrix multiplication rule. This gives a direct factorisation technique called Doolittle's method.

If instead we require $u_{ii} = 1$ and we do not restrict the ℓ_{ii} to be unity, we obtain Crout's method. Again the factorisation, if it exists, is unique and the elements ℓ_{ij} and u_{ij} can be determined using the matrix multiplication rule; we have

$$\left.\begin{aligned} \ell_{ij} &= a_{ij} - \sum_{k=1}^{i-1} \ell_{ik}u_{kj} \ , \quad i = 1,\ldots,n \ , \ j \leq i \ , \\ u_{ij} &= \frac{1}{\ell_{ii}}\left(a_{ij} - \sum_{k=1}^{i-1} \ell_{ik}u_{kj}\right) \ , \quad i = 1 ,\ldots,n-1 \ , \ j > i. \end{aligned}\right\} \quad (1.17)$$

Note that the form of (1.17) allows the elements ℓ_{ij} , u_{ij} to be calculated to a high degree of accuracy if inner products are accumulated using double precision arithmetic. The operation count of each of the Crout, Doolittle and Gauss elimination algorithms is identical, namely $(\frac{1}{3} n^3 + O(n^2))$ to produce the LU factorisation of A.

Routines in the NAG Library implement Crout's method with partial pivoting, both with and without iterative refinement, to solve problem 1; the elements of L, U are determined in the order: first column of L, first row of U, second column of L, second row of U etc., with $u_{ii} = 1$, $i = 1,\ldots,n$. The elements a_{ij} of A are overwritten by ℓ_{ij} , $j = 1,\ldots n$, $i \geq j$ and by u_{ij} , $i = 1,\ldots,n-1$, $j > i$, and the pivoting strategy is such that $|\ell_{ii}|$ is maximised using row interchanges, in order to guard against zero or small values of ℓ_{ii} . The solution of $A\underline{x} = \underline{b}$ is then obtained by solving the two triangular systems:

$$\left.\begin{aligned} L\underline{y} &= P\underline{b} \\ U\underline{x} &= \underline{y} \end{aligned}\right\} \quad (1.18)$$

where P = I, the identity matrix, if no interchanges are required.

If the matrix A is symmetric positive definite, i.e.

$$A = A^T \quad \text{and} \quad \underline{x}^T A\underline{x} > 0 \text{ for all } \underline{x} \neq \underline{0} ,$$

then A has a unique Cholesky factorisation given by

$$A = LL^T \tag{1.19}$$

where L is a real nonsingular lower triangular matrix with positive diagonal elements. Then the special form of A may be used to computational advantage; the elements of ℓ_{ij} are given by

$$\left.\begin{aligned} \ell_{ii} &= (a_{ii} - \sum_{k=1}^{i-1} \ell_{ik}^2)^{1/2} , \quad i = 1,\ldots,n , \\ \ell_{ij} &= \frac{1}{\ell_{jj}} (a_{ij} - \sum_{k=1}^{j-1} \ell_{ik}\ell_{jk}) , \quad j = 1,\ldots n-1 , \quad i > j \end{aligned}\right\} \tag{1.20}$$

and the cost of factorising A is $(\frac{n^3}{6} + O(n^2))$ operations together with n square root evaluations. The algorithm is numerically stable and the solution of $A\underline{x} = \underline{b}$ is found in the form: $L\underline{y} = \underline{b}$, $L^T\underline{x} = \underline{y}$. If required, the Cholesky algorithm can be easily modified in order to dispense with the extraction of square roots, by factorising A as

$$A = \tilde{L}D\tilde{L}^T \tag{1.21}$$

where $\tilde{L}$ is a real nonsingular unit lower triangular matrix and D is a diagonal matrix with positive diagonal elements.

Implementations of the Cholesky algorithm, for example, in the NAG Library, use double precision arithmetic for the accumulation of inner products. The NAG Library does not cater for the situation where A is symmetric but not positive definite; however, the LINPACK routine SSIFA implements a stable diagonal pivoting factorisation algorithm which expresses A in the form

$$A = LDL^T \tag{1.22}$$

where D is a symmetric block-diagonal matrix with blocks of order one or two, and L is unit lower triangular with $L_{i+1,i} = 0$ if $D_{i+1,i} \neq 0$. For further details, the reader is referred to Bunch (1971), Bunch and Parlett (1971), Barwell and George (1976).

2.5 Solution of rectangular systems

The least-squares problem is usually solved by the method of orthogonal factorisation; the method operates on the original coefficient matrix A and the accuracy of the solution then depends on the condition number of the problem. This technique may also be applied to the solution of square systems; however the associated cost renders it an unpopular choice when compared with Gauss elimination.

Any m × n matrix can be decomposed in the form

$$A = QR \tag{1.23}$$

where Q is an m × m orthogonal matrix and R is an m × n matrix of the form

$$R = \begin{bmatrix} U_R \\ 0 \end{bmatrix} \quad \text{or} \quad R = [U_R \quad V_R] \tag{1.24}$$

according as m > n or m < n, with U_R an n × n or m × m upper triangular matrix. Now the least-squares problem of determining $\underline{x}$ such that

$$||\underline{r}||_2 = ||\underline{b} - A\underline{x}||_2 \tag{1.25}$$

is minimised, is equivalent to minimising $||\tilde{\underline{r}}||_2$ where

$$\tilde{\underline{r}} = Q^T\underline{r} \ . \tag{1.26}$$

We shall consider in detail the case where m > n. If we partition $Q^T\underline{b}$ as $[\underline{c} \quad \underline{d}]^T$ where $\underline{c}$, $\underline{d}$ are vectors of length n, (m−n) respectively, then $\tilde{\underline{r}} = [\underline{c}-U_R\underline{x} \quad \underline{d}]^T$. If A has linearly independent columns then U_R is nonsingular and we can choose $\underline{x}$ so that

$$U_R\underline{x} = \underline{c} \ . \tag{1.27}$$

The vector $\underline{d}$ is independent of $\underline{x}$ and the solution of (1.27) provides the value of $\underline{x}$ which minimises $||\tilde{\underline{r}}||_2$ and hence $||\underline{r}||_2$. Then $||\underline{r}||_2 = ||\underline{d}||_2$. If however A has linearly dependent columns, U_R will be singular. We note that the singular value decompositions of A and U_R are related as follows. Since $A = Q \begin{bmatrix} U_R \\ 0 \end{bmatrix} = U \begin{bmatrix} \Sigma \\ 0 \end{bmatrix} V^T$, we have $\begin{bmatrix} U_R \\ 0 \end{bmatrix} = Q^T U \begin{bmatrix} \Sigma \\ 0 \end{bmatrix} V^T$.

Hence A and U_R have the same singular values and right singular vectors, but different left singular vectors. Now suppose A has rank $k < n$. Then only the first k singular values of A will be non-zero, that is,

$$A = U \begin{bmatrix} \Sigma \\ 0 \end{bmatrix} V^T \tag{1.28}$$

where

$$\Sigma = \begin{bmatrix} S & 0 \\ 0 & 0 \end{bmatrix}$$

and $S = \text{diag}\,[\sigma_i]$ is a $k \times k$ nonsingular matrix with

$$\sigma_1 \geqslant \sigma_2 \geqslant \dots \geqslant \sigma_k > 0.$$

If

$$U_R = \tilde{U} \tilde{\Sigma} \tilde{V}^T \tag{1.29}$$

is the singular value decomposition of U_R it follows that

$$A = Q \begin{bmatrix} \tilde{U} \tilde{\Sigma} \tilde{V}^T \\ 0 \end{bmatrix} = U \begin{bmatrix} \Sigma \\ 0 \end{bmatrix} V^T \tag{1.30}$$

where $U = Q \begin{bmatrix} \tilde{U} & 0 \\ 0 & I \end{bmatrix}$, $\Sigma = \tilde{\Sigma}$, $V = \tilde{V}$. This result may be used (see Hammarling (1983)) and the SVD of U_R computed (taking advantage of the upper triangular form of U_R) if U_R proves to be singular; we show below how this decomposition facilitates the determination of $\underline{x}$.

Now $\tilde{\underline{r}} = \begin{bmatrix} \underline{c} - \tilde{U}\Sigma V^T\underline{x} \\ \underline{d} \end{bmatrix}$ and since $||\tilde{\underline{r}}||_2 = ||\tilde{U}^T\tilde{\underline{r}}||_2$ it follows that a least-squares solution of $A\underline{x} = \underline{b}$ is given by minimising $||\hat{\underline{r}}||_2$ where

$$\hat{\underline{r}} = \tilde{U}^T\tilde{\underline{r}} \ . \tag{1.31}$$

If we partition $\tilde{U}^T\underline{c}$ and $V^T\underline{x}$ as $[\underline{e} \quad \underline{f}]^T$ and $[\underline{g} \quad \underline{h}]^T$ respectively, where $\underline{e}$, $\underline{g}$ are vectors of length k, and $\underline{f}$, $\underline{h}$ are vectors of length (n-k) , then $\hat{\underline{r}} = \begin{bmatrix} \hat{\underline{r}}_1 \\ \hat{\underline{r}}_2 \end{bmatrix}$ with $\hat{\underline{r}}_1 = \begin{bmatrix} \underline{e} - S\underline{g} \\ \underline{f} \end{bmatrix}$. $||\hat{\underline{r}}||_2$ is minimised by choosing $\underline{g}$ such that

$$S\underline{g} = \underline{e} \ . \tag{1.32}$$

Recall that $\underline{e}$ consists of the first k elements of $\tilde{U}^T\underline{c}$ and $\underline{c}$ consists of the first n elements of $Q^T\underline{b}$. Finally it follows that a least-squares solution of $A\underline{x} = \underline{b}$ is given by

$$\underline{x} = V \begin{bmatrix} \underline{g} \\ \underline{h} \end{bmatrix} \tag{1.33}$$

with $||\underline{r}||_2^2 = ||\underline{d}||_2^2 + ||\underline{f}||_2^2$. However, $\underline{x}$ is not unique; the particular solution for which $||\underline{x}||_2$ is also a minimum is called the minimal least-squares solution and this clearly occurs when $\underline{h} = \underline{0}$.

Note that we may formally write the minimal least-squares solution as $\underline{x} = A^+\underline{b}$ where

$$A^+ = V \, [\, \Sigma^+ \quad 0 \,] \, U^T$$

$$\Sigma^+ = \begin{bmatrix} S^{-1} & 0 \\ 0 & 0 \end{bmatrix}, \quad S = \begin{bmatrix} \sigma_1 & & \\ & \ddots & \\ & & \sigma_k \end{bmatrix}, \quad k < n$$

$$U^T = \begin{bmatrix} \tilde{U}^T & 0 \\ 0 & I \end{bmatrix} Q^T .$$

The pseudo-inverse A^+, and hence $\underline{x}$, are not continuous functions of the elements of A; when σ_i is small but not zero, S^{-1} will contain the element $\frac{1}{\sigma_i}$, but when $\sigma_i = 0$, this term will be replaced by zero. It is therefore necessary to decide when a computed σ_i should be regarded as zero; the effect of neglecting or not neglecting small singular values is examined in section 4.1 of Chapter 2.

Algorithms for the solution of the least-squares problem 2 should provide stable techniques for calculation of the various matrix factors required above. They should also, if required, be able to determine the rank of a matrix.

Routines exist in both NAG and LINPACK for the solution of least-squares problems. In particular the NAG Fortran routine F04JGF allows the least-squares solution to be computed from the QR factorisation of A, if U_R is sufficiently well conditioned, by means of orthogonal Householder transformations. A test is incorporated on the computed condition number of U_R in order to determine whether U_R is singular. In particular, if $K_E(U_R) \times \text{TOL} > 1.0$, where TOL is a user-specified tolerance parameter which reflects the largest relative error in the elements of A, and the suffix E denotes the Euclidean norm, then U_R is regarded as singular; the least-squares solution is then obtained via the SVD of U_R which is computed using the stable method of Golub and Reinsch (1970). This method reduces U_R to upper bidiagonal form via Householder transformations and then reduces this form to diagonal form using a variant of the symmetric QR algorithm for eigenvalues (see Chapter 2). The rank of A is determined as the largest integer k for which

$$\sigma_k > \text{TOL} \times \sigma_1$$

holds, where $\{\sigma_i\}$ are the computed singular values of U_R (and A). The cost of performing the Householder factorisation of A and the SVD of U_R is approximately $(mn^2 - \frac{n^3}{3})$ and $6n^3$ operations respectively.

2.6 Iterative refinement of least-squares solutions of overdetermined systems

The following iterative technique, analysed by Bjorck (1967, 1968) generalises the method of iterative refinement for improving the accuracy of the computed solution of problem 1 to the simultaneous improvement of both the computed residual vector and least-squares solution of problem 2(i). The method is based on the observation that the least-squares solution $\underline{x}$ satisfies the normal equations

$$A^T A\underline{x} = A^T\underline{b}$$

or

$$A^T\underline{r} = \underline{0} \tag{1.34}$$

where $\underline{r} = \underline{b} - A\underline{x}$ is the residual vector. Then the (m+n)-vector $[\underline{r}\ \ \underline{x}]^T$ satisfies the nonsingular linear system

$$\begin{bmatrix} I & A \\ A^T & 0 \end{bmatrix} \begin{bmatrix} \underline{r} \\ \underline{x} \end{bmatrix} = \begin{bmatrix} \underline{b} \\ \underline{0} \end{bmatrix} \tag{1.35}$$

where I is the m × m identity matrix. Given the QR factorisation of A, the solution of square systems of order (m+n), whose coefficient matrix is that of (1.35), may be readily obtained in approximately $(4mn-n^2)$ operations (see Golub and Van Loan (1983), Ch. 6).

We can associate with the system (1.35) the residual vector $[\underline{f}\ \ \underline{g}]^T$ defined by

$$\begin{bmatrix} \underline{f} \\ \underline{g} \end{bmatrix} = \begin{bmatrix} \underline{b} \\ \underline{0} \end{bmatrix} - \begin{bmatrix} I & A \\ A^T & 0 \end{bmatrix} \begin{bmatrix} \underline{r} \\ \underline{x} \end{bmatrix} \tag{1.36}$$

and the error vector $[\underline{p} \quad \underline{z}]^T$ which satisfies

$$\begin{bmatrix} I & A \\ A^T & 0 \end{bmatrix} \begin{bmatrix} \underline{p} \\ \underline{z} \end{bmatrix} = \begin{bmatrix} \underline{f} \\ \underline{g} \end{bmatrix} . \qquad (1.37)$$

The method of iterative refinement for square systems, given in section 2.2, can now be applied to improve the accuracy of the solution of (1.35).

2.7 Vector and parallel processing

Recent years have witnessed increased activity in the field of application of parallel processing techniques to the design of algorithms which will run efficiently (with respect to computation time) on parallel computers. In particular, Dongarra, Gustavson and Karp (1984) have examined the effect of restructuring the basic algorithm of Gauss elimination (implemented in Fortran) for the square system $A\underline{x} = \underline{b}$, with respect to execution speed on a Cray-like machine. The basic algorithm uses three loops, and six permutations are thus possible for arranging the three loop indices. Although each of the permutations will yield the same operation count for the LU factorisation of A, their performance varies greatly because of the way information is accessed. For a survey of parallel algorithms in linear algebra, the reader is referred to Heller (1978).

3. Sparse matrix techniques

Methods for sparse problems are sought which are efficient with respect to computation time and storage requirements; such methods exploit the sparsity of the coefficient matrix by avoiding unnecessary arithmetic operations on zero elements and by storing only the non-zero elements and their positions in the matrix. The application of the Gauss elimination algorithm to a sparse problem of type 1 inevitably leads to 'fill-in'. That is, elements which are initially zero become non-zero, thus destroying the sparsity pattern in the coefficient matrix. Various graph reordering strategies, as they apply to so-called adjacency graphs of sparse matrices, will be described, their aim being to reduce the problem of fill-in. The various forms of frontal solution, developed primarily for the solution of finite-element equations, will not be discussed here; the interested reader may consult Duff and Reid (1983) for an account of one such method.

Iterative methods appear to be ideal for sparse problems because only the non-zero elements need to be stored, and if the matrix is the result of a finite-difference or finite-element discretisation on a regular grid it is possible to use an even more compact form. We restrict our discussion to iterative methods based on the conjugate gradient or Lanczos method. These methods, after being out of favour for many years, are now coming back into prominence.

There have been a number of recent conferences on sparse matrix methods and the reader is referred to the following published proceedings: Barker (1976), Duff and Stewart (1979) and Duff (1982a).

3.1 Matrix reordering

For symmetric matrices, the reordering problem is to find a permutation matrix P such that factorisation of P^TAP involves less fill-in than the original ordering. If A is positive definite, then P^TAP is also positive definite and it is possible to proceed with a Cholesky factorisation, pivoting on the diagonal components, without being troubled by round-off errors. There are a number of approaches to the problem, the choice often being governed by the sophistication of the data structures that are available. The simplest approach assumes that a matrix is stored as a profile, within which the fill-in is automatically confined; then the most popular method is a version of the Cuthill-McKee ordering.

Algorithms for reordering a symmetric matrix are most easily described in terms of graph theory. The undirected graph G(A) associated with an $n \times n$ matrix A has n nodes x_i , $i = 1,\ldots,n$. Corresponding to each non-zero component a_{ij} $(i \neq j)$ in A there exists an edge $x_i - x_j$ in G(A).

Thus G(A) contains all the information on the structure of A, but no information concerning the numerical values a_{ij} . Nodes x_i and x_j are said to be <u>neighbours</u> if $a_{ij} \neq 0$ and we define

$$\mathrm{Adj}(x) = \{\text{neighbours of } x\} \quad .$$

The degree of a node x is the number of nodes in Adj(x) and is denoted by $|\mathrm{Adj}(x)|$. For any set X of nodes,

$$\mathrm{Adj}(X) = \{ \bigcup_{x \in X} \mathrm{Adj}(x) \} \setminus X .$$

If the sparse matrix is banded, then in general there will be many zeros within the band and the fill-in will be extensive. An algorithm for restructuring A is equivalent to an algorithm for reordering the nodes of G(A).

Cuthill-McKee (CM) algorithm

(i) Determine a suitable starting node x_1 .

(ii) For i = 1,...,n, number the unnumbered neighbours of x_i in order of increasing degree .

It has been shown that this can often be improved, provided a variable band (profile, envelope, skyline, staircase) solver is available. The modified algorithm is known as the Reverse-Cuthill-McKee (RCM) algorithm and contains the additional step

(iii) Invert the order.

It has been proved that this step cannot increase the size of the envelope.

A problem with these orderings is in performing step (i). In general terms a node on the edge of the graph, preferably at the end of a long thin arm, would be preferable to one in the middle of the graph. In order to define an algorithm for the starting node it is necessary to introduce additional graph theoretic notation and to view the RCM algorithm from a different direction. A rooted level structure $\Lambda(x) = \{L_0(x),\ldots,L_{\ell(x)}(x)\}$ is a partitioning of the nodes of the graph into sets $L_i \equiv L_i(x)$, i = 0, ...,$\ell(x)$, where

$$L_0 = \{x\} \ , \ L_{-1} = \emptyset \ ,$$

$$L_{i+1} = \mathrm{Adj}(L_i) \setminus L_{i-1} \ , \ i = 0,\ldots,\ell(x)-1 \ .$$

With this notation, step (ii) can be reformulated as

(ii) Form $\Lambda(x_1)$; order the nodes level-by-level in order of increasing degree.

Gibbs-Poole-Stockmeyer (GPS) algorithm

(i) Select x, a node of minimum degree.

(ii) Construct $\Lambda(x)$.

(iii) For each $y \in L_{\ell(x)}$, construct $\Lambda(y)$ and if $\ell(y) > \ell(x)$ replace x by y and return to (ii).

Nested dissection

If it is possible to adopt a more sophisticated storage scheme than profile storage, a number of alternative methods are available. The basic philosophy behind the nested dissection algorithm relies on the assumption that the matrix can be split into loosely connected sub-structures so that the penalty for saving storage, by recomputing certain components of the factorised matrix every time they are required, is not unacceptably high. Consider the symmetric block matrix

$$A = \begin{bmatrix} A_{11} & A_{12} \\ A_{12}^T & A_{22} \end{bmatrix}$$

where the A_{ii} are banded submatrices and A_{12} is very sparse. It is possible to devise a scheme for the factorisation of A in which the non-zero envelopes in the diagonal blocks are stored explicitly and the non-zeros only of A_{12} are stored column-by-column. Thus if

$$L = \begin{bmatrix} L_{11} & \\ W_{12}^T & L_{22} \end{bmatrix},$$

the submatrix W_{12} is not stored; it is computed as and when necessary. The factorisation is carried out as

(a) $L_{11}L_{11}^T = A_{11}$: L_{11} stored in place of A_{11}

(b) Since $L_{11}W_{12} = A_{12}$, and $L_{22}L_{22}^T = A_{22} - W_{12}^T W_{12}$,

$$L_{22}L_{22}^T = A_{22} - A_{12}^T\,(L_{11}^{-T}\,(L_{11}^{-1}\,A_{12})) \equiv A_{22}' .$$

Hence A'_{22} is computed column-by-column, i.e. A_{22} is modified column-by-column; then L_{22} is stored explicitly. The idea of nested dissection is to structure the blocks A_{11} and A_{22} in a similar manner proceeding down until the diagonal blocks are either 1×1 or 2×2.

The savings obtainable by using nested dissection appear quite dramatic in theory, but there are considerable overheads in such a complex method. It is found in practice that a one-way dissection, with a few partitions, within each of which the RCM algorithm is used, is often a highly efficient compromise on all but the largest problems.

For general sparse matrices, the minimum degree algorithm, due to Markowitz (1957), when combined with some threshold pivoting to retain stability, is widely used. An additional technique that can be usefully applied to sparse matrices with a more-or-less random structure is to check for reducibility, i.e. try to reorder the matrix into block-triangular form (Duff and Reid 1978).

3.2 Conjugate gradients and the Lanczos method

Iterative techniques such as successive over-relaxation (SOR) and alternating direction implicit (ADI) methods are widely used to solve finite-difference or finite-element discretisations on regular grids. The convergence can be very rapid, but for systems with less regularity there is the problem of finding the best acceleration parameters for a particular situation. The conjugate gradient (CG) method has the advantage that no parameters are required; in the classical form the method is:

(i) given $\underline{x}_1$ (= $\underline{0}$) set $\underline{d}_1 = \underline{r}_1 = \underline{b}$

(ii) for k = 1,...,n

$$\alpha_k = \underline{r}_k^T \underline{r}_k / \underline{d}_k^T A\underline{d}_k \tag{1.38}$$

$$\underline{x}_{k+1} = \underline{x}_k + \alpha_k\underline{d}_k \tag{1.39}$$

$$\underline{r}_{k+1} = \underline{r}_k - \alpha_k A\underline{d}_k \tag{1.40}$$

$$\beta_k = \underline{r}_{k+1}^T \underline{r}_{k+1} / \underline{r}_k^T \underline{r}_k \tag{1.41}$$

$$\underline{d}_{k+1} = \underline{r}_{k+1} + \beta_k \underline{d}_k \ . \qquad (1.42)$$

If the matrix A is symmetric and positive definite, this is equivalent to a descent method (see Chapter 12) for minimising

$$f(\underline{x}) = \underline{x}^T A \underline{x} - 2\, \underline{x}^T \underline{b} \ . \qquad (1.43)$$

Hence with exact arithmetic the true solution is achieved in at most n iterations. If exact arithmetic is not used then the conjugacy conditions

$$\underline{d}_\ell^T A \underline{d}_k = 0 \ , \quad \ell \neq k \qquad (1.44a)$$

and

$$\underline{r}_\ell^T \underline{r}_k = 0 \ , \quad \ell \neq k \qquad (1.44b)$$

do not hold and finite termination is no longer guaranteed. A more important result from a computational viewpoint is that the initial decrease in the residuals depends on the clustering of the eigenvalues: the bigger the cluster, the more rapid the decrease in the residual.

It can be shown (Paige and Saunders 1975) that the CG method is equivalent to the <u>Lanczos method</u> defined by:

(i) $\underline{v}_1 = \underline{b} / \gamma_1 \ , \quad \gamma_1^2 = \underline{b}^T \underline{b}$ (1.45)

$\underline{v}_0 = \underline{0}$

(ii) for $k = 1,\ldots,n$

$$\gamma_{k+1} \underline{v}_{k+1} = \underline{r} = A\underline{v}_k - \delta_k \underline{v}_k - \gamma_k \underline{v}_{k-1} \ , \qquad (1.46)$$

$$\delta_k = \underline{v}_k^T A \underline{v}_k \quad \text{with} \quad \gamma_{k+1}^2 = \underline{r}^T \underline{r} \ .$$

If we write

$$V_k = [\underline{v}_1 \ \ldots \ \underline{v}_k]$$

$$T_k = \begin{bmatrix} \delta_1 & \gamma_2 & & \\ \gamma_2 & \delta_2 & \gamma_3 & \\ & . & . & . \\ & & \gamma_k & \delta_k \end{bmatrix} \tag{1.47}$$

then

$$V_k^T V_k = I \equiv [\underline{e}_1 \quad \ldots \quad \underline{e}_k]$$

and the recurrence relation (1.46) can be written as

$$AV_k = V_k T_k + \gamma_{k+1} \underline{v}_{k+1} \underline{e}_k^T \ . \tag{1.48}$$

The solution is then

$$\underline{x}_n = V_n \underline{y}_n \tag{1.49}$$

where

$$T_n \underline{y}_n = \gamma_1 \underline{e}_1 \quad (\text{as } V_n^T \underline{b} = \gamma_1 \underline{e}_1) \ . \tag{1.50}$$

The connection between the two methods is clear when A is positive definite, since, for any intermediate $k \leqslant n$,

$$T_k = V_k^T A V_k$$

is then also positive definite and there exists a factorisation

$$T_k = L_k D L_k^T \ .$$

We define

$$\underline{p}_k = L_k^T \underline{y}_k \tag{1.51}$$

and

$$C_k = V_k L_k^{-T} \; . \tag{1.52}$$

Then the k × k system

$$T_k \underline{y}_k = \gamma_1 \underline{e}_1$$

is solved (by forward substitution) as

$$\left.\begin{aligned} L_k D \underline{p}_k &= \gamma_1 \underline{e}_1 \\ L_k C_k^T &= V_k^T \\ \underline{x}_k &= C_k \underline{p}_k \; . \end{aligned}\right\} \tag{1.53}$$

If the starting approximations for both algorithms are consistent as stated then (1.52) and (1.42) are equivalent as are (1.53) and (1.39), together with (1.40) and (1.46). It is then possible to identify

$$\underline{p}_k^T = [\alpha_1 \; \ldots \; \alpha_k] \; ,$$

$$C_k = [\underline{d}_1 \; \ldots \; \underline{d}_k]$$

and $\underline{r}_k$ is equivalent to $\underline{v}_k$ without the unnecessary normalisation.

Having established the connection between conjugate gradients and the Lanczos method, it is possible to develop methods based on the latter for the case when A is indefinite, when the former may break down. The conjugate gradient method assumes an LDL^T factorisation of T_k , but for an indefinite matrix a QR (or LQ) factorisation leads to a very stable algorithm and this is the basis of the SYMMLQ algorithm of Paige and Saunders (1975). An alternative Lanczos algorithm using selective re-orthogonalisation to preserve the conjugacy conditions (1.43), has been proposed by Parlett and Scott (1979). The Lanczos method is also known as bi-conjugate gradients when applied to indefinite or asymmetric matrices (Jacobs 1982); it is equivalent to minimising $\underline{r}^T\underline{r}$ rather than $\underline{r}^T A^{-1}\underline{r}$ as in (1.43).

3.3 Preconditioned conjugate gradient (PCG) method

Algorithms based on the Lanczos method can be used to solve symmetric indefinite systems, but the convergence of the basic conjugate gradient method can be very slow (or non-existent) if the eigenvalues of the positive definite matrix are not sufficiently clustered.

In general the eigenvalues of A will be evenly spread throughout the spectrum and it is necessary to precondition problem 1 (section 1) in order to cluster the eigenvalues. Meijerink and van der Vorst (1979) give a good illustration of the bunching effect of preconditioning. The procedure is to find a matrix M close to A such that $M\underline{x} = \underline{b}$ is easy to solve. Then, if the Cholesky factorisation of M is $M = NN^T$, problem 1 is replaced by

$$(N^{-1}AN^{-T})N^T\underline{x} = N^{-1}\underline{b} \,, \tag{1.54}$$

i.e. the numerical algorithm is equivalent to solving

$$A_0\,\underline{y} = \underline{b}_0 \,,$$

$$N^T\underline{x} = \underline{y} \,,$$

where $A_0 = N^{-1}AN^{-T}$ and $\underline{b}_0 = N^{-1}\underline{b}$. If M is close to A, A_0 is close to the identity matrix and has eigenvalues clustered around unity. If the Cholesky factors of M can be found, then it is possible to apply the method of conjugate gradients to the solution of (1.54) to give the algorithm:

(i) given $\underline{x}_1$ $(= M^{-1}\underline{b})$ set $\underline{r}_1 = \underline{b} - A\underline{x}_1$, $\underline{d}_1 = M^{-1}\underline{r}_1$

(ii) for k = 1,..., until $\underline{r}_k^T\,\underline{r}_k < \epsilon$ (given)

$$\alpha_k = \underline{r}_k^T\, M^{-1}\underline{r}_k \,/\, \underline{d}_k^T\, A\underline{d}_k \tag{1.55}$$

$$\underline{x}_{k+1} = \underline{x}_k + \alpha_k\underline{d}_k \tag{1.56}$$

$$\underline{r}_{k+1} = \underline{r}_k - \alpha_k\, A\underline{d}_k \tag{1.57}$$

$$\beta_k = \underline{r}_{k+1}^T M^{-1} \underline{r}_{k+1} / \underline{r}_k^T M^{-1} \underline{r}_k \qquad (1.58)$$

$$\underline{d}_{k+1} = M^{-1} \underline{r}_{k+1} + \beta_k \underline{d}_k \,. \qquad (1.59)$$

Clearly $\underline{a} = M^{-1}\underline{c}$ is found by solving $NN^T\underline{a} = \underline{c}$.

There are a number of different strategies for selecting the preconditioning matrix M; the two most popular are incomplete Cholesky factorisation and iteration. The incomplete Cholesky factorisation method (the ICCG method, Meijerink and van der Vorst 1977) generates the factors NN^T of the preconditioner M to have the same sparsity pattern as A (the ICCG(0) method) or to allow additional selected fill-in (Munksgaard 1979).

The components of N are determined by applying the Cholesky factorisation algorithm to the designated non-zeros only and defining all other components as zero. The algorithm is known to be stable if A is an 'M-matrix' (Meijerink and van der Vorst 1977), but in other situations it is necessary to modify the ICCG. Manteuffel (1979) uses a shifted incomplete factorisation as preconditioner; he has applied the method to three-dimensional finite-element analysis of elasticity. The technique is to reduce the size of the off-diagonal terms used in the factorisation. Thus if

$$A = I - B \quad \text{(after diagonal scaling)}$$

then the preconditioner is an incomplete factorisation of

$$I - \frac{1}{1+\alpha} B \,,$$

where $\alpha \approx 0.015$ is used by Manteuffel. An alternative modification for the diagonal terms in finite-difference approximations has been suggested by Gustaffson (1979) who has tested the MICCG(0) and MICCG(2) versions in the reference.

All the preconditioners based on Cholesky factorisation rely on the matrix A being diagonally dominant. This is not true however for finite-element (and finite-difference) approximation other than the most basic discretisations. Alternative strategies have been developed specifically for this application; see for example, Axelsson (1979), Crisfield (1979, 1984), Jennings and Malik (1978). Conjugate gradient methods have been found to be well suited to computation on array processors such as the ICL DAP, and examples of preliminary studies can be found in Wait and Martindale (1984).

An alternative strategy is not to define the preconditioning matrix M or its factors NN^T explicitly but to replace the terms $M^{-1}\underline{r}_k$, etc. in (1.55), (1.58) and (1.59) by a subsidiary vector $\underline{z}_k$ which is computed by an inner iteration. If m steps of an iterative method are used this is known as an m-step PCG method (Adams 1983). Alternatively a few steps of a multigrid method can be used (Markham 1983). Jennings and Malik (1978) compare the incomplete factorisation approach with a symmetric SOR-CG method. Preconditioned CG-like methods for nonsymmetric matrices are given by Axelsson (1979), van der Vorst (1981), Elman (1981), Eisenstat, Elman and Schultz (1983), Concus and Golub (1976). Other preconditioners have been suggested for computation on vector machines (Dubois, Greenbaum and Rodrigue 1979) for which incomplete factorisation would not be the most efficient approach.

3.4 Linear least squares

The Lanczos bi-conjugate gradient algorithm of section 3.2 solves problem 2(i) (of section 1) for the case m = n and it forms the basis of one of the popular methods for overdetermined systems. The solution of the overdetermined problem 2(i) satisfies the normal equations

$$A^TA\underline{x} = A^T\underline{b} \tag{1.60}$$

but solving the problem via (1.60) leads, in general, to numerical instabilities and to the destruction of the sparsity structure. In multiple regression calculations, the solution is often found via (1.60) in order to obtain the variance matrix $(A^TA)^{-1}$. On the other hand, if a numerically stable QR factorisation of A is used (see section 2.5), the same information can be derived efficiently, but once again with the loss of the sparse structure. In the notation of section 2.5, the QR factorisation of A yields $\underline{x}$ as the solution of the square system

$$U_R\,\underline{x} = \underline{c}$$

with $R = [U_R \quad 0]^T$.

If the rows and columns of A are permuted and the matrix P_1AP_2 is reduced to upper triangular form, then the sparsity of R depends only on P_2 (Gentleman 1976) but the amount of computation is affected by P_1. George and Ng (1983) provide a nested dissection algorithm for finding a good P_2 that guarantees a good P_1.

The sparsity can be used to an even greater extent if the variances are not required. The least-squares solution $\underline{x}$ and the residual vector $\underline{r} = \underline{b} - A\underline{x}$ satisfy equation (1.35), a symmetric indefinite system of size m+n, known as the residual equations. This is the basis of the LSQR algorithm of Paige and Saunders (1982a, b) which is similar in style to SYMMLQ mentioned in section 3.2.

An alternative is the method of Peters and Wilkinson (Duff and Reid 1976, or Bjorck and Duff 1980) which is to perform an LU factorisation of the matrix A and then to update the solution of the first n equations by solving a set of normal equations for which the coefficient matrix is L^TL and for which a stable Cholesky factorisation exists. In their numerical experiments, Duff and Reid (1976) found the method of Peters and Wilkinson to be preferable to a QR factorisation for sparse systems; for highly rectangular systems, the normal equations were satisfactory.

4. Software

The F04 chapter of the NAG Fortran Library and the package LINPACK both provide an excellent source of routines for the solution of non-sparse simultaneous linear equations and least-squares problems. In addition the NAG F04 chapter caters for some sparse problems.

The solution of problem 1 (see section 1) may be obtained using one of the NAG routines which implements Crout's method with partial pivoting, both with and without iterative refinement. Alternatively the LINPACK routine SGECO (see section 2.3) uses the method of Gauss elimination with partial pivoting; the method of iterative refinement is not used in this machine-independent routine.

Implementations of the Cholesky algorithm, for example in the NAG Library, use double precision arithmetic for the accumulation of inner products. The NAG Library does not cater for the situation where A is symmetric but not positive definite; however the LINPACK routine SSIFA implements a stable diagonal pivoting factorisation algorithm (see section

2.4) in this case.

The NAG routine F04JGF computes the least-squares solution of problem 2 (section 1) using the method of orthogonal factorisation described in section 2.5.

A useful survey of current sparse matrix software has been provided by Duff (1982b). The CM+GPS algorithm has been implemented by Gibbs, Poole and Stockmeyer (1976a, b), Crane, Gibbs, Poole and Stockmeyer (1976) and in the package SPARSPAK (George and Liu 1981). Improved versions of all these codes have been produced recently by Lewis (1982a, b).

Routines based on the minimum degree algorithm appear in the Harwell sparse matrix package MA28 (Duff 1977), YSMP (Eisenstat, Gursky, Schultz and Sherman 1982), SSLEST (Zlatev and Thompsen 1982) and in a symmetric form in SPARSPAK.

The SYMMLQ algorithm of Paige and Saunders (1975), described in section 3.2, is implemented in the NAG routine F04MBF which allows for preconditioning. The ITPACK routines (Kincaid, Respess, Young and Grimes 1982) include symmetric SOR-CG, Jacobi-CG and other CG variants.

Finally, an implementation of the LSQR algorithm may be found in the NAG routine F04QAF.

CHAPTER 2

Eigenvalue Problems

J.J. Du Croz and S.J. Hammarling

1. Introduction

We discuss first the standard form of algebraic eigenvalue problems: that is, given a real or complex matrix A, to compute a scalar λ and a non-null vector $\underline{x}$ such that

$$A\underline{x} = \lambda\underline{x} . \tag{2.1}$$

We may wish to compute only eigenvalues, λ , or to compute eigenvalues and their associated eigenvectors, $\underline{x}$. We may also wish to compute just a few selected eigenvalues or the complete eigen-spectrum.

First we shall discuss algorithms for dense matrices, and then algorithms for sparse matrices. Then we extend the discussion to the singular value problem, which is closely connected to the particular eigenvalue problem

$$A^T A\underline{x} = \lambda\underline{x} , \tag{2.2}$$

and to the generalised eigenvalue problem

$$A\underline{x} = \lambda B\underline{x} . \tag{2.3}$$

2. Algorithms for dense eigenvalue problems

We take as baseline the algorithms for dense eigenvalue problems which were extended and published as a collection of Fortran subroutines with the title 'EISPACK'; for references to EISPACK and other software, see section 6.

Ten years' experience has revealed no serious weaknesses in the EISPACK algorithms, but has shown the need for additional capabilities; many related algorithms have been published, some of which we review here. At the heart of most of the computations in EISPACK is some variant of the QR algorithm. This remains the best algorithm for computing all the eigenvalues of a matrix of moderate size. For a good description of how and why it works, see Golub and Van Loan (1983, Chs. 7, 8), or Watkins (1982).

2.1 Symmetric dense problems

We concentrate on the computation of all the n eigenvalues of A, so instead of (2.1) we take as the defining equation:

$$AQ = Q\Lambda \text{ , or equivalently, } Q^T AQ = \Lambda \text{ ,} \tag{2.4}$$

where $\Lambda = \text{diag}\ [\lambda_1 \ldots \lambda_n]$ (the eigenvalues) and Q is the orthogonal matrix of eigenvectors. The EISPACK approach is to compute Λ in two stages: first A is reduced to tridiagonal form T (using (n-2) Householder transformations):

$$Q_a^T AQ_a = T \text{ ,} \tag{2.5}$$

and then the QL algorithm (a variant of the QR algorithm) is applied iteratively to T, T being transformed by a sequence of Givens rotations:

$$Q_b^T TQ_b = \Lambda \text{ .} \tag{2.6}$$

(A may be complex Hermitian, in which case Q_a is unitary, but T can always be made real.) The eigenvalues are always well conditioned, the QR algorithm is globally (and rapidly) convergent (at least in exact arithmetic), and the exclusive use of orthogonal transformations ensures good numerical behaviour. For details, see the excellent book by Parlett (1980).

To compute all the eigenvectors of A, the straightforward approach is to form Q_a explicitly from (2.5) and then to compute $Q = Q_a Q_b$ at the same time as the QR algorithm is applied to T. Parlett points out that this can be speeded up if (2.6) is first computed without accumulating $Q_a Q_b$, and then (2.6) is computed again on a second copy of T using the now-known eigenvalues as shifts, and this time computing Q_aQ_b ; the number of iterations is reduced at this second attempt, theoretically to one per eigenvalue, and so Q_aQ_b can be computed more cheaply. An alternative approach is to use inverse iteration to compute the eigenvectors of T one by one; conventional advice is to use inverse iteration only when less than 25% of the eigenvectors are required, but in fact for sufficiently large matrices inverse iteration is the fastest method for computing all eigenvectors; the price is that it requires more storage and that the computed eigenvectors are less accurately orthogonal. See Peters and Wilkinson (1979) for a discussion of the practical behaviour of inverse iteration.

On parallel computers the essentially serial nature of the tridiagonal QR algorithm becomes a serious disadvantage, and Jacobi's algorithm has proved highly competitive. This is a single phase algorithm which applies a sequence of plane rotations to A, each of which reduces one of the off-diagonal elements to zero; such zeros may be lost in subsequent rotations, but the overall effect is to reduce the size of the off-diagonal elements until they are negligible. More arithmetic is involved than in computing (2.5) and (2.6), but on parallel machines it is possible to generate and apply n/2 rotations simultaneously. The details and the efficiency depend very much on the architecture of the machine (e.g. Modi and Parkinson (1982)).

2.2 Unsymmetric dense problems

For many purposes the appropriate analogue of (2.4) is the Schur factorisation

$$AQ = QU \ , \quad \text{or equivalently,} \quad Q^H AQ = U \tag{2.7}$$

where U is an upper triangular matrix, with the eigenvalues of A along the diagonal; they can be made to appear in any order. Q is a unitary matrix, the columns of which are called the Schur vectors. If A is real, it may have complex eigenvalues but to avoid complex arithmetic as much as possible, we can arrange that in (2.7) Q is a real orthogonal matrix, but U is not quite upper triangular: it is block upper triangular, with blocks of order one or two along the diagonal. The real eigenvalues of A are the 1 × 1 diagonal blocks, and the complex eigenvalues of A, which occur as complex conjugate pairs, are the eigenvalues of the 2 × 2 blocks. In this section, for brevity and simplicity, we describe algorithms in terms of complex matrices; the analogues for real matrices are immediate except that the presence of 2 × 2 blocks on the diagonal of U causes some complications in detail.

The Schur factorisation (2.7) may be computed in two stages: first A is reduced to upper Hessenberg form H (using (n−2) Householder transformations):

$$Q_a^H AQ_a = H , \tag{2.8}$$

and then the QR algorithm is applied iteratively to H:

$$Q_b^H HQ_b = U . \tag{2.9}$$

If eigenvectors of A are wanted, the eigenvectors of U may be computed by back-substitution:

$$UX = X\Lambda , \tag{2.10}$$

giving altogether

$$A (Q_aQ_bX) = (Q_aQ_bX) \Lambda . \tag{2.11}$$

One of the advantages of the Schur factorisation is that it can be computed using only unitary transformations, which ensures good numerical behaviour. A further consequence of the fact that Q is unitary is that this factorisation leads to a stable and reliable method for computing invariant subspaces - or more precisely for computing orthonormal bases of them. If the Schur factorisation of A is partitioned, we have

$$A [Q_1 \quad Q_2] = [Q_1 \quad Q_2] \begin{bmatrix} U_{11} & U_{12} \\ & U_{22} \end{bmatrix} \tag{2.12}$$

and hence:

$$AQ_1 = Q_1U_{11} . \tag{2.13}$$

The columns of Q_1 are mapped by A onto linear combinations of the columns of Q_1 ; that is, they span an invariant subspace of A, and, being mutually orthogonal, constitute an orthonormal basis of the subspace; moreover, this is the eigenspace corresponding to those eigenvalues of A which lie on the diagonal of U_{11} . Sometimes we may wish to compute the invariant subspace corresponding to a few eigenvalues (often a cluster of close eigenvalues, which may well correspond to a multiple eigenvalue in exact arithmetic); in other applications the desired subspace may correspond to a large subset of the eigenvalues (e.g. those with negative real part, or those inside the unit circle in the complex plane). In any case it is necessary to ensure that the relevant eigenvalues appear at the top of the diagonal of U. This can be achieved by the algorithm EXCHNG of Stewart (1976a), which computes the orthogonal transformations needed to interchange the adjacent diagonal blocks (whether 1 × 1 or 2 × 2) in the real Schur factorisation. To interchange two 1 × 1 blocks, U is pre- and post-multiplied by a single plane rotation; when 2 × 2 blocks are involved, a sequence of plane rotations is required. Repeated applications of EXCHNG can achieve any desired ordering of the eigenvalues.

An application of this procedure is found in the method of Laub (1979) for solving the algebraic Riccati equation which arises in control theory:

$$B + A^HX + XA - XCX = 0 \tag{2.14}$$

where B and C are Hermitian (all matrices are n × n). For simplicity we concentrate on the special case in which B and C are positive definite, for then X is also Hermitian and positive definite. We form the 2n × 2n matrix

$$M = \begin{bmatrix} A & -C \\ -B & -A^H \end{bmatrix} \tag{2.15}$$

and compute its Schur factorisation, ordered so that the eigenvalues with negative real part appear on the diagonal of U_{11} (it can be proved that M must have n such eigenvalues):

$$\begin{bmatrix} Q_{11}^H & Q_{21}^H \\ Q_{12}^H & Q_{22}^H \end{bmatrix} \begin{bmatrix} A & -C \\ -B & -A^H \end{bmatrix} \begin{bmatrix} Q_{11} & Q_{12} \\ Q_{21} & Q_{22} \end{bmatrix} = \begin{bmatrix} U_{11} & U_{12} \\ & U_{22} \end{bmatrix}. \quad (2.16)$$

Then X is given by $Q_{21}Q_{11}^{-1}$. The unsatisfactory feature of this method is that the Schur factorisation takes no advantage of the structure of M. Paige and Van Loan (1981) have shown that such a matrix (known as a Hamiltonian matrix) must have a special form of the Schur factorisation (2.16) in which $U_{22} = -U_{11}^H$, $Q_{22} = Q_{11}$ and $Q_{12} = -Q_{21}$, and an algorithm to compute this would be valuable (see Van Loan (1982a)).

Other applications of the Schur factorisation lie in the stable reduction (via unitary transformations) of other problems to a more manageable form. An important example is the algorithm of Bartels and Stewart (1972) for solving the Sylvester equation

$$AX + XB = C \quad (2.17)$$

where C and X may be rectangular. (A solution X exists if and only if A and -B have no eigenvalues in common.) If the Schur factorisations of A and B are $A = QUQ^H$ and $B = RVR^H$, then (2.17) reduces to

$$U\tilde{X} + \tilde{X}V = \tilde{C} \quad (2.18)$$

where $\tilde{X} = Q^H XR$ and $\tilde{C} = Q^H CR$. Since U and V are upper triangular, (2.18) can be solved by a back-substitution process. Bartels and Stewart also include an algorithm for solving the Lyapunov equation which results from (2.17) by setting $B = A^H$, C and X then being Hermitian:

$$AX + XA^H = C\ . \quad (2.19)$$

This reduces to

$$U\tilde{X} + \tilde{X}U^H = \tilde{C}\ . \quad (2.20)$$

If $\tilde{C}$ is positive definite, then so is $\tilde{X}$, and then Hammarling (1982) proposes an alternative back-substitution process to compute the Cholesky factors of $\tilde{X}$ rather than $\tilde{X}$ itself.

Golub, Nash and Van Loan (1979) have shown that (2.17) can be solved more efficiently (at a cost of extra workspace) if one of A or B is only reduced to upper Hessenberg form as in (2.8), and there are other problems where the Hessenberg decomposition can be used as a cheaper alternative to the Schur factorisation (see, for example, Van Loan (1982b)).

Sometimes it is useful to go beyond the Schur factorisation and reduce U to <u>block-diagonal</u> form

$$Y^{-1}UY = \begin{bmatrix} U_{11} & 0 & \cdots & 0 \\ & U_{22} & \cdots & 0 \\ & & \cdots & \cdot \\ & & & U_{ss} \end{bmatrix} \tag{2.21}$$

where the upper triangular blocks U_{ii} do not have any eigenvalues in common. This cannot be achieved by unitary transformations, but Bavely and Stewart (1979) have developed an algorithm in which a user-supplied tolerance controls the trade-off between the stability of Y and the size of the blocks. The essential idea is that

$$\begin{bmatrix} I & Y_{12} \\ & I \end{bmatrix}^{-1} \begin{bmatrix} U_{11} & U_{12} \\ & U_{22} \end{bmatrix} \begin{bmatrix} I & Y_{12} \\ & I \end{bmatrix} = \begin{bmatrix} U_{11} & \\ & U_{22} \end{bmatrix} \tag{2.22}$$

if

$$U_{11}Y_{12} - Y_{12}U_{22} = -U_{12} \tag{2.23}$$

and ill-conditioning in the transformation is indicated by large elements in Y_{12}. (2.23) has the same form as (2.20). A block-diagonal form with blocks as small as possible (consistent with stability) is useful, in particular, for computing functions of matrices (Kågström 1977, Golub and Van Loan 1983, Ch. 11). The ultimate step in this direction is to attempt to compute the Jordan normal form of U; although this form is of great theoretical value, its numerical determination raises many questions (Golub and Wilkinson 1976).

2.3 Condition numbers

A valuable addition to EISPACK would be routines for computing or estimating the condition numbers of eigenvalues and eigenvectors. The most popular measure of the sensitivity of a simple eigenvalue λ is (Wilkinson 1965, Ch. 2)

$$\text{cond}(\lambda) = \frac{||\underline{y}||_2 \, ||\underline{x}||_2}{|\underline{y}^H \underline{x}|} \tag{2.24}$$

where $\underline{x}$ and $\underline{y}$ are corresponding eigenvectors of A and A^H respectively; $\underline{y}$ is also called a left eigenvector of A. If λ is a multiple eigenvalue, it is usual to take cond(λ) as infinity. If A is Hermitian (or even normal), $A\underline{x} = \lambda\underline{x}$ implies also $A^H\underline{x} = \lambda\underline{x}$, and so cond(λ) = 1: all eigenvalues are well conditioned. For unsymmetric A, (2.24) is invariant under unitary transformations, and so can be computed from the matrix U in the Schur factorisation of A. Indeed if

$$U = \begin{bmatrix} U_{11} & \underline{u} & U_{13} \\ & \lambda & \underline{v}^H \\ & & U_{33} \end{bmatrix}$$

then $\underline{x} = Q\,[\underline{s} \;\; 1 \;\; \underline{0}]^T$ and $\underline{y} = Q\,[\underline{0} \;\; 1 \;\; \underline{t}]^T$ where

$$(U_{11} - \lambda I)\,\underline{s} = -\underline{u}\ ,$$
$$(U_{33} - \lambda I)^H\,\underline{t} = -\underline{v}\ .$$

Then $\text{cond}(\lambda) = \{(1 + \underline{s}^H\underline{s}).(1 + \underline{t}^H\underline{t})\}^{1/2}$.

Eigenvectors may be more ill-conditioned than the corresponding eigenvalues. We start from a measure of the sensitivity of an invariant subspace. Using the notation of (2.12), the sensitivity of the subspace spanned by the columns of Q_1 is given by the reciprocal of:

$$\text{sep}(U_{11}, U_{22}) = \min_{||X||_F=1} ||U_{11}X - XU_{22}||_F \quad (2.25)$$

(Stewart 1973, Varah 1979). In the special case where U_{11} reduces to a simple eigenvalue λ and

$$U = \begin{bmatrix} \lambda & \underline{c}^H \\ & B \end{bmatrix}, \quad (2.26)$$

the sensitivity of the eigenvector corresponding to λ is given by

$$\frac{1}{\text{sep}(\lambda, B)} = ||(\lambda I - B)^{-1}||_2 \quad . \quad (2.27)$$

The LINPACK condition estimator (see section 2.3 of Chapter 1) can be applied to estimate (2.27), and a generalisation of it to estimate (2.25) has been implemented by Byers (1984).

A totally different approach is to extend the idea of iterative refinement, as used in the solution of linear equations, to improve a first approximation to λ and $\underline{x}$ (Dongarra, Moler and Wilkinson 1983); as usual with iterative refinement, the computation of the residuals must be done in extended precision.

3. Algorithms for sparse symmetric problems

The methods described so far are not suitable for large sparse problems, principally because any application of unitary transformations to reduce the problem, as in (2.5) or (2.8), tends to cause a large amount of fill-in, and hence the amount of computation and storage needed becomes excessive; also those methods are designed for finding all eigenvalues, whereas with large sparse problems we often only want a small number of eigenvalues. Quite different algorithms are required. We describe two classes of methods for symmetric problems: simultaneous iteration, and Lanczos methods. The theory behind both methods is described by Parlett (1980, Ch. 10-14). Simultaneous iteration is better established; Lanczos methods have been studied intensively over the last ten years, but the development of robust general-purpose algorithms is still a subject of research. Lanczos methods can be much more efficient than simultaneous iteration for certain problems, and can be used to compute all the eigenvalues of a large sparse matrix. Both methods require a procedure for computing $A\underline{x}$ for any given $\underline{x}$, rather than access to the elements of A. If an algorithm is designed to find the largest eigenvalues (in absolute value), then it can easily be used to find the smallest eigenvalues if the procedure returns $A^{-1}\underline{x}$, i.e. returns $\underline{y}$ such that $A\underline{y} = \underline{x}$; an algorithm for solving a sparse symmetric system of equations may be used in conjunction with one for solving the eigenvalue problem (see for example Lewis (1977)). Similarly, to compute the eigenvalues closest to σ, the user should solve $(A - \sigma I)\underline{y} = \underline{x}$. In this section we consider only real A.

3.1 Simultaneous iteration

Simultaneous iteration is a method for obtaining a few of the largest eigenvalues (in absolute value) and their corresponding eigenvectors. The basic idea is that if Y is a p-dimensional subspace of R^n, then the sequence

$$Y,\ AY,\ A^2Y,\ A^3Y,\ \ldots$$

converges, under suitable conditions, to the eigenspace corresponding to the p largest eigenvalues. (When p = 1, this is simply the power method for finding the dominant eigenvalue.) In practice the method works with an orthonormal basis of A^iY at each stage: what is the best basis? If the columns of V_{i-1} are an orthonormal basis for $A^{i-1}Y$, we could simply form a QR factorisation of AV_{i-1} :

$$AV_{i-1} = Q_iR_i\ ; \quad V_i = Q_i\ . \tag{2.28}$$

(Incidentally, this is what is done in the QR algorithm, which can be regarded as simultaneous iteration on the whole n-dimensional space.) However, we achieve faster convergence if we compute:

$$\begin{aligned} H_i &= Q_i^T AQ_i \\ &= G_i\theta_iG_i^T \quad \text{(eigenvalue-eigenvector factorisation of } H_i) \\ V_i &= Q_iG_i\ . \end{aligned} \tag{2.29}$$

This is the Rayleigh-Ritz procedure, which yields:

$$V_i^T A V_i = \theta_i \; . \tag{2.30}$$

(This is not the same as (2.4): the V_i here are rectangular.) θ_i and V_i are in a certain sense the best set of approximations obtainable from the subspace $A^i Y$ to p eigenvalues and eigenvectors of A. (When $p = 1$, θ_i reduces to the Rayleigh quotient $\theta = \underline{v}^T A\underline{v}$, which minimises the residual norm $||A\underline{v} - \theta\underline{v}||_2$.) The diagonal elements of θ_i are called the Ritz values and

$$|\theta_j^{(i)} - \lambda_j| = O(|\lambda_{p+1} / \lambda_j|^i) \; , \quad j = 1,\ldots,p \; . \tag{2.31}$$

The columns of $V_i = Q_i G_i$ are called the Ritz vectors and they converge at the same time to the eigenvectors of A.

A very careful implementation of this method - including many refinements to save work, and to accelerate and measure convergence - was published by Rutishauser (1970) as an Algol 60 algorithm RITZIT. The only decision left to the user is the choice of p, which, as can be seen from (2.31), should be somewhat larger than the number of eigenvalues actually required, to avoid poor convergence if eigenvalues are closely clustered.

The method has been extended to unsymmetric problems in various ways, e.g. the 'lopsided' iteration of Stewart and Jennings (1981). An alternative approach, suggested by Stewart (1976b), is to use the Schur vectors rather than the eigenvectors.

3.2 Lanczos methods

Whereas simultaneous iteration works with subspaces of a fixed dimension p, the basic Lanczos method works with subspaces whose dimension increases by one at each iteration. Starting with an arbitrary vector $\underline{q}$, we work with a sequence of subspaces K_i (known as Krylov subspaces), where

$$K_i = \text{span}\,\{\underline{q}, A\underline{q}, \ldots, A^{i-1}\underline{q}\} \; . \tag{2.32}$$

We represent K_i by the orthonormal columns of an $n \times i$ matrix Q_i ; this can be done in such a way that $Q_i^T AQ_i$ ($= T_i$) is tridiagonal, and so it is economical to apply the Rayleigh-Ritz procedure. If the eigenvalue-eigenvector factorisation of T_i is

$$T_i = G_i \theta_i G_i^T$$

then the diagonal elements of θ_i are the Ritz values and the columns of $Y_i = Q_i G_i$ are the Ritz vectors corresponding to the subspace K_i . Moreover, the computation of Q_i and T_i is extremely cheap: it is based on the three-term recurrence:

$$\beta_{i+1} \underline{q}_{i+1} = A\underline{q}_i - \alpha_i \underline{q}_i - \beta_i \underline{q}_{i-1} \tag{2.33}$$

where by the orthogonality of the $\underline{q}_j$,

$$\alpha_i = \underline{q}_i^T A\underline{q}_i \tag{2.34}$$

and β_{i+1} is a normalising factor for $\underline{q}_{i+1}$. It is essential to use a stable form of this recurrence (Paige 1972, 1976). Then Q_i has columns $\underline{q}_1$, $\underline{q}_2$, ..., $\underline{q}_i$, and T_i has diagonal elements α_1 , α_2 ,..., α_i , and off-diagonal elements β_2 ,..., β_i . The superiority of Lanczos methods over simultaneous iteration stems in part from the fact that the Q_i and T_i retain information from all previous iterations. A only enters into (2.33) where $A\underline{q}_i$ is computed, and only five vectors of length n are needed to compute and store the T_i ; the previous $\underline{q}_j$ $(j < i)$ can be written to backing store and recalled when needed. (It is not necessary to compute all the Ritz vectors at each iteration.)

Originally (2.33) was proposed as a method for reducing A to tridiagonal form, which is achieved when $\beta_{i+1} = 0$, certainly when $i = n$. However, in finite-precision arithmetic the computed quantities resulting from (2.33) start to deviate drastically from the exact values: in particular the Q_i lose their orthogonality and the algorithm fails to terminate. The most obvious remedy - to reorthogonalise the Q_i at each stage - is hideously expensive. The alternative viewpoint presented above makes it plausible to expect, however, that even for $i \ll n$, some of the eigenvalues of T_i will converge to extreme eigenvalues of A, and this is observed in practice (when $i \simeq \sqrt{n}$, say). Rounding errors do cause problems: indeed loss of orthogonality in the Q_i is intimately associated with convergence of some of the Ritz values. The overall effect is not wholly disastrous: what happens in practice is that the algorithm starts to generate Ritz values which converge to spurious copies of eigenvalues that have already converged; this makes the algorithm inefficient and difficult to use.

Practical Lanczos algorithms include many refinements to overcome these difficulties. Selective orthogonalisation (Parlett and Scott 1979) is based on the observation that sufficient orthogonality among the q_i can be maintained if only occasional q_{i+1} are reorthogonalised against those Ritz vectors which have converged; the cost is not significant, especially if only a few eigenvalues are sought, and the method is effective in suppressing spurious copies of eigenvalues. Parlett and Reid (1981) have developed a very different algorithm: they keep a record of intervals within which relevant Ritz values are known to lie; these intervals are updated at each iteration by 'slicing the spectrum' of T_i (see Parlett (1980), Ch. 3), without ever computing the eigenvalues of any T_i ; spurious copies of eigenvalues of A can be detected when they appear. This algorithm has the disadvantage that it cannot detect the multiplicity of any multiple eigenvalue (though it can compute its value without difficulty). Indeed it is a theoretical limitation of the Lanczos method in exact arithmetic that it cannot detect multiplicities; even in the presence of rounding errors, although genuine copies of multiple eigenvalues may appear, they may do so out of sequence and hence be missed if the algorithm is terminated too soon. In the block Lanczos method, (2.33) is generalised to iterate on a block of p orthonormal vectors (usually $p \leqslant 4$); this is more expensive, and the T_i are now block tridiagonal or banded. However, convergence is improved, especially when eigenvalues are clustered, and multiple eigenvalues (with multiplicity $\leqslant p$) can be recognised as soon as they appear.

Attempts have been made to extend the Lanczos method to unsymmetric problems, but so far this work has not led to practical algorithms.

4. The singular value decomposition

The singular value decomposition (SVD) of an $m \times n$ matrix A is given by

$$A = UDV^H \tag{2.35}$$

where U is an $m \times n$ unitary matrix, V is an $n \times n$ unitary matrix and D is an $m \times n$ diagonal matrix with real nonnegative diagonal elements. The SVD always exists (Wilkinson 1977, 1978a) and since a permutation matrix is unitary, the diagonal elements may be chosen to be in order of descending magnitude down the diagonal. The case of most usual practical interest has A real and $m \geq n$, and for simplicity of discussion we shall concentrate on this case; the general case presents no essential additional complication.

The use of the SVD as an economical means of finding a least-squares solution to an overdetermined system of equations is described in Chapter 1, section 2.5. If D has the form $D = [\Sigma \quad 0]^T$, where $\Sigma = \text{diag}\,[\sigma_i]$, and $\sigma_1 \geq \sigma_2 \geq \ldots \geq \sigma_n \geq 0$ are the singular values of A, it follows that

$$A^TA = V\Sigma^2V^T , \tag{2.36}$$

which is the classical spectral factorisation of A^TA. Thus σ_j^2 is an eigenvalue of A^TA and $\underline{v}_j$, the jth column of V, is the corresponding eigenvector.

The SVD is often used in conjunction with, or as an alternative to, the QR (or QU) factorisation:

$$A = Q\begin{bmatrix} U_R \\ \\ 0 \end{bmatrix} , \tag{2.37}$$

where Q is an $m \times m$ orthogonal matrix and U_R is an $n \times n$ upper triangular matrix. Both QR factorisation and the SVD provide a means of representing A^TA and are important tools for solving problems that have been traditionally associated with A^TA, without having to take the numerically damaging step of explicitly forming A^TA. The reasons for avoiding the formation of A^TA are well documented (Golub 1965, Gentleman 1974b, Chambers 1977, Hammarling 1985) and we shall not elaborate on them here; instead we shall illustrate the use of the SVD in some statistical applications.

4.1 Numerical rank

The SVD is important because it provides the most reliable method of determining the numerical rank of a matrix and can be a great aid in analysing near linear dependence in the columns of A. If A is exactly of rank $k < n$ then it follows that

$$\sigma_{k+1} = \sigma_{k+2} = \ldots = \sigma_n = 0$$

and

$$A\underline{v}_j = \underline{0}\ , \quad j = k+1, k+2, \ldots, n\ ,$$

so that the last $(n-k)$ columns of V form an orthonormal basis for the null space of A.

Of course, in the presence of data and/or rounding errors A is unlikely to have exactly zero singular values and so it is important to understand the effect of neglecting or not neglecting small singular values. Let G be the $m \times n$ diagonal matrix

$$G = \begin{bmatrix} \text{diag}\,[g_i] \\ 0 \end{bmatrix}, \qquad g_i = \begin{cases} 0\ , & i = 1,2,\ldots,r \\ -\sigma_i\ , & i = r+1, r+2, \ldots, n\ , \end{cases}$$

so that $(D+G)$ is of rank r, and let E be the matrix $E = UGV^T$. Then the matrix $(A+E)$ is also of rank r and $||E||_2 = \sigma_{r+1}$. Hence regarding σ_{r+1} as zero corresponds to making a perturbation in A of order σ_{r+1} . Conversely if A is of rank n, but E is a matrix such that $(A+E)$ is of rank $r < n$ then it can readily be shown (Wilkinson 1978a) that

$$||E||_F^2 \geq \sum_{i=r+1}^{n} \sigma_i^2 ,$$

so that if the elements of E are small then the singular values σ_{r+1} , σ_{r+2} , ..., σ_n <u>must</u> also be small. Thus near rank-deficiency in A will be clearly exposed by the singular values.

Because the SVD can be computed by very stable numerical methods the above results also essentially hold for the computed singular values (Wilkinson 1978a, Hammarling 1985).

Of course it is not always easy to decide whether or not singular values are negligible (Golub, Klema and Stewart 1976, Stewart 1979, 1984) (for example, scaling is important), nevertheless the results of this section tell us about the effects of neglecting small singular values. Furthermore since $||A\underline{v}_j||_2 = \sigma_j$, the columns of V corresponding to small singular values give valuable information on near linear dependence in A.

4.2 Computation of the SVD

The SVD can be computed by first reducing A to bidiagonal form by a finite sequence of orthogonal transformations, akin to reducing the symmetric matrix A^TA to tridiagonal form, and then applying a variant of the QR algorithm to reduce the bidiagonal form to the diagonal matrix Σ (Golub and Kahan 1965, Golub and Reinsch 1970, Wilkinson 1977, 1978a).

We know from section 2.5 of Chapter 1 that the singular values and right singular vectors of A and U_R (see (2.37)) are identical. For many applications it is convenient to compute the QR factorisation of A, and only if U_R is not sufficiently well conditioned, to proceed to compute the SVD of U_R. We can take advantage of the upper triangular form of U_R in computing the SVD and when $m >> n$ the time taken will be dominated by the QR factorisation (Chan 1982a). This approach to the computation of the SVD allows much more flexibility because we can take advantage of the data structure in computing the QR factorisation and compute the SVD of what is usually a much smaller $n \times n$ upper triangular matrix. Generally the large matrix Q is not explictly required, but instead a vector of the form $Q^T\underline{b}$ is required and this can be obtained as the computation proceeds without computing Q.

For example, we can process A one row, or block of rows, at a time using standard updating procedures (Golub 1965, Gentleman 1974a, Gill and Murray 1977, Dongarra, Bunch, Moler and Stewart 1979, Hammarling 1985), which can be useful in real time situations, or when A cannot be held in store; when A has band or augmented band form, such as in spline applications, we can take advantage of the band form (Cox 1981) in computing the QR factorisation (we can actually take advantage through to the bidiagonal form; Golub, Luk and Overton 1981); or in more general sparse situations we can take advantage of sparsity in the QR factorisation (George and Heath 1980, Heath 1982).

As with the standard symmetric eigenvalue problem, in parallel environments Jacobi-type methods are often suitable (see for example Brent and Luk (1985)), and recently such an algorithm due to Kogbetliantz (1955) has received attention for both the SVD and for the generalised SVD discussed in section 4.5. This algorithm has the merit that it can take advantage of triangular form (Heath, Laub, Paige and Ward 1985).

Singular values, like the eigenvalues of a symmetric matrix, are always well conditioned, but singular vectors, like eigenvectors, may be sensitive to small changes in A. If high accuracy in the vectors is important they may be refined by an iterative refinement method analogous to that for the eigenvalue problem (Dongarra 1983).

4.3 The SVD and linear least-squares problems

The SVD is important in a number of multivariate statistical applications because, as mentioned earlier, it can be a great aid in the analysis of near multicollinearities in statistical data. A common application is that of linear least-squares problems (see Chapter 1, section 2.5), where we wish to determine the vector $\underline{x}$ to

$$\text{minimise } \underline{r}^T\underline{r}, \quad \text{where } \underline{b} = A\underline{x} + \underline{r} , \tag{2.38}$$

and $\underline{r}$ is the residual vector frequently assumed to come from a Normal distribution with

$$E(\underline{r}) = 0 \quad \text{and} \quad E(\underline{r}\underline{r}^T) = \sigma^2 I . \tag{2.39}$$

When A is rank-deficient, it is not always the minimal length solution that is desired (Cox 1982), but, with additional computation, other solutions can readily be obtained.

The SVD applied to the Jacobian can also be used in the solution of nonlinear least-squares problems and allows both small- and large-residual problems to be solved reliably (see Chapter 14).

When the second of (2.39) is replaced by

$$E(\underline{r}\underline{r}^T) = \sigma^2 W , \tag{2.40}$$

where W is symmetric nonnegative definite, then $\underline{x}$ is usually required to be the solution of the weighted least-squares problem

$$\text{minimise } \underline{r}^T W^{-1} \underline{r} , \quad \text{where} \quad \underline{b} = A\underline{x} + \underline{r} . \tag{2.41}$$

Of course, (2.41) reduces to (2.38) when W = I. Problem (2.41) is not defined when W is singular and solving (2.41) explicitly is numerically unstable unless W is well conditioned.

Let B be any matrix such that

$$W = BB^T \tag{2.42}$$

and let $\underline{e}$ satisfy

$$B\underline{e} = \underline{r} . \tag{2.43}$$

Then (2.41) is equivalent to the constrained least-squares problem

$$\begin{aligned} &\text{minimise} && \underline{e}^T \underline{e} \\ &\text{subject to} && \underline{b} = A\underline{x} + B\underline{e} , \end{aligned} \tag{2.44}$$

but now W is not required to be nonsingular. This is called the <u>generalised linear least-squares problem</u>. When $\underline{e}$ comes from a Normal distribution with

$$E(\underline{e}) = 0 , \quad E(\underline{e}\underline{e}^T) = \sigma^2 I \tag{2.45}$$

then Kourouklis and Paige (1981) have shown that the solution of (2.44) is still a best linear unbiased estimator. Note that (2.43) and the second of (2.45) imply (2.40). Methods for the solution of (2.44) have additionally been discussed by Paige (1978, 1979a, 1979b). Here we just indicate how the SVD may be used to solve (2.44).

Partitioning the matrix U of the SVD of A, as

$$U = [U_1 \;\; U_2] , \tag{2.46}$$

and using a similar splitting for vectors, the linear constraints in (2.44) become

$$\begin{bmatrix} U_1^T \underline{b} \\ U_2^T \underline{b} \end{bmatrix} = \begin{bmatrix} S & 0 \\ 0 & 0 \end{bmatrix} \begin{bmatrix} \underline{q} \\ \underline{h} \end{bmatrix} + \begin{bmatrix} U_1^T B\underline{e} \\ U_2^T B\underline{e} \end{bmatrix}$$

so that

$$S\underline{q} = U_1^T \underline{b} - U_1^T B\underline{e} \quad \text{and} \quad U_2^T \underline{b} = U_2^T B\underline{e} \quad . \tag{2.47}$$

We can see that $\underline{h}$ is arbitrary, the first of (2.47) determines $\underline{q}$ once $\underline{e}$ is given and $\underline{e}$ is wholly determined by the second of (2.47). An SVD of $U_2^T B$ either enables $\underline{e}$ to be found or shows that the equations are inconsistent, as would be the case for example when $U_2^T \underline{b} \neq \underline{0}$, but $U_2^T B = 0$. Further details and a discussion of inconsistency are given in Hammarling, Long and Martin (1983).

Paige (1978) has discussed the problem of updating the solution to (2.44) and the ideas there enable the Kalman filtering problem to be solved in a numerically stable manner even when the variance-covariance matrix W is ill-conditioned, without the need to add in arbitrary noise to the system.

The linear least-squares problem assumes that $\underline{b}$, the dependent variable, contains experimental error. In many applications A also contains experimental error and it may then be more appropriate to determine $\underline{x}$ to

$$\text{minimise} \quad \| [E \quad \underline{r}] \|_F^2 \quad , \text{ where} \quad \underline{r} = (A + E)\underline{x} + \underline{b} \ . \tag{2.48}$$

This is called the <u>total least-squares problem</u> (Golub and Van Loan 1980). Put

$$B = [A \quad \underline{b}] \ , \qquad G = [E \ \ -\underline{r}] \tag{2.49}$$

and suppose that B has full rank. Problem (2.48) can be written as

$$\text{minimise } \|G\|_F^2, \quad \text{where} \quad (B + G)\begin{bmatrix} \underline{x} \\ 1 \end{bmatrix} = \underline{0} \tag{2.50}$$

and so we require the minimum perturbation that makes B rank-deficient with $[\underline{x}^T \; 1]^T$ in the null space of (B+G). Let the SVD of B be

$$B = U\begin{bmatrix} \Sigma & 0 \\ 0 & \sigma_{n+1} \\ 0 & 0 \end{bmatrix}V^T \text{ and put } G = U\begin{bmatrix} 0 & 0 \\ 0 & -\sigma_{n+1} \\ 0 & 0 \end{bmatrix}V^T .$$

Then the matrix G makes (B+G) rank-deficient and is of minimum norm (see for example Golub and Van Loan (1983), corollary 2.3-3). If we partition V as

$$V = \begin{bmatrix} \tilde{V} & \underline{v} \\ \underline{r}^T & \rho \end{bmatrix}$$

then, if $\rho \neq 0$, for this choice of G it is easily verified that $\underline{x}$ is given by

$$\underline{x} = (1/\rho)\, \underline{v} \; .$$

A fuller discussion is given by Golub and Van Loan (1980, 1983) and by Van Huffel, Vandewalle and Staar (1984). The latter reference includes further discussion of the case where $\rho = 0$ and a comparison of linear least squares with total least squares.

The SVD can be used to solve many other problems in multivariate statistics (for example, Chambers (1977), Banfield (1978)) and allows proper judgement of whether or not quantities, such as principal components, are significant.

4.4 The SVD of a large sparse matrix

We mentioned in section 4.2 the possibility of using a sparse technique to obtain the QR factorisation of A, followed by the SVD of the resulting $n \times n$ matrix U_R. Unless A has a special structure we cannot usually take advantage of any sparsity that may be left in U_R. When n is large, or only some of the singular values and singular vectors are required, then such an approach may be unacceptable. Instead we may seek an iterative method to solve the problem.

Let the matrix U of the SVD of A be partitioned as in (2.46) with U_1 an $m \times n$ matrix, and define B and R as

$$B = \begin{bmatrix} 0 & A \\ A^T & 0 \end{bmatrix}, \qquad R = \frac{1}{\sqrt{2}} \begin{bmatrix} U_1 & U_1 \\ V & -V \end{bmatrix}. \tag{2.51}$$

The $(m+n) \times 2n$ matrix R satifies $R^T R = I$ and

$$BR = R \begin{bmatrix} \Sigma & 0 \\ 0 & -\Sigma \end{bmatrix}. \tag{2.52}$$

Thus B has as eigenvalues (m−n) zeros together with the 2n values $\pm\sigma_1$, $\pm\sigma_2$, ..., $\pm\sigma_n$. The eigenvector of B corresponding to the singular value σ_j is $[\underline{u}_j^T \;\; \underline{v}_j^T]^T$ and that corresponding to $-\sigma_j$ is $[\underline{u}_j^T \; -\underline{v}_j^T]$. Hence we can apply an eigenvalue routine designed for large sparse symmetric eigenvalue problems to B in order to find the singular values of A. In practice, only the method of Lanczos seems to have been considered.

If we apply the basic Lanczos scheme to B then

$$\delta_{j+1}\, \underline{w}_{j+1} = B\underline{w}_j - \gamma_j \underline{w}_j - \delta_j \underline{w}_{j-1}, \qquad \gamma_j = \underline{w}_j^T B \underline{w}_j \tag{2.53}$$

where $\underline{w}_0 = \underline{0}$ and δ_{j+1} is chosen so that $\underline{w}_{j+1}^T \underline{w}_{j+1} = 1$. If we choose $\underline{w}_1$ to have the form

$$\underline{w}_1 = \begin{bmatrix} \tilde{\underline{w}}_1 \\ \underline{0} \end{bmatrix}$$

then it is readily seen that $\gamma_j = 0$, $j = 1,2,\ldots$ and $\underline{w}_j$ has the form

$$\underline{w}_j = \begin{bmatrix} \underline{0} \\ \tilde{\underline{w}}_j \end{bmatrix}, \quad j \text{ even} ; \qquad \underline{w}_j = \begin{bmatrix} \tilde{\underline{w}}_j \\ \underline{0} \end{bmatrix}, \quad j \text{ odd}.$$

Thus, let us put

$$\underline{y}_j = \tilde{\underline{w}}_{2j-1}, \quad \underline{z}_j = \tilde{\underline{w}}_{2j}, \quad \beta_j = \delta_{2j-1}, \quad \alpha_j = \delta_{2j}.$$

Then the Lanczos scheme (2.53) becomes

$$\left.\begin{aligned} \alpha_j\underline{z}_j &= A^T\underline{y}_j - \beta_j\underline{z}_{j-1} \\ \beta_{j+1}\underline{y}_{j+1} &= A\underline{z}_j - \alpha_j\underline{y}_j \end{aligned}\right\} \qquad (2.54)$$

where $\underline{z}_0 = \underline{0}$ and α_j and β_j are chosen so that $\underline{z}_j^T\,\underline{z}_j = \underline{y}_j^T\,\underline{y}_j = 1$. Since the γ_j are all zero the original matrix associated with the kth step of this Lanczos process is

$$T = \begin{bmatrix} 0 & \alpha_1 & 0 & \ldots & 0 & 0 \\ \alpha_1 & 0 & \beta_2 & \ldots & 0 & 0 \\ 0 & \beta_2 & 0 & \ldots & 0 & 0 \\ . & . & . & \ldots & . & . \\ 0 & 0 & 0 & \ldots & 0 & \alpha_k \\ 0 & 0 & 0 & \ldots & \alpha_k & 0 \end{bmatrix}. \qquad (2.55)$$

Introducing the permutation matrix $P = [\underline{e}_1\ \underline{e}_{k+1}\ \underline{e}_2\ \underline{e}_{k+2}\ \ldots\ \underline{e}_k\ \underline{e}_{2k}]$ we find that

$$PTP^T = \begin{bmatrix} 0 & S^T \\ S & 0 \end{bmatrix}, \quad \text{where } S = \begin{bmatrix} \alpha_1 & \beta_2 & 0 & \dots & 0 \\ 0 & \alpha_2 & \beta_3 & \dots & 0 \\ 0 & 0 & \alpha_3 & \dots & 0 \\ \cdot & \cdot & \cdot & \dots & \cdot \\ 0 & 0 & 0 & \dots & \alpha_k \end{bmatrix}. \quad (2.56)$$

Hence, we need only compute the singular values of S, by the usual bidiagonal QR algorithm, rather than compute the eigenvalues of T. This Lanczos process is described in Paige (1974) and a block Lanczos version is described in Golub, Luk and Overton (1981).

This Lanczos process is also the basis of algorithm LSQR, for solving large sparse linear least-squares problems (see Chapter 1, section 3.4).

4.5 The generalised singular value decomposition

Here we briefly consider a generalisation of the SVD that applies to the matrix pair (A, B). Let A be an $m \times n$ matrix and B a $p \times n$ matrix. Then the generalised singular value decomposition (GSVD) is given by

$$A = U_A D_A [W^H S \quad 0] V^H, \qquad B = U_B D_B [W^H S \quad 0] V^H, \qquad (2.57)$$

where U_A, U_B, W and V are unitary, S is a real nonsingular diagonal matrix with positive diagonal elements and D_A and D_B are real diagonal matrices of the form

$$D_A = \begin{bmatrix} I & 0 & 0 \\ 0 & \Sigma_A & 0 \\ 0 & 0 & 0 \end{bmatrix}, \qquad D_B = \begin{bmatrix} 0 & 0 & 0 \\ 0 & \Sigma_B & 0 \\ 0 & 0 & I \end{bmatrix}$$

with $\Sigma_A = \text{diag}[\alpha_i]$, $\Sigma_B = \text{diag}[\beta_i]$ satisfying

$$\alpha_i^2 + \beta_i^2 = 1, \quad \alpha_i \geqslant 0, \quad \beta_i > 0.$$

The α_i and β_i can be chosen so that the α_i are in descending order and the β_i in ascending order. The pairs (α_i , β_i) are called the generalised singular values (Paige and Saunders 1981). It is straightforward to show that the diagonal elements of S are the non-zero singular values of the matrix $[A^H \; B^H]$.

When A and B are real, $p \geq n$ and B has full rank, (2.57) specialises to

$$A = U_A D_A X^{-1} , \quad B = U_B D_B X^{-1} , \quad \text{where } X = VS^{-1}W \tag{2.58}$$

with

$$D_A = \begin{bmatrix} \Sigma_A & 0 \\ 0 & 0 \end{bmatrix} \quad \text{and} \quad D_B = \begin{bmatrix} \Sigma_B & 0 \\ 0 & I \end{bmatrix} .$$

Furthermore

$$A^T A = X^{-T} \begin{bmatrix} \Sigma_A^2 & 0 \\ 0 & 0 \end{bmatrix} X^{-1} , \quad B^T B = X^{-T} \begin{bmatrix} \Sigma_B^2 & 0 \\ 0 & I \end{bmatrix} X^{-1}$$

so that X is a congruence matrix that simultaneously diagonalises $A^T A$ and $B^T B$. Hence the columns of X are the eigenvectors of the generalised eigenvalue problem $A^T A\underline{x} = \gamma B^T B\underline{x}$ and the (α_i^2 / β_i^2) are the non-zero eigenvalues (Van Loan 1976).

Analogously to the SVD, the GSVD provides a means of representing the pair $(A^T A, B^T B)$ and allows us to solve problems associated with the pair without having to take the numerically damaging step of explicitly forming $A^T A$ and $B^T B$. The GSVD was expressed in the form (2.57) rather than (2.58) by Paige and Saunders (1981) in order to encourage its computation by numerically stable algorithms. Efficient algorithms for computing the GSVD are beginning to emerge (Stewart 1983, Paige 1985b, Van Loan 1984).

The GSVD is also a useful tool in the analysis of linearly constrained and generalised linear least-squares problems (Van Loan 1983, Golub and Van Loan 1983, Paige 1985a).

5. Generalised eigenvalue problems

In this section we concentrate on the matrix pair (A, B) and problems associated with the <u>matrix pencil</u>

$$P(\lambda) = A - \lambda B . \tag{2.59}$$

We concentrate in particular on the case where $P(\lambda)$ is square and consider the generalised eigenvalue problem

$$A\underline{x} = \lambda B\underline{x} . \tag{2.60}$$

When B is nonsingular (2.60) is equivalent to the standard problem

$$(B^{-1}A)\,\underline{x} = \lambda\underline{x} , \tag{2.61}$$

but, unless B is well conditioned, the solution of (2.60) via (2.61) is numerically unstable. We prefer to perform well-conditioned (ideally unitary) equivalence transformations on A and B. When B is singular we have added features which can be illustrated by the three symmetric pencils

$$P_1(\lambda) = \begin{bmatrix} 1 & 1 \\ 1 & 1 \end{bmatrix} - \lambda \begin{bmatrix} 1 & 2 \\ 2 & 4 \end{bmatrix}, \quad P_2(\lambda) = \begin{bmatrix} 1 & 1 \\ 1 & 0 \end{bmatrix} - \lambda \begin{bmatrix} 1 & 2 \\ 2 & 4 \end{bmatrix},$$

$$P_3(\lambda) = \begin{bmatrix} \frac{1}{2} & 1 \\ 1 & 2 \end{bmatrix} - \lambda \begin{bmatrix} 1 & 2 \\ 2 & 4 \end{bmatrix},$$

for which $\det(P_1(\lambda)) = -\lambda$, $\det(P_2(\lambda)) \equiv -1$, $\det(P_3(\lambda)) \equiv 0$. Thus $P_1(\lambda)$ has only one finite eigenvalue at zero and $P_2(\lambda)$ has no finite eigenvalues. By considering the pencils $(\mu A - B)$ we see that $P_1(\lambda)$ and $P_2(\lambda)$ have respectively one and two infinite eigenvalues. The case $P_3(\lambda)$ illustrates a <u>singular pencil</u> for which

$$\det(A - \lambda B) \equiv 0 .$$

Non-square pencils are also said to be singular. If a pencil is not singular it is said to be regular and in the next three sections we consider such pencils.

5.1 Dense generalised symmetric problems

When A and B are symmetric and B is positive definite and well conditioned then the standard approach is to factorise B by the Cholesky factorisation:

$$B = U^T U , \tag{2.62}$$

where U is upper triangular, and then transform (2.60) to the standard symmetric problem

$$C\underline{y} = \lambda \underline{y} , \tag{2.63}$$

where

$$C = U^{-T} A U^{-1} , \quad \underline{y} = U\underline{x} .$$

The factorisation (2.62) can be replaced by the spectral factorisation and in this case it can be advantageous to arrange the eigenvalues of B in ascending order on the diagonal (see for example, Golub and Van Loan (1983), section 8.6). These approaches are akin to formulating the problem as in (2.61), except that symmetry is maintained, and so are not stable when B is ill-conditioned.

There are currently no algorithms available for symmetric problems that both use orthogonal transformations and maintain symmetry and so, if we wish to use such tranformations, we must forego symmetry and use the QZ algorithm (see section 5.3). If we give up orthogonality then algorithms do exist to simultaneously diagonalise A and B by means of congruence transformations which rely only on positive semi-definiteness of B.

Jacobi's method may be generalised to the pair (A, B) (Bathe and Wilson 1976, Parlett 1980), and recently Bunse-Gerstner (1984) has described a QR-like algorithm in which efforts are made to control the condition of the congruence transformations; the pair is first reduced to (T, D), with T symmetric tridiagonal and D diagonal and this form is maintained by the iterative procedure.

There currently appear to be no effective algorithms that maintain symmetry for the case where A and B are indefinite.

5.2 Sparse generalised symmetric problems

The approach to the large sparse generalised eigenvalue problem depends very much upon whether or not one is prepared to perform and retain a factorisation and whether or not B is well conditioned.

When B is positive definite and well conditioned, and we are prepared to perform a sparse Cholesky factorisation (see for example, George and Liu (1981)) then we can use the form (2.63) and use a sparse technique for the standard symmetric problem. Of course, we do not form C explicitly, but form products of the form $C\underline{v}$ by first solving $U\underline{w} = \underline{v}$ and then solving $U^T\underline{z} = A\underline{w}$ to give $\underline{z} = C\underline{v}$. For example, for the Lanczos process we have

$$\beta_{i+1}\underline{q}_{i+1} = C\underline{q}_i - \alpha_i\underline{q}_i - \beta_i\underline{q}_{i-1}$$

and we need to form $C\underline{q}_i$ at each step. This approach was first discussed by Golub, Underwood and Wilkinson (1972) in the context of a band symmetric pencil. We note that band symmetric pencils may also be solved by a method due to Crawford (1973), in which the problem is reduced to a standard symmetric problem, with the band structure preserved. (See also Wilkinson (1977).)

Alternatively, we can use the form (2.61) and in this case, when products of the form $\underline{z} = B^{-1}A\underline{v}$ are required, we obtain $\underline{z}$ by solving $B\underline{z} = A\underline{v}$. For example, after multiplying through by B, the Lanczos process corresponding to this form is

$$\tilde{\beta}_{i+1}B\tilde{\underline{q}}_{i+1} = A\tilde{\underline{q}}_i - \tilde{\alpha}_i B\tilde{\underline{q}}_i - \tilde{\beta}_i B\tilde{\underline{q}}_{i-1} .$$

In many practical sparse problems it is the small, rather than the large, eigenvalues that are required, and in this case we reverse the roles of A and B. Interior eigenvalues may be obtained by expressing (2.60), for some appropriate μ, as

$$B\underline{x} = (\lambda - \mu)^{-1}\ (A - \mu B)\underline{x} \qquad (2.64)$$

and now $(A - \mu B)$ plays the role of B. If B is well conditioned, but not positive definite, then we can still use this approach provided we replace the Cholesky factorisation by some other appropriate sparse factorisation. We can often safely use the $U^T DU$ factorisation, where D is diagonal and U is unit upper triangular, but it is then important to monitor the stability of the factorisation (Parlett 1980, section 3.2; Ericsson 1983a, b). An alternative formulation to (2.64) that is commonly used in practice is to express (2.60) as

$$B(a - \mu B)^{-1} B\underline{x} = (\lambda - \mu)^{-1} B\underline{x} . \tag{2.65}$$

If we are not prepared to factorise then we must either use an iterative method to solve the linear equations that arise, or use a method that does not give rise to linear equations (Scott 1981).

Alternative methods, particularly when B is ill-conditioned, tend to be based upon Rayleigh quotient iteration (Golub and Van Loan 1983, equation (8.6-3)) and/or inverse iteration (Peters and Wilkinson 1979). Szyld (1983) describes an effective algorithm that switches between Rayleigh quotient and inverse iteration in order to find an eigenvalue in a given interval, together with the corresponding eigenvector. The resulting indefinite equations are solved using algorithm SYMMLQ with preconditioning (Paige and Saunders 1975).

Some possibilities for indefinite symmetric pencils are discussed in Ericsson (1983c) and a bibliographical tour of the large sparse generalised eigenvalue problem is given in Stewart (1976c).

5.3 The dense generalised unsymmetric problem

The Schur factorisation of a matrix A can be generalised to the Stewart factorisation (Stewart 1972) for the matrix pair (A, B) as

$$A = QUZ^H , \quad B = QRZ^H , \tag{2.66}$$

where Q and Z are unitary and U and R are upper triangular matrices. Analogously to the Schur factorisation, when A and B are real, Q and Z can be chosen to be orthogonal if we allow U and R to be block upper triangular, with 1×1 and 2×2 blocks; the complex conjugate eigenvalues correspond to the 2 × 2 blocks. The factorisation can be obtained, using unitary transformations throughout, by the QZ algorithm of Moler and Stewart (1973). (See also Ward (1975), Kaufman (1977).) The problem of balancing (2.60), with the aim of improving the accuracy of the eigenvalues computed by the QZ algorithm, is discussed in Ward (1981).

As with the Schur factorisation the eigenvalues can be chosen to be in any order in (2.66) (Van Dooren 1981a). In particular, this enables orthonormal bases for deflating subspaces to be computed by stable methods. If we partition Q, U, R and Z conformably as

$$Q = [Q_1 \quad Q_2], \quad U = \begin{bmatrix} U_{11} & U_{12} \\ 0 & U_{22} \end{bmatrix}, \quad R = \begin{bmatrix} R_{11} & R_{12} \\ 0 & R_{22} \end{bmatrix},$$

$$Z = [Z_1 \quad Z_2]$$

then

$$AZ_1 = Q_1U_{11}, \qquad BZ_1 = Q_1R_{11}$$

and the columns of Z_1 form a deflating subspace for the pair (A, B) corresponding to the eigenvalues defined by the pencil $(U_{11} - \lambda R_{11})$.

5.4 Singular pencils

When the pencil $(A - \lambda B)$ is singular, then in exact arithmetic (2.66) will have at least one diagonal pair (u_{ii}, r_{ii}) such that $u_{ii} = r_{ii} = 0$. However, in the presence of rounding errors we are unlikely to obtain an exactly zero pair and so great caution must be exercised before assuming that $\lambda_i = (u_{ii} / r_{ii})$ has any meaning. Indeed Moler and Stewart (1973) recommended that the pairs (u_{ii}, r_{ii}), rather than the ratios λ_i, be output by implementations of their algorithm. It should also be appreciated that in the presence of a zero pair, other apparently reasonable pairs may also yield meaningless ratios (Wilkinson 1979).

The appropriate generalisation of the Jordan canonical form for a matrix pencil is the Kronecker canonical form (Gantmacher 1959), and Wilkinson (1978b) and Van Dooren (1979) have shown how to expose the Kronecker structure of a pencil by a numerically stable method. In essence the pencil is factorised as

$$P(\lambda) = Q \begin{bmatrix} A_r - \lambda B_r & X & X & X \\ 0 & A_f - \lambda B_f & X & X \\ 0 & 0 & A_i - \lambda B_i & X \\ 0 & 0 & 0 & A_c - \lambda B_c \end{bmatrix} Z^H , \quad (2.67)$$

where Q and Z are unitary and X denotes a non-null matrix; the pencil $(A_f - \lambda B_f)$ contains only the finite eigenvalues of $P(\lambda)$, $(A_i - \lambda B_i)$ contains only the infinite eigenvalues of $P(\lambda)$ and the non-square pencils $(A_r - \lambda B_r)$ and $(A_c - \lambda B_c)$ contain only the Kronecker row and column structure respectively. The two middle pencils may be interchanged, as may the two corner pencils. The algorithms utilise the SVD in order to make reliable decisions about rank.

Following (2.67) we can safely apply the QZ algorithm to the regular pencil $(A_f - \lambda B_f)$. Both Van Dooren and Wilkinson recommend the use of an algorithm to extract the singular part of a pencil prior to the use of the QZ algorithm. For further discussion and applications see Van Dooren (1981b, 1983). The computation of the Kronecker canonical form itself is discussed in Kågström (1983).

5.5 Other generalised problems

Here we give some references to the somewhat sparse literature on problems of the form $P(\lambda)\underline{x} = \underline{0}$, where $P(\lambda)$ is a nonlinear polynomial in λ, or a general nonlinear function of λ. In practice such problems are often associated with vibration analysis of some form or another and are often symmetric. The most common such problem is the quadratic problem which arises in damped vibration; this can frequently be converted to a generalised symmetric problem $A\underline{x} = \lambda B\underline{x}$, with B positive definite (see for example Lancaster (1966; 1977, section 3)); but algorithms also exist to deal with the quadratic problem directly (Scott 1980, Scott and Ward 1982).

The above Lancaster references also treat the more general nonlinear problems and important additional references include Wilkinson (1965), Peters and Wilkinson (1970), Kublanovskya (1970), Wittrick and Williams (1971) and Ruhe (1973). The SVD of $P(\lambda_i)$ is a useful tool in verifying whether or not λ_i is an eigenvalue of $P(\lambda)$ and in obtaining the eigenvectors corresponding to λ_i. (See for example Ferris, Hammarling, Martin and Warham (1983), appendix 3.)

6. Software

(All the software referred to here is in Fortran.)

For dense eigenvalue problems, including generalised problems, comprehensive sets of Fortran subroutines can be found in EISPACK (Smith, Boyle, Dongarra, Garbow, Ikebe, Klema and Moler 1976; Garbow, Boyle, Dongarra and Moler 1977) and in the NAG Library and IMSL. A third edition of the EISPACK software (Dongarra and Moler 1983) includes minor improvements in portability and efficiency, especially on vector-processing machines; additional improvements for vector-processing machines have been discussed by Dongarra, Kaufman and Hammarling (1985), and by Kaufman (1984) for banded problems and some of these improvements have been incorporated into the NAG Library (Du Croz 1983). Regrettably and inconveniently there are no routines in EISPACK, nor in the NAG Library or IMSL, which return the Schur factorisation of an unsymmetric matrix to the user, although only a trivial modification of the source-text is required to suppress the computation of eigenvectors: such an adaptation of the EISPACK routine HQR2 has been published by Stewart (1976a), among others.

Software for computing the SVD is included in both EISPACK and LINPACK (Dongarra et al. 1979) as well as in the NAG Library. The NAG routines compute the SVD via the QR factorisation, as does the algorithm of Chan (1982b); the NAG routine F02WDF explicitly allows the user to stop at the QR factorisation if U_R is not too ill-conditioned, and this is the basis of the NAG routine F04JGF for solving linear least-squares problems. Algorithm LSQR for sparse least-squares problems is available in the ACM algorithms (Paige and Saunders 1982b) and as the NAG routine F04QAF.

Many ACM algorithms provide extensions to the capabilities of EISPACK to re-order the eigenvalues in the Schur factorisation (Stewart 1976a) or in the Stewart factorisation of the generalised problem (van Dooren 1982); to solve the Sylvester and Lyapunov equations (Bartels and Stewart 1972); to compute condition numbers of eigenvalues (Chan, Feldman and Parlett 1977); to perform iterative improvement of eigenvalues and eigenvectors (Dongarra 1982); and for the complex QZ algorithm (Garbow 1978; also in the NAG Library, F02GJF). Crawford's method for banded symmetric generalised problems is in the NAG Library (F02FHF). A routine to compute the Jordan normal form is given by Kågström and Ruhe (1980) and a routine which determines the Kronecker structure of a matrix pencil is in the NPL linear algebra subroutine library (Hammarling, Kenward, Symm and Wilkinson 1981).

For sparse problems, there is less software available. The method of simultaneous iteration for standard symmetric problems is implemented in various Fortran descendants of Rutishauser's algorithm RITZIT: in the Harwell Library (EA12), in the ACM algorithms (Nikolai 1979), and in the NAG Library (F02FJF); the last two also cater for generalised problems. The Lanczos method of Parlett and Reid (1981) is in the Harwell Library (EA15); Scott (1982) has published a block Lanczos code with selective orthogonalisation, and a software package for the spectral transformation Lanczos method is described in Ericsson and Ruhe (1982). Stewart and Jennings (1981) give a routine for unsymmetric sparse problems based on a 'lopsided' simultaneous iteration.

CHAPTER 3

ORDINARY DIFFERENTIAL EQUATIONS: INITIAL-VALUE PROBLEMS

G. Hall and J. Williams

1. Introduction

In this chapter on the initial-value problem, we discuss some of the significant developments in the area of algorithms. It has of course been necessary to restrict the topics and in selecting our material it has not been possible to include such areas as systems of coupled differential and algebraic equations. We restrict ourselves to algorithms for treating the standard problem

$$\left.\begin{aligned} &\underline{y}' = \underline{f}(x, \underline{y}) \ , \quad a \leq x \leq b \ , \quad \underline{y} \in R^m \ , \\ &\underline{y}(a) = \underline{y}_0 \ . \end{aligned}\right\} \qquad (3.1)$$

This topic is treated fully up to 1975 in Hall and Watt (1976), further in Chapters 7 and 8 in Baker and Phillips (1981), and Hall and Williams (1982). We begin with a broad introduction to some standard material (see for example Lambert (1973) for full discussion and definitions), and in particular we identify those approaches and classes of formulae which form the basis of the most successful software for (3.1).

The most widely used software is based on the idea of a discrete variable method. These methods proceed step by step and advance the numerical solution from x_n to $x_{n+1} = x_n + h_n$, $x_0 = a$, by computing $\underline{y}_{n+1}$ as an approximation to the true solution $\underline{y}(x_{n+1})$. They attempt to cover the range [a,b] in the most efficient way possible whilst satisfying the user's accuracy requirements and possibly other conditions, e.g. printing out approximations at specified points. This often involves at some stage changing both the stepsize h_n and the underlying formula being used. The combination of <u>variable step</u> and <u>variable formula</u> has led to some of the most efficient and sophisticated software.

2. Runge-Kutta and linear multistep methods

The step-by-step methods can be conveniently classified in terms of the amount of information they require to pass from x_n to x_{n+1}. The one-step Runge-Kutta (RK) methods only require y_n and are therefore self-starting. For simplicity consider the scalar form of (3.1); then

$$y(x) = y(x_n) + \int_{x_n}^{x} f(t, y(t))\, dt , \tag{3.2}$$

and with $x = x_{n+1} = x_n + h$ a quadrature rule replacement yields the formula

$$y_{n+1} = y_n + h \sum_{i=1}^{s} b_i\, f(x_n + c_i h, y_{ni}) \tag{3.3a}$$

where $y_{ni} \approx y(x_n + c_i h)$. These 'intermediate' approximations can be generated similarly from (3.2) with $x = x_n + c_i h$ by quadrature formulae using the same abscissae, to give

$$y_{ni} = y_n + h \sum_{j=1}^{s} a_{ij}\, f(x_n + c_i h, y_{nj}) \quad , \quad i = 1,2,\ldots,s \; . \tag{3.3b}$$

Equations (3.3) define the general s-stage RK formula for determining y_{n+1} from y_n ; it is completely specified by the abscissae $\{c_j\}$ and weights $\{b_i\}$ and $\{a_{ij}\}$, conveniently represented as follows:

$$\begin{array}{c|cccc} c_1 & a_{11} & a_{12} & \ldots & a_{1s} \\ c_2 & a_{21} & a_{22} & \ldots & a_{2s} \\ . & . & & & . \\ . & . & & & . \\ . & . & & & . \\ c_s & a_{s1} & a_{s2} & & a_{ss} \\ \hline & b_1 & b_2 & \ldots & b_s \end{array} \quad \equiv \quad \begin{array}{c|c} \underline{c} & A \\ \hline & \underline{b}^T \end{array}$$

Changing the stepsize h presents no difficulty. Some workers write (3.3a) as

$$y_{n+1} = y_n + h \sum_{i=1}^{s} b_i k_i$$

where now the $\{k_i\}$ are regarded as 'intermediate' approximations to the values of the derivative y'; so (3.3b) is replaced by

$$k_i = f(x_n + c_i h, \; y_n + h \sum_{j=1}^{s} a_{ij} k_j) \quad , \quad i = 1,2,\ldots,s.$$

In either form, if A is strictly lower triangular ($a_{ij} = 0$, $i \leq j$) the formula is explicit. The same formulae in vector form apply to the system (3.1). Implicit methods are significantly more expensive to use, since in general (3.3b) is a system of sm nonlinear equations for the s vectors $\{\underline{k}_i\}$. Their use is restricted to stiff systems, discussed in section 7.

To derive useful linear multistep methods (LMM's) suppose that the k back values y_n , y_{n-1} ,..., y_{n-k+1} have been computed; then from (3.2) with $x = x_{n+1}$, the next approximation y_{n+1} can be obtained from

$$y_{n+1} = y_n + \int_{x_n}^{x_{n+1}} P(t)\, dt$$

where P(t) is a polynomial interpolating f-values. If P(t) is of degree k-1 and satisfies

$$P(t_{n-i}) = f_{n-i} \equiv f(x_{n-i} \, , \, y_{n-i}) \quad , \quad i = 0,1,2,..,k-1 \quad ,$$

the resulting formula is the <u>explicit</u> Adams-Bashforth (AB) method. If in addition P(t) satisfies

$$P(t_{n+1}) = f_{n+1} \equiv f(x_{n+1} \, , \, y_{n+1}) \, ,$$

and so is of degree k, then the resulting formula is implicit since y_{n+1} is not yet available; this is the Adams-Moulton (AM) method.

The actual form of the Adams formulae depends on the way the interpolating polynomial P(t) is represented. For example, in terms of f-values, the AB method is

$$y_{n+1} = y_n + h \sum_{i=0}^{k} \beta_{ki} f_{n-i} \qquad (3.4)$$

if the back values are assumed to have equal spacing h. This is the <u>fixed-coefficient formula</u>, while unequal spacing yields the <u>variable-coefficient formula</u>. Changing the accuracy of the formula or stepsize is not straightforward and for this reason a divided difference or Nordsieck form is preferred. To obtain the Nordsieck form, regard (3.4) as using the saved information

$$\underline{Y}_n = [y_n \;\; hf_n \;\; hf_{n-1} \;\; \ldots \;\; hf_{n-k}]^T .$$

These values uniquely determine a polynomial $\Pi_{k+1}(x)$ of degree k+1 satisfying

$$\Pi_{k+1}(x_n) = y_n \;\; , \;\; \Pi'_{k+1}(x_{n-i}) = f_{n-i} \;\; , \quad i = 0,1,\ldots,k.$$

This polynomial can be <u>represented</u> in the normalised Taylor form

$$\Pi_{k+1}(x) = \sum_{i=0}^{k+1} \gamma_{ni} \left(\frac{x - x_n}{h} \right)^i$$

and now the saved information $\{\gamma_{ni}\}$ is interpreted as

$$\underline{\gamma}_n = [y_n \;\; hf_n \;\; \frac{h^2}{2} f'_n \;\; \ldots \;\; \frac{h^{k+1}}{(k+1)!} f_n^k \;]^T .$$

Formula (3.4) can be written via a linear transformation of $\underline{Y}_n$ in terms of $\underline{\gamma}_n$. If now $h \to \alpha h$, then the saved information can be adjusted to the next stepsize via

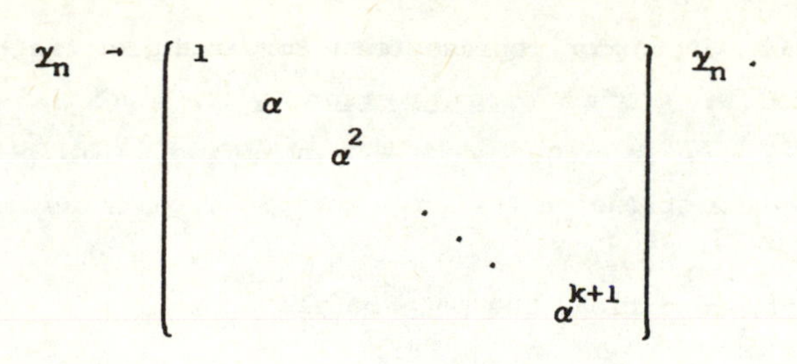

Being multistep formulae, the Adams formulae all require k starting values.

Another most important class of LMM can be derived from (3.1) directly. Taking the scalar form, let P(t) now satisfy the interpolatory conditions

$$P(t_{n+1-i}) = y_{n+1-i}\ , \quad i = 0,1,\ldots,k\ .$$

Then using (3.1), the backward differentiation formulae (BDF) are defined by

$$\left[\frac{d}{dt}P(t)\right]_{t=x_{n+1}} = f_{n+1}\ .$$

In their simplest form they become

$$\sum_{j=0}^{k} \alpha_j y_{n+1-j} = h\beta_k f_{n+1}\ ,$$

but they are usually implemented in some other form, for example, Nordsieck.

3. Accuracy and stability

As a first measure of the accuracy of a formula we define the <u>local truncation error</u> $\underline{\tau}_n$; this is the amount by which the true solution $\underline{y}(x)$ of (3.1) fails to satisfy the formula. For the general k-step LMM (constant stepsize form)

$$\sum_{i=0}^{k} \alpha_i \underline{y}_{n+i} = h \sum_{i=0}^{k} \beta_i \underline{f}_{n+i}\ , \tag{3.5}$$

$$\underline{\tau}_{n+k} = \sum_{i=0}^{k} \alpha_i\ \underline{y}(x_{n+i}) - h \sum_{i=0}^{k} \beta_i\ \underline{y}'(x_{n+i})\ .$$

The formula (3.5) is of <u>order p</u> if when applied to <u>sufficiently smooth</u> solutions of (3.1) the <u>global error</u> $\underline{y}_n - \underline{y}(x_n)$ satisfies $||\underline{y}_n - \underline{y}(x_n)|| = O(h^p)$ as $h \to 0$, $x_n = a + nh$ fixed. This pth order convergence requires restrictions on the $\{\alpha_i\}$ and $\{\beta_i\}$, given as follows.

(i) Zero stability condition. The polynomials

$$\rho(z) = \sum_{i=0}^{k} \alpha_i z^i \quad , \quad \sigma(z) = \sum_{i=0}^{k} \beta_i z^i$$

satisfy $\rho(1) = 0$, $\rho'(1) = \sigma(1)$ and no zero of $\rho(z)$ has modulus exceeding one; those of modulus one are simple.

(ii) $||\underline{T}_{n+k}|| = O(h^{p+1})$ as $h \to 0$.

The above AB and AM formulae are of orders k and k+1 respectively. For the general RK formula (k_i-form),

$$\underline{T}_{n+1} = \underline{y}(x_{n+1}) - \underline{y}(x_n) - h \sum_{i=1}^{s} b_i \bar{\underline{k}}_i \quad ,$$

where $\bar{\underline{k}}_i = \underline{f}(x_n + c_i h,\ \underline{y}(x_n) + h \sum_{j=1}^{s} a_{ij} \bar{\underline{k}}_j)$; it can be shown that the formula is of order p when $||\underline{T}_{n+1}|| = O(h^{p+1})$. This order condition places nonlinear constraints on the $\{b_i\}$, $\{c_i\}$ and $\{a_{ij}\}$.

For a general formula advancing the numerical solution from x_n to x_{n+1} , we may also regard the value of $\underline{y}_{n+1}$ as an approximation to $\underline{u}(x_{n+1})$, the <u>local solution</u> at x_{n+1} , defined by

$$\underline{u}' = \underline{f}(x, \underline{u}) \quad , \quad x \geqslant x_n$$

$$\underline{u}(x_n) = \underline{y}_n \ .$$

The error $\underline{y}_{n+1} - \underline{u}(x_{n+1})$ made in this one step is called the <u>local error</u>. Many software packages actually control the size of this quantity as a means of controlling the overall global error.

Although convergence is necessary for a formula to be of practical value, it is far from sufficient. What is also of vital importance is the behaviour of the approximations $\{\underline{y}_n\}$ for stepsizes not tending to zero. An acceptable situation is that, in the presence of local truncation and rounding errors, the computed $\underline{y}_n$-values have the same qualitative behaviour as the true solution $\underline{y}(x)$; that is, errors do not grow and swamp the approximations. The actual accuracy of the $\{\underline{y}_n\}$ is determined by the stepsizes. In general this analysis can be carried out only for a very small class of differential equations, the simplest of which is

$$y' = \lambda y \quad , \quad x \geq 0 \quad , \quad \mathrm{Re}(\lambda) \leq 0 \; . \tag{3.6}$$

All solutions of (3.6) are bounded and so we demand that our formula also yields bounded $\{y_n\}$ when applied to this problem. The analysis is simplest for constant stepsize h. A formula is absolutely stable for a given h and λ if it yields bounded approximations when applied to (3.6). The set of values of $h\lambda$ forms a region S in the left half of the complex plane; S is called the region of absolute stability.

Applying the general s-stage RK formula (3.3) to (3.6) gives

$$y_{n+1} = R(h\lambda)y_n \quad , \quad R(h\lambda) = 1 + \lambda\underline{b}^T(I - h\lambda A)^{-1}\underline{e} \quad ,$$

where $\underline{e} = [1\ 1\ \ldots\ 1]^T$ has s components. $R(h\lambda)$ is in general a rational function of numerator and denominator degrees s (which can be regarded as an approximation to $e^{h\lambda}$). For bounded $\{y_n\}$ we require $|R(h\lambda)| \leq 1$; this condition defines the stability region S. If the formula is explicit, $R(h\lambda)$ reduces to a polynomial.

When the general k-step LMM (3.5) is applied to (3.6) the $\{y_n\}$ satisfy

$$\rho(E)y_n - h\lambda\sigma(E)y_n = 0$$

where E denotes the shift operator. Therefore for bounded solutions we require that the k zeros $\{z_i(h\lambda)\}$ of the stability polynomial $\rho(z) - h\lambda\sigma(z)$ satisfy $|z_i(h\lambda)| \leq 1$ and those of modulus one are simple; this defines the stability region S.

For Runge-Kutta and linear multistep methods the test equation can be extended to a special form of (3.1),

$$\underline{y}' = A\,\underline{y} + \underline{\varphi}(x) \quad , \quad \underline{y} \in R^m \quad , \tag{3.7}$$

where A is a constant $m \times m$ matrix which has distinct eigenvalues $\{\lambda_i\}$ with non-positive real parts (bounded solutions). The application of RK and LM methods leads to difference equations for $\{\underline{y}_n\}$, which can be uncoupled since A is diagonalizable, and the situation reduces to that of test equation (3.6) where each λ_i , $i = 1,2,\ldots,m$, takes the role of λ. Hence for the classical fourth-order RK formula, absolute stability requires $h\lambda_i \in S$, $i = 1,2,\ldots,m$ (see Fig. 1).

The further extension to general nonlinear problems is not possible. However, some insight is gained by a local linearisation of (3.1) about a known solution $\underline{z}(x)$, yielding

$$\underline{y}' = J(x, \underline{z})\, \underline{y} + \underline{\Psi}(x) ,$$

where the Jacobian matrix $J = \left[\dfrac{\partial f_i}{\partial y_j}\right]$ is evaluated on the solution curve $\underline{z}(x)$. This is the variable-coefficient form of (3.7), so if we are prepared to accept a 'frozen Jacobian', we have a 'local stability' condition of the form $h\lambda_i(x) \in S$ where $\{\lambda_i(x)\}$ are the eigenvalues of $J(x, \underline{z})$; we are therefore essentially back to (3.6).

Stability regions have been useful in helping to select and understand the behaviour of numerical methods. It is remarkable how formulae selected on this basis via the test equation (3.6) have proved so successful when applied to general nonlinear systems. Special nonlinear test equations have also been proposed (Dahlquist 1976, Butcher 1975), but the resulting formulae have not yet led to general-purpose software which is significantly better than that developed on the basis of test equation (3.6). Some stability regions are illustrated in Figs. 1 - 3 .

In Fig. 2, P_k refers to the kth order AB formula, which being explicit, can be used to provide a first approximation y^P_{n+1} (P ≡ predicted value) to y_{n+1} . An evaluation (E ≡ evaluation) $\underline{f}(x_{n+1} , y^P_{n+1})$ is carried out and inserted in the implicit kth order AM formula to obtain y^C_{n+1} (C ≡ corrected value); a further evaluation $\underline{f}(x_{n+1} , y^C_{n+1})$ can now be made. If no further corrections are made then $y_{n+1} \equiv y^C_{n+1}$, $\underline{f}_{n+1} \equiv \underline{f}(x_{n+1} , y^C_{n+1})$; this is referred to as the P_kEC_kE mode of the Adams formulae. Used in a predictor-corrector mode (possibly with further corrections), the Adams formulae form the basis of very successful and widely used software for nonstiff problems.

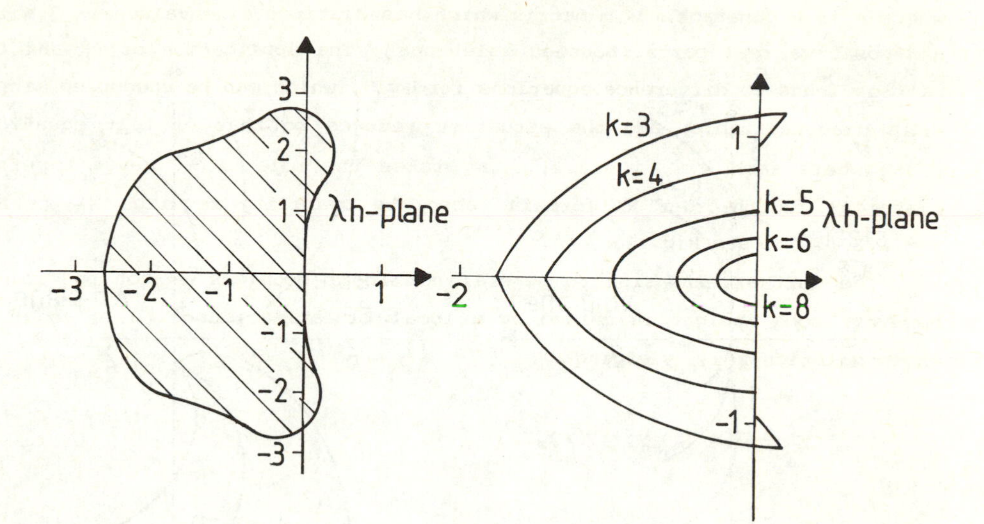

FIG. 1 Stability region for fourth-order explicit RK

FIG. 2 Stability regions for AB-AM (P_kEC_kE form)

In Fig. 3 the stability regions of the backward differentiation formulae are shown; as will be clear for k up to 6, they are well suited to solving stiff problems. Both the BDF and Adams formulae are implemented in variable-step, variable-order form.

4. Use of Runge-Kutta methods

We now consider recent developments affecting the standard variable-step fixed-order Runge-Kutta code for the initial-value problem

$$\underline{y}' = \underline{f}(x, \underline{y}) \quad , \quad \underline{y}(a) = \underline{y}_0 \quad ; \quad \underline{y} \ , \ \underline{f} \in R^m \ . \tag{3.8}$$

Such codes have always been popular and a heavily used part of any program library. It has become clear that the choice of basic formulae and some implementation details have a significant effect on both efficiency and reliability. The termination criteria for (3.8) vary in practical problems.

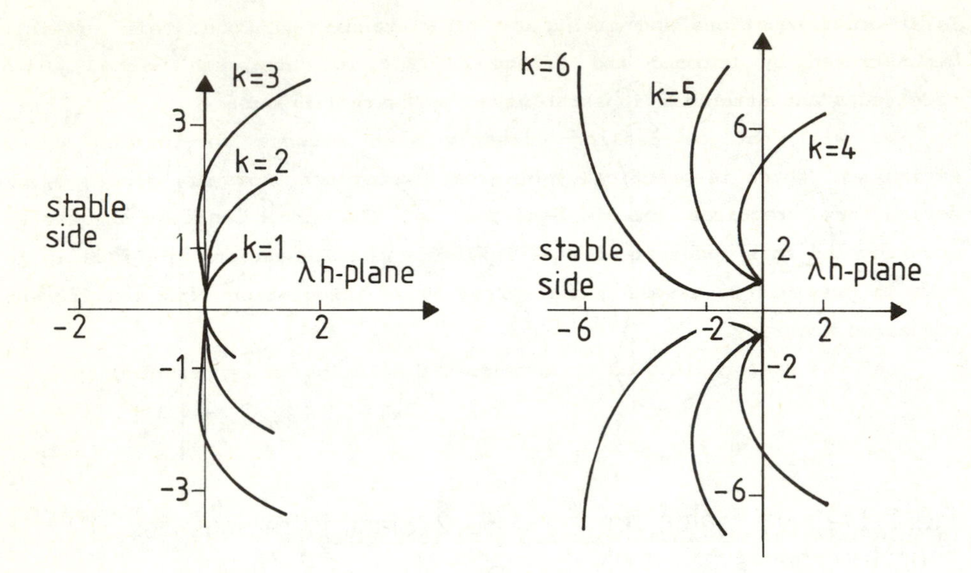

FIG. 3 Stability boundaries for BDF formulae

Output of the solution may be required at specified points in $a \leqslant x \leqslant b$: in some cases the integration is to be continued to find a value of x for which an equation of the following form holds

$$\underline{g}(x, \underline{y}(x), \underline{y}'(x)) = \underline{0} . \tag{3.9}$$

We concentrate here on RK formulae of orders 4 and 5. It was clearly established in the tests of Hull, Enright, Fellen and Sedgwick (1972) that codes based on these formulae can be the most efficient and reliable for moderate accuracy requirements on problems for which evaluations of the derivatives $\underline{f}(x, \underline{y})$ are not too expensive. It is valid to compare costs in codes of this type by the number of $\underline{f}$-evaluations. RK codes take more evaluations per step than an Adams code, but they are able to take larger steps and overheads are a much less significant part of the cost. For severe accuracy requirements a higher-order RK code will be more efficient, but our discussion will also be relevant to them. It is also the case that RK formulae specially derived for second-order equations can be far superior, on suitable problems, to re-writing the problem as an equivalent system of first-order equations and making use of a standard RK code. This is amply demonstrated by Dormand and Prince (1978). To date such methods have received scant attention in established software libraries.

We will not be dealing directly with methods for global error estimation, which is practically of great importance. However, global error estimation requires smooth behaviour of the error against requested accuracy and this requirement will influence the discussion; the idea is to make it possible to assess global error by re-integration with a different requested accuracy.

Let the RK formula used to advance the solution be represented by

$$\underline{y}_{n+1} = \underline{y}_n + h_n \underline{G}(x_n , \underline{y}_n) . \tag{3.10}$$

In this interval the local solution $\underline{u}(x)$ is defined by

$$\underline{u}'(x) = \underline{f}(x, \underline{u}(x)) \quad , \quad \underline{u}(x_n) = \underline{y}_n . \tag{3.11}$$

The stepsizes are chosen to control the local error $\underline{y}_{n+1} - \underline{u}(x_{n+1})$. If the formula is of order p we can write, for sufficiently smooth problems,

$$\underline{y}_{n+1} - \underline{u}(x_{n+1}) = \underline{\varphi}(x_n)h_n^{p+1} + O(h_n^{p+2}) . \tag{3.12}$$

The code requires an asymptotically correct, computable error estimate $\underline{\epsilon}_{n+1} \simeq \underline{y}_{n+1} - \underline{u}(x_{n+1})$ so that

$$\underline{\epsilon}_{n+1} = \underline{\varphi}(x_n)\, h_n^{p+1} + O(h_n^{p+2}) \tag{3.13}$$

also holds. The stepsize can be controlled by requiring that

$$||\underline{\epsilon}_{n+1}|| \leqslant \text{TOL} \tag{3.14}$$

where TOL represents the user-specified accuracy request. The step is rejected if the test (3.14) fails. The assumption that the principal truncation error dominates in (3.13) easily leads to the formula

$$h_{opt} = \left[\frac{TOL}{||\underline{\epsilon}_{n+1}||} \right]^{\frac{1}{p+1}} h_n \tag{3.15}$$

for the locally optimal stepsize, aiming to use the largest stepsize that will satisfy (3.14). To avoid the inefficiency of many rejected steps the next stepsize h_{n+1} (or h_n in case of a failure) is taken to be βh_{opt}, $0 < \beta < 1$; $\beta = 0.9$ is typical of many well tested codes. Finally it is regarded as best to modify the accepted result by including the error estimate, thereby improving the order of global accuracy. This is termed local extrapolation and amounts to continuing the integration with $\underline{y}^E_{n+1} = \underline{y}_{n+1} - \underline{\epsilon}_{n+1}$. From (3.12), (3.13) we have

$$\underline{y}^E_{n+1} - \underline{u}(x_{n+1}) = O(h_n^{p+2}) . \tag{3.16}$$

There is no estimate available for the local error in $\underline{y}^E_{n+1}$. In this mode the use of $\underline{\epsilon}_{n+1}$ in (3.14) is to ensure a reasonable stepsize strategy which can be used to produce accuracy close to that requested. For some purposes later we still require that $\underline{\epsilon}_{n+1}$ be a good estimate for the local error in $\underline{y}_{n+1}$.

This brief summary will suffice for our discussion in the next two sections. However, for completeness we indicate some variations and essential modifications. Some authors have preferred the error control $||\underline{\epsilon}_{n+1}|| \leq h_n$ TOL, local error per unit step, in place of (3.14). For nonstiff problems the case between these has never been very strongly made. Error per step, (3.14), with local extrapolation is used in D02PAF (NAG routine), DERKF (DEPAC, see Shampine and Watts (1980)) and DVERK (Hull, Enright and Jackson 1976) and is superior in the opinion of the present writer. The actual value of TOL used in (3.14) is constructed by a device such as

$$TOL = RTOL * ||\underline{y}_n|| + ATOL ,$$

where the user inputs RTOL and ATOL, thus permitting absolute, relative or mixed error control. (The norm may include component-wise weighting.)

The stepsize $h_{n+1} = \beta h_{opt}$ may be further modified before proceeding to the next step. This can be to limit, for safety considerations, the increase h_{n+1} / h_n or to hit (or smoothly approach) output points. A more conservative strategy still is adopted after stepsize failures. Some care is worth devoting to the choice of initial stepsize, which can affect smoothness of solutions. The general idea is to produce a smooth sequence of stepsizes to avoid inefficiencies and obtain smooth error behaviour with respect to TOL. For general-purpose codes it is difficult to know how far to go in this direction. Special codes, see Vu (1983), have successfully taken such investigations much further.

5. Derivation of Runge-Kutta formulae

The general s-stage explicit formula is determined by the parameters

$$\begin{array}{c|cccccc} 0 & 0 & & & & & \\ c_2 & a_{21} & & & & & \\ c_3 & a_{31} & a_{32} & & & & \\ \cdot & \cdot & \cdot & \cdot & & & \\ \cdot & \cdot & \cdot & & \cdot & & \\ \cdot & \cdot & \cdot & & & \cdot & \\ c_s & a_{s1} & a_{s2} & \cdot & \cdot & \cdot & a_{s,s-1} \\ \hline & b_1 & b_2 & \cdot & \cdot & \cdot & b_{s-1} \quad b_s \end{array} \tag{3.17}$$

which represent (in the scalar case) the formulae

$$\left.\begin{aligned} k_1 &= f(x_n\ ,\ y_n) \\ k_i &= f\Big(x_n + c_i h\ ,\ y_n + h \sum_{j=1}^{i-1} a_{ij} k_j\Big)\ ,\quad i = 2,3,\ldots,s \\ y_{n+1} &= y_n + h \sum_{i=1}^{s} b_i k_i\ . \end{aligned}\right\} \tag{3.18}$$

Early work concentrated on the derivation of a single formula, such as the well-known classical fourth-order, four-stage method

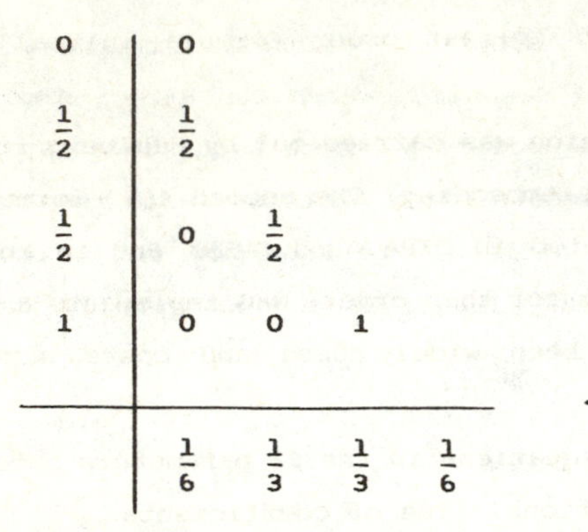

0	0			
$\frac{1}{2}$	$\frac{1}{2}$			
$\frac{1}{2}$	0	$\frac{1}{2}$		
1	0	0	1	
	$\frac{1}{6}$	$\frac{1}{3}$	$\frac{1}{3}$	$\frac{1}{6}$

Error estimation could be achieved by the procedure now known as doubling, one view of which is that each step is repeated with two equal half-steps. For an s-stage method the cost is 3s-1 evaluations, $f(x_n, y_n)$ being common to the two steps beginning at x_n. If the full step gives an approximation $\hat{y}_{n+1}$, and two half-steps give y_{n+1}, we have asymptotically

$$\hat{y}_{n+1} - u(x_{n+1}) = \varphi(x_n)\, h_n^{p+1} + \ldots ,$$

$$y_{n+1} - u(x_{n+1}) = 2\varphi(x_n) \left[\frac{h_n}{2}\right]^{p+1} + \ldots .$$

Hence a local error estimate for the better approximation, y_{n+1}, is

$$\epsilon_{n+1} = \frac{\hat{y}_{n+1} - y_{n+1}}{2^p - 1} = 2\varphi(x_n) \left[\frac{h_n}{2}\right]^{p+1} + \ldots .$$

The locally extrapolated value y^E_{n+1}, (3.16), is

$$y^E_{n+1} = \frac{2^p\, y_{n+1} - \hat{y}_{n+1}}{2^p - 1} = u(x_{n+1}) + O(h_n^{p+2}) .$$

It is clear that the formula for y^E_{n+1} is just an RK formula of the form (3.18), and this procedure can be viewed as a way of constructing a (p+1)st order formula from a pth order formula. A more general derivation would be to derive simultaneously a pth and (p+1)st order formula initially from a common set of k_i values.

Such a systematic investigation was carried out by Fehlberg for various orders. The technique is known as embedding. The second 4(5) formulae given in Fehlberg (1970) were implemented in DEPAC and DVERK and in independent tests came to be regarded as best for this order; NAG implements a different pair. The Fehlberg pair has been widely used and costs six function evaluations per step.

The derivation involves 30 equations in the 32 parameters obtained from (3.17) with s = 6, where an additional line of coefficients

$$\left| \; b_1^E \quad b_2^E \quad . \; . \; . \quad b_6^E \right.$$

is included and the following equation is added to (3.18):

$$y_{n+1}^E = y_n + h \sum_{i=1}^{6} b_i^E k_i \; .$$

We then have $\epsilon_{n+1} = y_{n+1} - y_{n+1}^E$. The local error expansions have the form

$$\left.\begin{aligned} y_{n+1} - u(x_{n+1}) &= h^5 \sum_{j=1}^{9} T_{5,j} D_{5,j} + h^6 \sum_{j=1}^{20} T_{6,j} D_{6,j} + \dots \\ y_{n+1}^E - u(x_{n+1}) &= h^6 \sum_{j=1}^{20} T_{6,j}^E D_{6,j} + \dots \; . \end{aligned}\right\} \quad (3.19)$$

Here the $T_{r,j}$, $T_{r,j}^E$ are nonlinear combinations of the parameters and the $D_{r,j}$ are the so-called elementary differentials of f(x, y) of a given order. As a measure of the size of these terms we use $T_M = ||(T_{M,j})||_2$, $T_M^E = ||(T_{M,j}^E)||_2$, as in Dormand and Prince (1980) and Shampine (1984b); also let $E_M = ||(T_{M,j} - T_{M,j}^E)||_2$. Having achieved orders 4, 5 for y_{n+1} and y_{n+1}^E ($T_M = T_M^E = 0$, M = 1, 2, 3, 4, $T_5^E = 0$) Fehlberg's primary aim was to make T_5 small. This has the effect of improving the accuracy of y_{n+1}.

In view of the fact that local extrapolation is generally advantageous it now appears likely that better formulae are possible. We may state the objectives as follows.

(i) Minimise T_6^E - improving the accuracy of y_{n+1}^E <u>and</u> the accuracy of the error estimate for y_{n+1}.

(ii) Ensure that the principal term dominates in the expansion of $\epsilon_{n+1} = y_{n+1} - y_{n+1}^E$; here, from (3.19), we must compare T_5 and E_6 - this is to ensure accurate stepsize selection, particularly at low accuracy. It is important to have a code that performs well for the commonly occurring low accuracy requests.

(iii) Maximise the size of the stability region of the formula used.

What is not clear is the relative weighting to be given to those objectives and there are certainly other, less well-defined, characteristics of methods that show up in tests. Stiffness detection is a feature one wants to incorporate in a standard code.

Dormand and Prince (1980) derive three 4(5) pairs. Two of these, RK5(4)7M and RK5(4)7S, also include $f(x_{n+1}, y_{n+1}^E)$ in constructing the formula for y_{n+1}, giving one extra parameter. The cost is still six function evaluations on successful steps since this is then the first evaluation on the next step. The third, RK5(4)6M, uses the same data as the Fehlberg formulae. They all emphasise objective (i) and RK5(4)7S pays attention to (iii). In their derivations one parameter not affecting (i) is left free for tuning on a set of test problems, which may improve the performance with respect to (ii). The Fehlberg pair would be judged poor from objective (ii) since the primary goal was to reduce T_5. On objective (i) they give

	T_6^E
Fehlberg	3.36×10^{-3}
RK5(4)7M	3.99×10^{-4}
RK5(4)7S	1.81×10^{-3}
RK5(4)6M	1.23×10^{-3}

Their tests show RK5(4)7M to be superior in achieving a specified global accuracy at minimum cost - on one problem 800 f-evaluations against 1450 for Felhberg in achieving accuracy 10^{-6}. It is clear that the Fehlberg pair can no longer be regarded as the best available.

6. Dense output in Runge-Kutta codes

The output requirements of the user force further additions or modifications to the RK formulae. Restricting the stepsize for dense output directly impairs efficiency, and may introduce consequent fluctuations in succeeding steps. Reliability also suffers and lack of smoothness with respect to requested accuracy is inevitable. Linear multistep codes have a natural interpolant as the basis for the method and the stepsize can be chosen independent of output. This feature is also of great value for problems involving root finding, as in (3.9). Recent work has begun to address this problem, at least for RK methods of orders 4 and 5. Higher orders will present more formidable difficulties.

Horn (1983) considers a scheme of the form, $0 < \sigma < 1$,

$$\left.\begin{aligned} k_1 &= f(x_n, y_n) , \\ k_i &= f(x_n + c_i^* \sigma h , \; y_n + \sigma h \sum_{j=1}^{i-1} a_{ij}^* k_j) , \\ &\qquad i = 2,3,\ldots,s^* , \\ y_{n+1} &= y_n + \sigma h \sum_{i=1}^{s^*} b_i^* k_i . \end{aligned}\right\} \quad (3.20)$$

One can write down the conditions on the parameters for this scheme to have order p for all σ.

The intention is to embed a given formula pair in (3.20) which may require more stages, $s^* > s$. This involves taking

$$c_i^* = c_i / \sigma \; , \quad a_{ij}^* = a_{ij} / \sigma \; , \quad i = 1,2,\ldots,s \; ,$$

where c_i, a_{ij} are the coefficients of the given formulae. In this way the evaluations k_i, $i = 1,2,\ldots,s$ are the usual stages of the embedded formulae. The full scheme (3.20) is called in only when interpolation for output is required on that step for some particular values of σ. The coefficients b_i^* are now low degree polynomials in σ. For the Fehlberg pair ($s = 6$) it was shown that with $s^* = 7$, giving one extra evaluation, (3.20) could be made of order 4 for all σ. Although we may be using order 5 after local extrapolation (y^E_{n+1}), the fact that the local error estimate for y_{n+1} is reliable (objective (i)) and that (3.14) is satisfied can be used to confirm that the fourth-order interpolated values are sufficiently accurate. It is, of course, possible to get fifth-order interpolated values by increasing s^* but this may not be worthwhile.

Shampine (1984a, b) analyses the question of interpolation fully and presents an alternative technique. Since interpolation is only considered after successful steps we have available y_n, $f(x_n, y_n)$, y^E_{n+1}, $f(x_n, y^E_{n+1})$. With one extra evaluation using the Horn (1983) result with $\sigma = \frac{1}{2}$ a value $y_{n+1/2}$ can be obtained. Then Hermite interpolation can be used to produce a polynomial interpolant for the whole interval. This approach guarantees a globally C^1 interpolant over the whole range of integration. It is shown that such a scheme can be written in the form of Horn's approach, of which it is a particular case. This is useful for assessing the quality of the interpolation polynomial approximation. Horn's scheme does not, in general, produce even a continuous interpolant over the whole range. Shampine investigates interpolation for several well-known RK schemes and develops a good interpolation scheme for the best Dormand and Prince formulae, RK5(4)7M.

7. Stiff systems

Here for system (3.1) the Jacobian matrix $J(x, \underline{y}) = \left[\dfrac{\partial f_i}{\partial y_j}\right]$ has eigenvalues $\{\lambda_i(x)\}$ for which $\text{Re}(\lambda_i(x))$, $i = 1,2,\ldots,m$, vary greatly in magnitude (and are usually negative for <u>most</u> of the range $[a,b]$); varying from -10^{-9} to -10^{-3} would not be unusual. Therefore, from section 3, it follows that the absolute stability condition for <u>explicit</u> RK methods and LMM's ,

$$h|\lambda_i(x)| = O(1) \quad , \quad x \in [a,b] \quad , \quad i = 1,2,\ldots,m \; ,$$

imposes a severe restriction on the stepsize <u>throughout the range</u>. Generally this renders the computation prohibitively expensive. This is because the nature of the solution $\underline{y}(x)$ is such that over $[a,b]$, h can be increased significantly (usually through several orders of magnitude), whilst still approximating the solution with sufficient accuracy. Ideally, the user's accuracy requirements should determine h, not stability conditions imposed by the formula.

General purpose software for stiff systems must therefore be based on a formula whose stability region S contains an unbounded part of the left half plane. It follows that RK methods and LMM's must be implicit. Suitable LMM's are the BDF; see Fig. 3. Certain implicit RK formulae up to high orders have S as the open left half plane (Butcher 1976) and such formulae are called A-stable (Dahlquist 1963). For a LMM to be A-stable the order must not exceed two. A-stability is desirable but not essential for stiff problems. A good discussion of codes for stiff problems is given by Shampine and Gear (1979).

In the next two sections we are mainly concerned with some recent developments for stiff problems. It is generally agreed that one of the outstanding problems in this area is the efficient solution of large systems (Gear 1981). Such problems arise directly in physical systems or as part of another problem, for example, the method of lines for parabolic partial differential equations (see Chapter 9).

The major computational difficulty arises from the fact that we have to use implicit formulae. For example, advancing from x_n to $x_{n+1} = x_n + h_n$ via the BDF

$$\sum_{j=0}^{k} \alpha_{jn}\underline{y}_{n+1-j} = h_n\beta_{kn}\underline{f}_{n+1} \quad ,$$

requires the solution of

$$\underline{y}_{n+1} = h_n\gamma_{kn}\,\underline{f}(x_{n+1}\,,\,\underline{y}_{n+1}) + \underline{\Psi}_n \quad , \quad \gamma_{kn} > 0 \quad ,$$

where $\underline{\Psi}_n$ is known. Rewriting (dropping the subscripts and x-variable), the system takes the form

$$\underline{F}(\underline{y}) \equiv \underline{y} - h\gamma\,\underline{f}(\underline{y}) - \underline{\Psi} = \underline{0} \quad , \quad \gamma > 0 \quad , \gamma = O(1) \quad . \qquad (3.21)$$

Now typically, throughout [a,b] $\max_i |\mathrm{Re}(\lambda_i)|$ is large, where $\{\lambda_i\}$ are the eigenvalues of the Jacobian $J = \left[\dfrac{\partial f_i}{\partial y_j}\right]$; hence $||J|| >> 1$. Solving (3.21) numerically by functional iteration

$$\underline{y}^{(\nu+1)} = h\gamma\, \underline{f}(\underline{y}^{(\nu)}) + \underline{\Psi} \;, \quad \nu = 0,1,2,\ldots,$$

requires, <u>throughout</u> [a,b], $h\gamma||J|| < 1$ for convergence. This is a prohibitive restriction on the stepsize similar to that imposed by an <u>explicit</u> formula. This restriction could be accepted during a fast transient stage where h must be small for accuracy requirements. The most widely used iteration for (3.21) is of Newton type (NT): if $\tilde{J}$ is an approximation to the Jacobian J,

$$(I - h\gamma\tilde{J})\,(\underline{y}^{(\nu+1)} - \underline{y}^{(\nu)}) = -\,\underline{F}(\underline{y}^{(\nu)}), \qquad \nu = 0,1,2,\ldots, \tag{3.22}$$

and in practice this can be convergent for values of h which are reasonably consistent with the user's accuracy requirements. A typical acceptance test would be to try to satisfy

$$||\underline{y}^{(\nu)} - \underline{y}^{*}|| \leqslant c.\,\mathrm{TOL} \;, \quad c > 0 \;, \tag{3.23}$$

where TOL is the user's requested global error, and $\underline{F}(\underline{y}^{*}) = \underline{0}$. For <u>large</u> systems it is now clear that a major part of the computation and storage is spent in solving systems of linear algebraic equations. A standard approach to handling (3.22) is to compute the L, U factors of the matrix $(I - h\gamma\tilde{J})$ and retain them for as many <u>integration</u> steps as possible - until lack of convergence (say after three iterations) forces either an update or a reduction in h. The factors may be reused if h is only slightly altered. The step h may be increased by an order change, when γ will also change, possibly for a further k steps - this could also force an update in the factors.

8. Treatment of the Newton equations

The larger the system (3.22) the more likely is the Jacobian to be <u>sparse</u> and so full use must be made of those direct methods which use sparse matrix techniques. This is the case in FACSIMILE (Curtis 1980); a variant of LSODE (see section 10), LSODES, also uses sparse Gauss elimination. Carver and Boyd (1979) illustrate that sparse matrix techniques can be efficient for even quite small systems (< 25 equations). Substantial savings can also be made by optimising the <u>evaluation</u> of the Jacobian by using finite-difference approximations $J \approx [\Delta f_i/\Delta y_j]$ and exploiting the sparsity structure of J (Curtis, Powell and Reid 1974). Carver and MacEwen (1981) give an example in which for 242 equations only seven f-evaluations were required to obtain $\tilde{J}$. Pivoting strategies for $(I - h\gamma\tilde{J})$ can often be reused in subsequent factorisations.

Other approaches to simplifying (3.22) are possible, for example partitioning. To be worthwhile such methods usually require assumptions about the distribution of eigenvalues of J. Typically it is assumed that the stiffness is due to a comparatively few 'stiff eigenvalues' for which $|h\gamma\lambda_i| >> 1$, and the remaining 'nonstiff eigenvalues' satisfy $|h\gamma\lambda_i| < 1$. The idea now is to partition (3.22) so as to correspond to NT iteration for the stiff part and functional iteration for the nonstiff part; e.g. see Robertson (1976) and more recently a sophisticated algorithm due to Bjorck (1982) which uses block QR factorisations.

Interestingly it is argued by Curtis (1983) that on the basis of many large stiff systems (illustrated by five test problems of orders 40 to 760) appearing in chemical kinetics, such an assumption about 'stiff' and 'nonstiff' eigenvalues is not valid. Partitioning methods are then not appropriate; similar remarks apply to exponential-fitting methods. Further, he makes the point that for large stiff problems both second-derivative methods (requiring y'' and hence J^2) and many implicit RK methods would be very inefficient. There is the possibility that application areas will arise in the future which lead to large systems with eigenvalue distributions for which partitioning methods are appropriate.

The majority of codes use Gauss elimination (G.E.) with partial pivoting for solving the Newton equations (3.22). Banded and sparse G.E. may also be options. Recently a good deal of effort has gone into an investigation of iterative methods for (3.22), for example, by Gear and Saad (1983), Chan and Jackson (1984). These try to use some of the developments made over the past decade in iterative methods for large sparse nonsymmetric systems of linear algebraic equations; for work up to 1981 see Hageman and Young (1981).

Some of the work of Chan and Jackson (CJ) will now be briefly described. To begin, it is clear that the linear system (3.22) need not be solved exactly. The predicted value $\underline{y}^{(0)}$ for the exact solution $\underline{y}^*$ is fairly accurate so the Newton corrections are small, and only a few significant digits are required. Indeed, the 'corrector equation' $\underline{F}(\underline{y}) = \underline{0}$, can be regarded as a 'stabiliser'. CJ give theoretical arguments concerning the inexact Newton-type method, to show that (under certain conditions on $\tilde{J}$ and

$\underline{y}^{(0)}$) if (3.22) is solved <u>approximately</u> so that $\{\underline{y}^{(\nu)}\}$ satisfies

$$||(I - h\gamma \tilde{J})(\underline{y}^{(\nu+1)} - \underline{y}^{(\nu)}) + \underline{F}(\underline{y}^{(\nu)})|| \leq \eta ||\underline{F}(\underline{y}^{(\nu)})|| \quad , \quad (3.24)$$

for constant $\eta < 1$, then $\underline{y}^{(\nu)} \to \underline{y}^{*}$ linearly. This is a relationship between the residual of the linear system (3.22) and the residual of the nonlinear system (3.21), which combined with (3.23) can be used as an acceptance test for the Newton correction stage of the <u>inexact</u> NT method.

Regard now the equation for the Newton corrections as

$$A\,\underline{z} = \underline{b} \quad , \quad A = I - h\gamma\tilde{J} \quad . \quad (3.25)$$

Techniques related to partial differential equations for the efficient iterative solution of this system, e.g. SOR, ADI, require A to have special properties and 'optimum' parameters are needed. For a general purpose ODE solver we cannot easily use such methods; A is not usually symmetric positive definite, and it lacks a regular structure.

From (3.24) we seek an approximation to the solution of (3.25) which has 'small residuals', and for this reason CJ consider the application of a generalised Conjugate Residual method. In its <u>original</u> form the method is defined for A symmetric positive definite. The iteration yields iterates $\underline{z}_0$, $\underline{z}_1$, $\underline{z}_2$, ... , where $\underline{z}_i$ is chosen to minimise over a certain subspace of R^m the Euclidean norm of the residual

$$\underline{r}_i = \underline{b} - A\,\underline{z}_i \quad .$$

The subspace is the translated Krylov subspace,

$$\underline{z}_0 + \{\underline{r}_0 , A\underline{r}_0 , \ldots , A^{i-1}\underline{r}_0\} \quad ,$$

where $\underline{z}_0$ is the initial guess for $\underline{z}$. The method only requires the user to supply a routine to compute $A\underline{v}$ for any $\underline{v}$; hence full advantage can be taken of sparsity. CJ quote the bound

$$||\underline{r}_i|| \leq 2 \left[\frac{1 - 1 / \sqrt{\kappa(A)}}{1 + 1 / \sqrt{\kappa(A)}} \right]^i ||\underline{r}_0|| \quad ,$$

where $\kappa(A)$ is the spectral condition number of A. For stiff problems we can expect A to be ill-conditioned as h increases, since

$$\kappa(A) = ||A||_2 \, ||A^{-1}||_2 \geq \max_i |1 - h\gamma\lambda_i| \, / \, \min_i |1 - h\gamma\lambda_i| \quad ,$$

where $\{\lambda_i\}$ are the eigenvalues of $\tilde{J}$, which suggests very slow convergence of the method. Hence the whole process is modified to treat a preconditioned form of (3.25),

$$\hat{A}\,\hat{\underline{z}} = \underline{b}$$

in which $\hat{A} = AM^{-1}$, $\hat{\underline{z}} = M\underline{z}$, and M is 'close' to A and inexpensive to invert.

The important feature of the above method is that it can be extended to <u>nonsymmetric</u> problems (Eisenstat, Elman and Schultz 1983), to yield the Preconditioned Generalised Conjugate Residual (PGCR) method. The method is convergent if $\hat{A}$ is positive real, that is $(\underline{x}, \hat{A}\underline{x}) > 0$ for all non-zero $\underline{x} \in R^m$, equivalently, if $\frac{1}{2}(\hat{A} + \hat{A}^T)$ is positive definite. <u>Prior</u> to preconditioning A can be positive real for practical stiff problems (through properties of the Jacobian related to stability properties of the differential system).

CJ carry out some extensive numerical testing with the PGCR method. Overall the results show that the method compares well with LSODE (see section 10) which uses direct methods to solve for the Newton correction. With increasing problem size PGCR is more economical in storage. Examples include the method of lines applied to a Convection-Diffusion equation in two and three dimensions, yielding problem sizes up to $m = 9^2$ and 9^3 respectively.

Alternative forms of Krylov subspace methods have also been considered by Gear and Saad (1983). An important feature here is that there is no need to compute and store approximate Jacobians $\tilde{J}$; all that is required is that $\tilde{J}\underline{v}$ can be computed for any $\underline{v}$ (preconditioning in CJ means that $\tilde{J}$ must be available explicitly). Now $\tilde{J}\underline{v}$ may be estimated from the approximation

$$\tilde{J}\underline{v} \simeq \frac{1}{\epsilon}\,[\underline{f}(\underline{y} + \epsilon\underline{v}) - \underline{f}(\underline{y})]\,,$$

employing one function evaluation (since $\underline{f}(\underline{y})$ is known). Hence the various matrix solvers required for different matrix structures in $\tilde{J}$ are not required. Preconditioning is not yet included in this approach. Gear and Saad carry out some numerical testing by incorporating their Krylov subspace iterative method as an option in LSODE; comparisons are then made by using LSODE with full Gauss elimination. They treat for example a system of the form

$$\underline{y}' = A\,\underline{y} + \underline{g}(\underline{y}),$$

for which LSODE forms the Jacobian automatically. As the system size m increases, the iterative approach is seen to be superior (Table 1).

Table 1

Results of Gear and Saad (1983)

	LSODE (G.E.)	LSODE (Krylov)
m = 50		
No. of f-evaluations	1382	643
No. of steps	151	180
Core memory	2972	1750
Run times	1'09	1'05
m = 80		
No. of f-evaluations	1935	640
No. of steps	149	172
Core memory	7142	2410
Run times	3'03	1'38

In summary the iterative treatment of the linear system for the Newton equations can lead to worthwhile gains in efficiency for large stiff problems.

9. Extrapolation methods

For completeness we report briefly on developments that have been made in extrapolation methods. The original method for nonstiff problems was proposed by Gragg (1965). It has always been widely used and has compared well with other methods (Hull et al. 1972). Recent improvements in implementation will enhance its reputation and some theoretical questions have also been clarified.

The basis of the method is several repeated integrations over a large step x_0 to $x_0 + H$ using the explicit mid-point rule.

Let $h_i = H/n_i$, n_i even. The sub-integrations take the form (for a scalar equation)

$$\left.\begin{aligned} & z_0 = y_0 \,, \\ & z_1 = z_0 + h_i\, f(x_0\,, z_0) \,, \\ & z_{j+1} = z_{j-1} + 2h_i\, f(x_0 + jh_i\,, z_j)\,,\ j = 1,2,\ldots,n_i-1\,, \\ & S(h_i) = \frac{1}{2}\,[z_{n_i} + z_{n_{i-1}} + h_i\, f(x_0+H,\, z_{n_i})]\,. \end{aligned}\right\} \quad (3.26)$$

$S(h_i)$ is a smoothed approximation to $y(x_0 + H)$ and the error has an expansion in powers of h_i^2. (3.26) may be regarded as an RK method of order 2. By using a sequence of sub-integrations for a sequence of integers $n_1 < n_2 < \ldots$, the values $S(h_i)$ can be used as a basis for an h^2-extrapolation tableau, producing progressively (RK) approximations of orders 4, 6, 8, ... until the required accuracy is achieved. Shampine (1983) proves that for any problem, the optimal sequence for the n_i-values is

$$\{2,\ 6,\ 8,\ 10,\ 12,\ 14,\ 16,\ 18,\ 20,\ 22\}.$$

The same questions are also considered for extrapolation methods for stiff problems.

A theory for order and stepsize selection is given by Deuflhard (1983). The results of this paper give an improvement in performance, most striking for stiff problems using an extrapolation method based on a semi-implicit mid-point rule presented in Bader and Deuflhard (1983). The extrapolation methods for stiff equations considered here are a significant advance for algorithms in this area.

10. **Software**

Algorithms fall into the two broad classes of stiff and nonstiff solvers. Perhaps the most widely used and highly successful software packages combine the Adams formulae for nonstiff problems and the backward differentiation formulae for stiff problems. Originally developed by Gear (1971a) in his famous DIFSUB (see also Gear (1971b)), the user selected the stiff/nonstiff option and the code used the BDF in variable-step, variable-order form via the Nordsieck representation. Extensive developments have since been made. For example, the variable-coefficient forms of the formulae are used in EPISODE (Byrne and Hindmarsh 1975). These developments have culminated in the 'state of the art' code called LSODE along with its variants. The standard version was developed by Hindmarsh (1980). The variant LSODA, due to Hindmarsh (1980) and Petzold (1980), has the facility to switch automatically between stiff and nonstiff, so the user does not have to decide whether the problem is stiff or not. A further code LSODI due to Painter and Hindmarsh (Hindmarsh 1980) treats linearly implicit systems of the form

$$A(x, \underline{y})\, \underline{y}' = \underline{g}(x, \underline{y}) \quad ,$$

where $A(x, \underline{y})$ is an $m \times m$ matrix.

Chaper D02 of the NAG Library (Mark 11) contains general routines for first-order systems implementing a fourth-order Runge-Kutta method (D02PAF), the Adams method (D02QAF), and the Backward Differentiation method (D02QBF). These routines provide a large number of options covering the initial stepsize, error control, intermediate output, and termination criteria, and the user interfaces are relatively complicated. They form the basis for a larger set of easy-to-use routines, which require less information from the user, and are intended to solve the problems which occur most frequently in practice.

CHAPTER 4

Ordinary Differential Equations: Boundary-Value Problems

J.L. Mohamed

1. Introduction

Boundary-value problems for ordinary differential equations appear in many forms and often occur naturally as high-order equations. Consider a first-order system of the following form

$$\underline{y}' = \underline{f}(x, \underline{y}, \underline{p}) , \qquad a \leqslant x \leqslant b \tag{4.1a}$$

subject to m multipoint boundary conditions

$$\left.\begin{aligned} &\underline{g}\{\underline{y}(a_1(\underline{p})), \ldots , \underline{y}(a_r(\underline{p}))\} = \underline{0} , \\ &a \leqslant a_1 < a_2 < \ldots < a_r \leqslant b , \quad r \geqslant 2 \end{aligned}\right\} \tag{4.1b}$$

where $\underline{y}$, $\underline{f}$, $\underline{g}$ are vectors of length n, $\underline{g}$ is of length m, and $\underline{p}$ is a vector of unknown parameters which, together with $\underline{y}$, is to be determined. If $\underline{p}$ is absent from (4.1) the parameter-independent problem consists of finding $\underline{y}$ such that both the differential equations and the boundary conditions (b.c.s) are satisfied. In this case we usually have m = n. We assume that a solution exists; Keller (1968) gives sufficient conditions for the existence of a solution in simple cases.

The problem of restating an nth order system over a given range [a,b] in the form (4.1), may require more than a simple conversion of the nth order differential equation to a system of n first-order equations. One or both of the end-points a, b may be infinite, with one or more of the boundary conditions given asymptotically. The treatment of these conditions may introduce further parameters into the problem, which need to be determined as part of the solution process.

The problem of classification of boundary-value problems (BVPs) has been discussed in depth in an article by Gladwell (1980). BVPs can be classified by the following properties, although this list is by no means exhaustive: linear or nonlinear, nonsingular or singular, parameter-dependent or parameter-independent, with two-point or multipoint b.c.s, with separated or non-separated b.c.s, with finite or infinite range of integration.

The examples given below may be classified as follows:

$$\text{(i)} \quad \underline{y}' = F(x)\underline{y} + \underline{g}(x) , \qquad a \leqslant x \leqslant b \qquad (4.2a)$$

$$A\underline{y}(a) = \underline{\alpha} , \quad B\underline{y}(b) = \underline{\beta} , \qquad -\infty < a,b < \infty \qquad (4.2b)$$

where F, A, B are matrices of order n. This is a parameter-independent system of n linear differential equations, to be solved subject to separated b.c.s at the two finite points a, b.

$$\text{(ii)} \quad \underline{y}' = \underline{f}(x, \underline{y}) , \qquad a \leqslant x \leqslant b \qquad (4.3a)$$

$$\underline{g}(\underline{y}(a), \underline{y}(b)) = \underline{0} , \qquad -\infty < a,b < \infty \qquad (4.3b)$$

where $\underline{y}$, $\underline{f}$, $\underline{g}$ are vectors of length n. This parameter-independent system of n nonlinear differential equations is to be solved subject to non-separated b.c.s at the two finite points a, b.

$$\text{(iii)} \quad y'' = \frac{y^3 - y'}{2x} , \qquad 0 \leqslant x \leqslant 16 \qquad (4.4a)$$

$$y(0) = 0.1 , \qquad y(16) = \frac{1}{6} . \qquad (4.4b)$$

This is a specific example of a second-order parameter-independent BVP, which is used in the NAG Library documentation for boundary-value routines. It is nonlinear with separated b.c.s; it is also singular in the sense that the right-hand side of (4.4a) is undefined at x = 0. However, it may be shown that y has the following truncated power series expansion near the origin

$$y(x) \simeq 0.1 + 0.1\, p\sqrt{x} + 0.001x ,$$

where p is an unknown parameter. This expansion can be used to provide a b.c. at $x = x_0$ where x_0 is close to the origin; the resulting parameter-dependent problem can then be solved over the reduced range $[x_0,16]$.

(iv)
$$\left.\begin{aligned} y' &= \tan\varphi \,, \\ v' &= -\frac{(\alpha\sin\varphi + 0.0002v^2)}{v\cos\varphi} \,, \\ \varphi' &= -\frac{\alpha}{v^2} \,, \end{aligned}\right\} \quad 0 \leqslant x \leqslant \beta \,. \tag{4.5a}$$

$$\left.\begin{aligned} &y(0) = 0 \,, \quad v(0) = 500 \,, \quad \varphi(0) = 0.5 \,, \\ &y(\beta) = 0 \,, \quad v(\beta) = 450 \,, \quad \beta < \infty \text{ is unspecified.} \end{aligned}\right\} \tag{4.5b}$$

This is an example of a parameter-dependent system of first-order differential equations, used in the NAG Library documentation for routine DO2HBF. It represents the equations of motion of a projectile fired at angle φ to the horizontal, towards a target which is at distance β from the point of projection. The aim is to determine the gravitational constant α and the range β over which the projectile must be fired in order to hit the target with a given velocity. The problem is nonlinear with separated b.c.s; five conditions are needed to determine the solution and the parameters α and β.

(v)
$$\left.\begin{aligned} y'' &= \frac{2(y'^2 - z)}{3y} \,, \\ z'' &= -yz' \,, \end{aligned}\right\} \quad 0 \leqslant x \leqslant \infty \,. \tag{4.6a}$$

$$\left.\begin{aligned} &y(0) = 0 \,, \quad z(0) = 1 \,, \\ &y'(x),\ z'(x) \to 0 \quad \text{as} \quad x \to \infty \,. \end{aligned}\right\} \tag{4.6b}$$

This illustrates a coupled system of second-order nonlinear differential equations with separated b.c.s, which is singular in a double sense: the problem is undefined at $x = 0$ and b.c.s are specified at infinity. It is possible (see Gladwell (1980)) to obtain the leading terms in power series expansions for y and z at small values of x, and also in asymptotic expansions for y and z at large values of x. These expansions are parameter-dependent and the problem is thus restated as one in which the solutions y, z and certain unknown parameters are required.

(vi)
$$(p(x)y')' + \{q(x) + \lambda r(x)\}y = 0 \,, \qquad a < x < b \tag{4.7a}$$

$$\left.\begin{aligned} a_1\, y(a) - a_2\, p(a)y'(a) = 0\ , \\ b_1\, y(b) - b_2\, p(b)y'(b) = 0\ , \end{aligned}\right\} \quad -\infty < a,b < \infty\ , \qquad (4.7b)$$

where p, q, r are continuous and p, r are positive throughout (a,b). This second-order linear two-point BVP with separated b.c.s is referred to as a <u>Sturm-Liouville eigenvalue problem</u>; the aim is to determine the values λ_n of λ (the eigenvalues) for which the differential equation (4.7a) has a nontrivial solution (eigenfunction) y_n that satisfies the boundary conditions (4.7b). This parameter-dependent problem is called singular if for instance a or b is infinite, or if $p(x) \to 0$ as $x \to a$ or b.

There is a wealth of literature associated with methods for the numerical solution of BVPs, in particular for two-point BVPs, written as a system of n first-order differential equations of the form

$$\underline{y}' = \underline{f}(x,\ \underline{y},\ \underline{p})\ , \qquad a \leqslant x \leqslant b$$

subject to n b.c.s

$$\underline{g}(\underline{y}(a(\underline{p})),\ \underline{y}(b(\underline{p}))) = \underline{0}\ .$$

The reader is referred to Chapters 15-18 of Hall and Watt (1976), Chapters 10, 12 of Baker and Phillips (1981) and to papers by Aktas and Stetter (1977), Walsh (1977), Fox (1980). The most widely used methods are shooting, finite differences and series expansions, and we shall restrict our attention to these. This is not to say that other methods are unworthy of consideration; a useful and detailed catalogue of methods, together with references to and descriptions of available software, may be found in Childs, Scott, Daniel, Denman and Nelson (1979) and in Gladwell (1980).

2. Shooting methods

Consider the solution of the nth order parameter-independent BVP, required over the finite interval [a,b],

$$y^{(n)}(x) = f(x,\ y,\ y',\ \ldots\ ,\ y^{(n-1)})\ , \qquad a \leqslant x \leqslant b \qquad (4.8a)$$

subject to n separated b.c.s of the form

$$\left.\begin{array}{ll} y^{(i-1)}(a) = \alpha_i \;, & 1 \leq i \leq n \\ y^{(j-1)}(b) = \beta_j \;, & 1 \leq j \leq n \end{array}\right\} \quad (4.8b)$$

where k b.c.s are prescribed at x = a and (n-k) b.c.s are prescribed at x = b. Then if we write

$$y_1 = y \,,$$

$$y_i = y'_{i-1} \,, \qquad i = 2,3,\ldots,n \,,$$

the nth order BVP is equivalent to the first-order system

$$\left.\begin{array}{ll} y'_i = y_{i+1} \,, & i = 1,2,\ldots,n-1 \,, \\ y'_n = f(x,\, y_1 \,,\, y_2 \,,\, \ldots \,,\, y_n) \,, & \end{array}\right\} \quad (4.9a)$$

with boundary conditions of the form

$$\left.\begin{array}{ll} y_i(a) = \alpha_i \,, & i = i_1 \,,\, i_2 \,,\ldots,\, i_k \\ y_j(b) = \beta_j \,, & j = j_1 \,,\, j_2 \,,\ldots,\, j_{n-k} \,. \end{array}\right\} \quad (4.9b)$$

Now if we introduce (n-k) parameters p_1 , p_2 ,..., p_{n-k} for the unknown boundary values at x = a and k parameters p_{n-k+1} ,..., p_n for the unknown boundary values at x = b, we obtain two initial-value problems with boundary conditions given in terms of $\underline{p}$.

Let $\underline{y}_a(x, \underline{p})$ satisfy (4.9a), with n boundary conditions at x = a, k given by (4.9b) and n-k given in terms of the parameters p_1 , p_2 ,..., p_{n-k} . Let $\underline{y}_b(x, \underline{p})$ be defined similarly, satisfying (4.9a) with n-k boundary conditions given by (4.9b), and k given in terms of p_{n-k+1} ,..., p_n . Then we can integrate forwards for $\underline{y}_a(x, \underline{p})$ and backwards for $\underline{y}_b(x, \underline{p})$ using guessed values of the parameters. If the parameters are chosen correctly, then at some 'matching' point x = m, where a ≤ m ≤ b, we should have

$$\underline{y}_a(m, \underline{p}) = \underline{y}_b(m, \underline{p}) \ .$$

The solution of the BVP (4.8) is thus equivalent to the problem of finding $\underline{p}$ such that the <u>matching equations</u>

$$\underline{F}(\underline{p}) \equiv \underline{y}_a(m, \underline{p}) - \underline{y}_b(m, \underline{p}) = \underline{0} \tag{4.10}$$

are satisfied. We can use Newton's method to provide a sequence of estimates $\{\underline{p}^{(k)}\}$ to the solution of (4.10); given an initial approximation $\underline{p}^{(0)}$ to $\underline{p}$, the iterative scheme is

$$\left.\begin{aligned} J(\underline{p}^{(k)})\,\underline{\delta}^{(k+1)} &= -\,\underline{F}(\underline{p}^{(k)}) \,, \\ \underline{p}^{(k+1)} &= \underline{p}^{(k)} + \underline{\delta}^{(k+1)} \,, \end{aligned}\right\} \tag{4.11}$$

where J is the Jacobian matrix of $\underline{F}$ with (i,j) element

$$\frac{\partial F_i}{\partial p_j} \,, \qquad i,j = 1,\ldots,n \ .$$

There is a unique solution if and only if J is nonsingular. The elements of J may be calculated either by forming and solving the <u>variational equations</u> of the differential system (see Hall and Watt (1976), Ch. 16), or by approximating the (i,j) element of J by a simple difference quotient,

$$\frac{\partial F_i}{\partial p_j} \simeq \frac{F_i(\underline{p} + \underline{e}_j \delta p_j) - F_i(\underline{p})}{\delta p_j} \ . \tag{4.12}$$

The latter is easier to implement; it requires the forward and backward solution of (n-k) and k differential systems respectively, but the differential equations are identical to those from which $\underline{y}_a$ and $\underline{y}_b$ are obtained, the only difference being that the initial conditions are perturbed.

The convergence of the Newton iteration depends critically on having a good initial approximation $\underline{p}^{(0)}$. For $\underline{p}^{(0)}$ in the neighbourhood of the solution of the matching equations (4.10), Newton's method displays second-order convergence if the solution is isolated. However, the convergence rate can be slow if a poor approximation $\underline{p}^{(0)}$ is chosen or if the initial-value problem is inherently unstable (see later). In practice, the Newton method (4.11) is modified slightly to increase the region of convergence. A correction $\underline{\delta}^{(k+1)}$ to the previous iterate $\underline{p}^{(k)}$ is calculated by solving the first equation of (4.11); the new iterate is then taken to be

$$\underline{p}^{(k+1)} = \underline{p}^{(k)} + \lambda_{k+1}\,\underline{\delta}^{(k+1)}$$

where λ_{k+1} is a damping factor in the range $0 < \lambda_{k+1} \leq 1$, which is chosen to ensure that some norm of $\underline{F}(\underline{p})$ is reduced at each step. The most reliable way of obtaining a good estimate for $\underline{p}^{(0)}$ is via continuation; a description of this technique is deferred to section 4. The choice of δp_j in (4.12) is also important, and it should be related to the accuracy of the integration.

Other numerical difficulties which can arise in the shooting method are inherent instability, relative instability, or stiffness in the initial-value problem. In the case of inherent instability, there is rapid growth of one or more solution components y_i , and overflow may occur. Relative instability arises when the solution contains a dominant component, which tends to produce a singular or ill-conditioned Jacobian matrix. Both forms of instability may be alleviated by splitting the range of integration into smaller intervals and applying the shooting method over each interval. This is the method of multiple shooting, where a series of initial-value problems is solved, one for each sub-interval $[x_i, x_{i+1}]$, with matching of the computed solutions at the points x_i . A detailed description of the method may be found, for example, in Keller (1968), and in Deuflhard (1980) which discusses recent advances in multiple shooting techniques; an implementation of the method is described in Gladwell (1980).

A more general formulation of a boundary-value problem allows for the specification of auxiliary differential equations, which are not involved in the matching process, but which provide variables needed to integrate the remaining equations. The problem may be written as follows, where (4.13b) are the auxiliary, or 'driving', equations:

$$\underline{y}' = \underline{f}_1(x, \underline{y}, \underline{z}, \underline{p}) , \qquad (4.13a)$$

$$\underline{z}' = \underline{f}_2(x, \underline{z}, \underline{p}) , \qquad (4.13b)$$

$$\underline{y}(a) = \underline{g}_1(\underline{p}) , \quad \underline{y}(b) = \underline{g}_2(\underline{p}) , \tag{4.13c}$$

$$\underline{z}(a) = \underline{g}_3(\underline{p}) , \tag{4.13d}$$

$$\underline{e}(\underline{p}) = \underline{0} , \tag{4.13e}$$

where

$$\underline{p} = [p_1 \; \dots \; p_m]^T ,$$

$$\underline{y} = [y_1 \; \dots \; y_\ell]^T , \qquad \ell \leq m ,$$

$$\underline{z} = [z_1 \; \dots \; z_{n-\ell}]^T ,$$

and $\underline{e}$ is a vector of length $m-\ell$. The shooting method is applied to the solution of this BVP with matching point taken at the end-point of the range of integration $x = b$. The differential system for $\underline{z}$ is not involved in the matching process, which is applied only to the solution of (4.13a), the 'active' equations. The computed solution of (4.13a), (4.13b), with initial conditions

$$\underline{y}(a) = \underline{g}_1(\underline{p}) , \qquad \underline{z}(a) = \underline{g}_3(\underline{p}) ,$$

is required to satisfy

$$\underline{y}(b) = \underline{g}_2(\underline{p}) .$$

If $\ell = m$ then equations (4.13e) are missing and the problem is equivalent to solving the following system for $\underline{p}$:

$$\underline{F}(\underline{p}) \equiv \underline{y}(b) - \underline{g}_2(\underline{p}) = \underline{0} .$$

If however $\ell < m$, we require $\underline{p}$ to satisfy the additional $(m-\ell)$ equations (4.13e), and the problem is equivalent to solving

$$\underline{F}(\underline{p}) \equiv \begin{bmatrix} \underline{y}(b) - \underline{g}_2(\underline{p}) \\ \underline{e}(\underline{p}) \end{bmatrix} = \underline{0} .$$

The parameters $\underline{p}$ need not be restricted to boundary values; they may include eigenvalues, parameters in the coefficients of the differential equations, coefficients in power series or asymptotic expansions for boundary values, or the length of the range of integration, etc. .

3. Shooting methods for Sturm-Liouville eigenvalue problems

A nonsingular Sturm-Liouville eigenvalue problem is defined by a self-adjoint second-order differential equation of the form

$$(p(x)\, y')' + Q(x, \lambda)\, y = 0\ , \qquad a < x < b\ , \tag{4.14a}$$

$$a_1\, y(a) - a_2\, p(a)y'(a) = 0\ , \tag{4.14b}$$

$$b_1\, y(b) - b_2\, p(b)y'(b) = 0\ , \tag{4.14c}$$

where a, b are finite, and p, Q satisfy certain regularity conditions. This may be solved by a shooting method, based on a Prüfer transformation of the differential equation.

The method of shooting described in the previous section can be applied directly to the solution of (4.14), but Pryce and Hargrave (1977) point out that although this method works well for low-order eigenvalues, it loses accuracy if high-order eigenvalues are required, because of the rapid oscillation of the associated eigenfunctions. The standard approach for solving Sturm-Liouville problems is based on a phase-amplitude representation of y (see Bailey (1966)); the second-order differential equation in (4.14) is transformed into two first-order nonlinear equations by writing

$$y(x) = \rho(x) \sin \theta(x) \tag{4.15a}$$

$$p(x)\, y'(x) = \rho(x) \cos \theta(x) \tag{4.15b}$$

where $\theta(x)$, $\rho(x)$ are the phase and amplitude of $y(x)$. Thus

$$\tan \theta(x) = \frac{y(x)}{p(x)y'(x)}$$

and on differentiating this expression with respect to x we find

$$\theta'(x) = \frac{1}{p(x)} \cos^2 \theta(x) + Q(x, \lambda) \sin^2 \theta(x)\ . \tag{4.16a}$$

The boundary conditions (4.14b), (4.14c) become

$$\left.\begin{aligned} a_1 \sin \theta(a) - a_2 \cos \theta(a) = 0 \,, \\ b_1 \sin \theta(b) - b_2 \cos \theta(b) = 0 \,, \end{aligned}\right\} \qquad (4.16b)$$

which define angles α and β such that

$$\left.\begin{aligned} &\theta(a) = \alpha = \tan^{-1} (a_2 / a_1) \,, \qquad 0 \leqslant \alpha < \pi \\ &\theta(b) = \beta + n\pi \,, \quad \beta = \tan^{-1} (b_2 / b_1) \,, \quad 0 < \beta \leqslant \pi \end{aligned}\right\} \qquad (4.16c)$$

where n is the eigenvalue index. (The eigenvalues are assumed to be ordered so that $0 \leqslant \lambda_0 < \lambda_1 < \lambda_2 < \ldots$.) In addition, we can obtain a first-order differential equation for $\rho(x) = (y(x) / \sin \theta(x))$, by differentiating this expression with respect to x, and substituting from equations (4.15b), (4.16a). We find

$$\frac{\rho'(x)}{\rho(x)} = \sin \theta(x) \cos \theta(x) \left[\frac{1}{p(x)} - Q(x, \lambda) \right] . \qquad (4.17)$$

In order to find the (n+1)st eigenvalue of (4.14), given an initial estimate $\lambda_n^{(0)}$, equation (4.16a) is integrated forwards from x = a and backwards from x = b to a matching point m in (a,b) using the boundary conditions (4.16c) to obtain $\theta_a(m, \lambda_n^{(0)})$ and $\theta_b(m, \lambda_n^{(0)})$. Then the eigenvalue λ_n is obtained as the root of

$$f(\lambda_n) = \theta_a(m, \lambda_n) - \theta_b(m, \lambda_n) \,,$$

which can be determined using the Newton-Raphson method

$$\lambda_n^{(k+1)} = \lambda_n^{(k)} - \frac{f(\lambda_n^{(k)})}{f'(\lambda_n^{(k)})} \,.$$

The equation (4.17) is required only for the calculation of the eigenfunctions of (4.14).

It is sometimes convenient to use a <u>scaled</u> Prüfer transformation (see Pryce and Hargrave (1977)), which is more appropriate for the determination of a specified eigenvalue and the corresponding eigenfunction of a <u>singular</u> problem on a finite or infinite range. Further discussion may be found in the papers by Bailey, Gordon and Shampine (1978) and Bailey and Shampine (1979) which consider the automatic solution of Sturm-Liouville eigenvalue problems.

4. Finite-difference methods

Shooting methods provide approximations to the solution of the BVP in a step-by-step fashion; in finite-difference methods we calculate the whole solution simultaneously, at a set of (not necessarily uniform) mesh-points in the range [a,b]. Derivatives in the differential equations are replaced by finite-difference approximations and the effect of this process is to reduce the problem to that of solving a set of linear or nonlinear equations.

We take as an example the algorithm developed by Lentini and Pereyra (see Pereyra (1979)) for the solution of a general two-point BVP of the form

$$\underline{y}' = \underline{f}(x, \underline{y}), \quad a \leqslant x \leqslant b, \tag{4.18a}$$

$$\underline{g}(\underline{y}(a), \underline{y}(b)) = \underline{0}, \tag{4.18b}$$

where $\underline{y}$, $\underline{f}$, $\underline{g}$ are n-vectors. The implicit trapezoidal rule is used to approximate the solution $\underline{y}$ at a set of (non-uniform) mesh-points x_j, $j = 1,\dots,m+1$ where

$$a = x_1 < x_2 < \dots < x_{m+1} = b .$$

Thus, if $\underline{w}_j \simeq \underline{y}(x_j)$, $j = 1,\dots,m+1$ we obtain the system of difference equations

$$\frac{\underline{w}_{j+1} - \underline{w}_j}{h_j} = \frac{1}{2}\left[\underline{f}(x_j, \underline{w}_j) + \underline{f}(x_{j+1}, \underline{w}_{j+1})\right], \quad j = 1,\dots,m, \tag{4.19a}$$

$$\underline{g}(\underline{w}_1, \underline{w}_{m+1}) = \underline{0}, \tag{4.19b}$$

where $h_j = x_{j+1} - x_j$. Following Pereyra, we separate $\underline{g}$ into three vector components as

$$\underline{g}(\underline{w}_1, \underline{w}_{m+1}) = \begin{bmatrix} \underline{g}^{(1)}(\underline{w}_1) \\ \underline{g}^{(2)}(\underline{w}_1, \underline{w}_{m+1}) \\ \underline{g}^{(3)}(\underline{w}_{m+1}) \end{bmatrix}, \tag{4.20}$$

where $\underline{g}^{(1)}$ corresponds to those b.c.s (a total of p say) given at x = a, $\underline{g}^{(2)}$ to r non-separated b.c.s involving a and b, and $\underline{g}^{(3)}$ to q = n-p-r b.c.s given at x = b. Then equations (4.19) may be written in vector form as

$$\underline{F}(\underline{w}) = \underline{0}, \tag{4.21}$$

where

$$\underline{F}(\underline{w}) = \begin{bmatrix} \underline{g}^{(1)}(\underline{w}_1) \\ \underline{w}_2 - \underline{w}_1 - \frac{h_1}{2}(\underline{f}_1 + \underline{f}_2) \\ \cdot \\ \cdot \\ \cdot \\ \underline{w}_{m+1} - \underline{w}_m - \frac{h_m}{2}(\underline{f}_m + \underline{f}_{m+1}) \\ \underline{g}^{(2)}(\underline{w}_1, \underline{w}_{m+1}) \\ \underline{g}^{(3)}(\underline{w}_{m+1}) \end{bmatrix},$$

$$\underline{w} = [\underline{w}_1\ \underline{w}_2\ \ldots\ \underline{w}_{m+1}]^T,$$

$$\underline{w}_j = [w_{1j}\ w_{2j}\ \ldots\ w_{nj}]^T.$$

Thus we have a system of (m+1)n nonlinear equations to solve for the unknown values w_{ij}, i = 1,...,n, j = 1,...,m+1. A modified Newton method applied to the solution of (4.21) requires the Jacobian matrix $J_F(\underline{w})$ of $\underline{F}$, and the partitioning in (4.20) has been chosen so as to give a matrix J_F which is as near as possible to block tridiagonal form, with blocks of order n :

$$J_F(\underline{w}) = \begin{bmatrix} A_1 & C_1 & & & & & \\ B_2 & A_2 & C_2 & & & 0 & \\ & \cdot & \cdot & \cdot & & & \\ & & \cdot & \cdot & \cdot & & \\ & 0 & & \cdot & \cdot & \cdot & \\ & & & & B_m & A_m & C_m \\ D_1 & & & & & B_{m+1} & A_{m+1} \end{bmatrix} .$$

Note that the first p rows of A_1 contain the elements of $\dfrac{\partial \underline{g}^{(1)}}{\partial \underline{w}_1}$ and the last

(r+q) rows of A_{m+1} contain the elements of $\left[\dfrac{\partial \underline{g}^{(2)}}{\partial \underline{w}_{m+1}} , \dfrac{\partial \underline{g}^{(3)}}{\partial \underline{w}_{m+1}} \right]^T$. The $n \times n$

blocks B_j , C_j have the following forms:

$$B_j = \begin{bmatrix} ////////// \\ \hline 0 \end{bmatrix} \begin{matrix} p \times n \\ (n-p) \times n \end{matrix} \quad , \quad C_j = \begin{bmatrix} 0 \\ \hline ////// \end{bmatrix} \begin{matrix} p \times n \\ (n-p) \times n \end{matrix}$$

and

$$D_1 = \begin{bmatrix} 0 \\ \hline ////////// \\ \hline 0 \end{bmatrix} \begin{matrix} p \times n \\ r \times n \\ q \times n \end{matrix} .$$

(The forms of B_j , C_j differ slightly from those given in Pereyra (1979).)

Thus equations (4.21) are solved using an iterative scheme as for (4.10):

$$J_F(\underline{w}^{(k)})\,\underline{\delta}^{(k+1)} = -\,\underline{F}(\underline{w}^{(k)})\,, \qquad (4.22a)$$

$$\underline{w}^{(k+1)} = \underline{w}^{(k)} + \lambda_{k+1}\,\underline{\delta}^{(k+1)}\,, \qquad 0 < \lambda_{k+1} \leq 1\,. \qquad (4.22b)$$

The solution of equation (4.22a) may be obtained via a stable LU factorisation of J_F, with practically no fill-in (see Keller (1974)), where

$$L = \begin{bmatrix} I_n & & & & & \\ \beta_2 & I_n & & & & \\ & \beta_3 & I_n & & & \\ & & \cdot & \cdot & & \\ & & & \cdot & \cdot & \\ & & & \cdot & \cdot & \\ & & & & \beta_m & I_n \\ \xi_1 & \xi_2 & \cdot & \cdot & \xi_{m-1} \;\; \beta_{m+1} & I_n \end{bmatrix}, \quad U = \begin{bmatrix} \alpha_1 & \gamma_1 & & & & \\ & \alpha_2 & \gamma_2 & & & \\ & & \alpha_3 & \gamma_3 & & \\ & & & \cdot & \cdot & \\ & & & & \cdot & \cdot \\ & & & & \cdot & \cdot \\ & & & & & \alpha_m \;\; \gamma_m \\ & & & & & \alpha_{m+1} \end{bmatrix}$$

are block matrices with blocks of order n, and are such that β_j, B_j have the same structure; so too do γ_j, C_j and ξ_j, D_1.

A good estimate for $\underline{w}^{(0)}$ may be obtained via the method of <u>continuation</u>, which proceeds as follows:

Given a sequence of (user-specified) values

$$0 = \epsilon_1 < \epsilon_2 < \ldots < \epsilon_p = 1\,,$$

where ϵ is a 'continuation parameter', a sequence of problems of the form

$$\left.\begin{aligned} \underline{y}' &= \underline{f}(x, \underline{y}, \epsilon_i)\,, \\ \underline{g}(\underline{y}(a), \underline{y}(b), \epsilon_i) &= \underline{0}\,, \end{aligned}\right\} \qquad (4.23)$$

is solved for i = 1,...,p. This sequence is such that when i = 1, the problem in (4.23) is easy to solve and when i = p, problem (4.23) yields the original BVP whose solution is required. If ϵ_{i+1} is close to ϵ_i the method proceeds on the assumption that the (i+1)st problem can be solved fairly easily using the final iterate of the ith problem as a starting approximation.

Finally we note that if $h = \max_j h_j$, then the error in the approximate solution $\underline{w}$ has an expansion in powers of h^2 and a <u>deferred correction</u> technique (see Hall and Watt (1976), Ch. 15) may be used to improve the accuracy of the computed solution. Following this, an estimate for the leading term in the local truncation error of the trapezoidal method can be obtained on each sub-interval $[x_j, x_{j+1}]$ and the mesh refined in an attempt to keep the error constant in norm on this mesh; this practice is referred to as 'equidistribution' of the error and further details may be found in Lentini and Pereyra (1977).

5. Collocation methods

Suppose we have a system of n nonsingular linear differential equations with multipoint b.c.s, where the ith differential equation in the system has order m_i and is of the form (see Gladwell (1980)):

$$\sum_{j=1}^{m_i+1} \sum_{k=1}^{n} f_{kj}^{i}(x)\, y_k^{(j-1)}(x) = r^i(x)\,, \qquad i = 1,\ldots,n\,. \tag{4.24}$$

The jth b.c. associated with (4.24) has the general form

$$\sum_{j=1}^{\ell_i+1} \sum_{k=1}^{n} f_{kj}^{ij}(x^{ij})\, y_k^{(j-1)}(x^{ij}) = r^{ij}(x^{ij}) \tag{4.25}$$

where $x^{ij} \in [a,b]$. Now if we assume that each $y_i(x)$ has a convergent Chebyshev series expansion, an approximation to each y_i is given by

$$y_i(x) \simeq \sum_{j=0}^{N}{}' a_{ij}\, T_j(t)\,, \qquad t = \frac{2x - a - b}{b - a}\,, \tag{4.26}$$

where $T_j(t)$ is the first-kind Chebyshev polynomial of degree j and the prime on the summation sign implies that the first term is halved. It follows that

$$y_i^{(r)}(x) \simeq \sum_{j=0}^{N-r}{}' \, a_{ij}^r \, T_j(t) \left(\frac{2}{b-a}\right)^r \tag{4.27a}$$

where

$$\left.\begin{array}{ll} 2j\, a_{ij}^r = a_{i,j-1}^{r+1} - a_{i,j+1}^{r+1}\,, & j \geq 1 \\ a_{ij} = 0\,, & j > n-r-1\,. \end{array}\right\} \tag{4.27b}$$

The collocation method determines the Chebyshev coefficients a_{ij}, $i = 1,\ldots,n$, $j = 0,1,\ldots,N$ by substituting the Chebyshev expansions of the solution components and their derivatives into the differential equation and b.c.s (4.24), (4.25); it is required that the b.c.s are satisfied exactly, while the differential equations are satisfied at a number of points, the so-called collocation points, in the range $[a,b]$. The number of collocation points, k say, must be specified by the user and these points are chosen at the extrema of a shifted Chebyshev polynomial of degree $k-1$. Then the remaining unknown Chebyshev coefficients are obtained by solving a dense linear system of equations. (If it is overdetermined, we use the method of least squares; see Chapter 1.)

The solution of a nonlinear system of BVPs may be found by this method if the system is first linearised (see Aktas and Stetter (1977), Albasiny (1979)), thereby reducing the original nonlinear system to a sequence of linear BVPs. This type of approach may be useful if the solution is required in a semi-analytical form such as (4.26) for later evaluation.

The spline collocation method of Ascher, Christiansen and Russell (1979) may also be used to determine the solution of a nonlinear system of BVPs. For convenience we write the system as

$$y_i^{(m_i)}(x) = F_i(x;\, \underline{z}(\underline{y}))\,, \qquad i = 1,\ldots,n\,, \tag{4.28}$$

where

$$m_1 \leq m_2 \leq \ldots \leq m_n$$

and

$$\underline{z}(\underline{y}) = (y_1\,,\, y_1'\,,\, \ldots\,,\, y_1^{(m_1-1)}\,,\, \ldots\,,\, y_n\,,\, y_n'\,,\, \ldots\,,\, y_n^{(m_n-1)})\,.$$

Equations (4.28) are to be solved subject to $\bar{m} = \sum_{i=1}^{n} m_i$ multipoint separated boundary conditions

$$g_j(x^j, \underline{z}(\underline{y})) = 0 , \qquad j = 1,\ldots,\bar{m} , \tag{4.29}$$

where $a \leqslant x^1 \leqslant x^2 \leqslant \ldots \leqslant x^{\bar{m}} \leqslant b$. Then if we define a mesh Π on $[a,b]$ by

$$\Pi : a = x_1 < x_2 < \ldots < x_{N+1} = b$$

where the sub-interval $[x_i, x_{i+1}]$ is of length h_i , $i = 1,\ldots,N$ and splines

$$P_{\ell,\Pi} = \{v : v|_{(x_i,x_{i+1})} \text{ is a polynomial of degree} < \ell,\ i = 1,\ldots,N\}$$

the method produces a vector approximation $\underline{v} = [v_1\ v_2\ \ldots\ v_n]^T$ to the required solution $y = [y_1\ y_2\ \ldots\ y_n]^T$, with

$$v_i \in P_{k+m_i,\Pi} \cap C^{m_i-1}[a,b] , \qquad i = 1,\ldots,n , \tag{4.30}$$

where $k \geqslant m_n$ is the number of collocation points in each open sub-interval of Π. The collocation approximation $\underline{v}$ is determined by requiring that $\underline{v}$ satisfies the b.c.s (4.29) exactly and the differential equations (4.28) at the collocation points.

A particular choice of the collocation points, defined in terms of the k Gauss-Legendre points on [-1,1] (see Ascher et al. (1979) for further details), then leads to the following theoretically attainable orders of accuracy:

$$||y_i^{(\ell)} - v_i^{(\ell)}||_\infty = O(h^{k+m_i+\ell}) , \quad \ell = 0,\ldots,m_i ; \quad i = 1,\ldots,n ,$$

$$|y_i^{(\ell)}(x_j) - v_i^{\ell}(x_j)| = O(h^{2k}) , \quad j = 1,\ldots,N ; \quad \ell = 0,\ldots,m_i-1 ;$$
$$i = 1,\ldots,n ,$$

where

$$h = \max_{1 \leqslant j \leqslant N} h_j .$$

6. Availability of software

The main methods discussed in this chapter are implemented in Mark 11 of the NAG Library, and a general description of the NAG routines for solving boundary-value problems is given by Gladwell (1979). It should be noted that for difficult nonlinear problems there is no guarantee that a particular method will work satisfactorily, and the user may have to try different approaches in order to obtain a solution.

For simple problems, routines based on the shooting method are generally easy to use, but for more general systems a rather complex calling sequence is required, which the user must study carefully. The basic routines are D02AGF, D02HAF, D02HBF; the extensions indicated in equations (4.13) are covered by D02SAF. All these routines use a fourth-order Runge-Kutta method with local error control to carry out the integration. The multiple shooting option is not available in the NAG routines at present.

The Sturm-Liouville eigenvalue problem is covered by the routines D02KAF, D02KDF, D02KEF, which use the phase-amplitude method discussed in section 3.

A finite-difference method is implemented in D02RAF on the lines described in section 4; this is adapted from the code PASVA3 developed by Lentini and Pereyra (Pereyra 1979). Chebyshev collocation methods (section 5) are implemented for linear problems in D02JAF, D02JBF, D02TGF.

The spline collocation method described in section 5 is implemented in COLSYS (Ascher, Christiansen and Russell 1981), which uses B-splines as basis functions.

CHAPTER 5

Numerical Integration
Part A. Extrapolation Methods for Multi-dimensional Quadrature

J.N. Lyness

1. Introduction

The principal purpose of this article is to draw the readers' attention to the existence of various expansions on which extrapolation may be based. Relatively few expansions in quadrature are known. These are usually simple to state and simple to use, but some require long involved proofs (which are not given here). We also discuss a simple way to use them.

A secondary purpose is to indicate how some of the elementary theory of numerical quadrature can be based on the Euler Maclaurin summation formula. If this is done, extrapolation in quadrature can be seen as a natural integrated branch of numerical quadrature and not (as is so often the case in textbooks) as an apparently independent theory.

The development of an extrapolation technique is conveniently considered in three phases. The first phase involves choosing a suitable expansion on which to base the extrapolation. The second consists of deciding on a quadrature rule and a mesh ratio sequence. A third phase involves choosing an efficient algorithm for the extrapolation calculation. For example, in the classical version of Romberg integration (see Bauer, Rutishauser and Stiefel (1963), Romberg (1955)) (in one dimension) these choices are: the Euler Maclaurin expansion ((5.1) below), the end-point trapezoidal rule and the geometric mesh ratio sequence ((5.8) below), and the Neville algorithm ((5.6) below).

The major limitation on use of extrapolation in quadrature seems to be the construction of a suitable expansion for the error functional. The next two sections are devoted exclusively to describing nearly all the expansions which are presently available. In section 2 we deal with an analytic integrand function and in section 3 with a function having a small number of recognizable and conveniently located singularities. In both we treat first the one-dimensional finite interval and then the N-dimensional hypercube. In section 4 we briefly treat phase 3, the algorithm. In section 6 we discuss phase 2, the choice of quadrature rule and mesh ratio sequence. It is in these sections that the difference of techniques because of dimensional effects comes into play. As a preliminary we treat numerical instability in section 5. Finally, in section 8 we discuss briefly the possible role of these methods in the present numerical quadrature environment. The emphasis in this article is on the underlying ideas. For proofs and theorems, the reader should refer to the cited references.

2. An analytic integrand function

A quite distinctive approach to numerical quadrature for the finite interval may be based on Richardson's deferred approach to the limit. One application of this is known as Romberg integration. From the present point of view, one may think of this in terms of considering in detail the convergence of a special Riemann sum to its limit. For convenience, we scale these results to the interval [0,1].

We define the end-point (m-panel) trapezoidal rule

$$R^{[m,1]}f = \frac{1}{m}\left\{ \frac{1}{2}\, f(0) + \sum_{j=1}^{m-1} f(j/m) + \frac{1}{2}\, f(1) \right\}. \qquad (5.1)$$

Here the positive integer m is the reciprocal of the more conventional steplength h. For all Riemann-integrable functions we can assert

$$\lim_{m\to\infty} R^{[m,1]}f \;=\; If \;\equiv\; \int_0^1 f(x)\,dx. \qquad (5.1a)$$

With no further information available, a possible procedure might be to calculate a sequence of trapezoidal rule approximations with increasing m, and to stop when it appears to have converged. (Incidentally, this 'sledgehammer approach' corresponds roughly with the correct one in the case that f(x) is analytic and periodic with period unity.) If one were to carry this out, one would almost certainly notice a pattern of regular behaviour as m becomes large. The nature of this pattern depends on the nature of f(x).

When $f(x)$ is analytic, it is governed by the Euler Maclaurin asymptotic expansion

$$R^{[m,1]}f = If + \frac{B_2}{2!}\frac{f'(1) - f'(0)}{m^2} + \frac{B_4}{4!}\frac{f^{(3)}(1) - f^{(3)}(0)}{m^4}$$

$$+ \dots + \frac{B_{2p}}{(2p)!}\frac{f^{(2p-1)}(1) - f^{(2p-1)}(0)}{m^{2p}}$$

$$+ \frac{1}{m^{2p+2}}\int_0^1 f^{(2p+2)}(t)\, K_{2p+2}^{[m,1]}(t)\, dt. \qquad (5.2)$$

Here B_{2j} are Bernoulli numbers and $B_{2j}/(2j!) \propto 1/(2\pi)^{2j}$. In view of this, if we had evaluated $R^{[m,1]}f$ for several values of m, we could try to predict the value of If by fitting these results to an expression of the form

$$R^{[m,1]}f = If + \frac{A_2}{m^2} + \frac{A_4}{m^4} + \dots \quad . \qquad (5.3)$$

A fitting technique of this sort is known as extrapolation. Note that to do this one needs to know the form of the expression (5.3) above but one does not need to know the numerical values of the coefficients

$$A_{2s} = \frac{B_{2s}}{(2s)!}\left(f^{(2s-1)}(1) - f^{(2s-1)}(0)\right).$$

Estimates of these quantities could become available if required by a by-product of the extrapolation process.

To carry out extrapolation, one chooses a mesh ratio sequence. This is a sequence of values of m, say $m_0, m_1, m_2, \dots$. Classically $m_j = 2^j$ but more efficient sequences exist. A convenient mechanism for extrapolation based on (5.3) involves the construction of the Romberg T-table

$$\begin{array}{lll} T_{0,0} & & \\ & T_{0,1} & \\ T_{1,0} & & T_{0,2} \\ & T_{1,1} & \\ T_{2,0} & & \end{array} \tag{5.4}$$

The initial column of this table is defined by

$$T_{j,0} = R^{[m_j,1]} f \, . \tag{5.5}$$

Elements of subsequent columns may be calculated from the two nearest neighbours in the previous column using the Neville Romberg algorithm, namely

$$T_{k,p} = \frac{m_{k+1}^2 T_{k+1,p-1} - m_k^2 T_{k,p-1}}{m_{k+1}^2 - m_k^2} \, . \tag{5.6}$$

Following the genesis of $T_{k,p}$ one finds that it is a linear combination of the form

$$T_{k,p} = \sum_{j=k}^{k+p} \gamma_{j,0}^{k,p} T_{j,0} \, , \tag{5.7}$$

the coefficients $\gamma_{j,0}^{k,p}$ depending on the mesh ratio sequence only. $T_{k,p}$ defined by (5.5) and (5.6) is, in fact, the approximation to If that one obtains by solving the p+1 linear equations in p+1 unknown coefficients of m^{-2s} obtained by disregarding the remainder term in (5.2) and letting m take the values m_j , j = k, k+1, ..., k+p.

In the classical theory (see Bauer et al. (1963)) one chooses mesh ratios

$$m_j = 2^j \, . \tag{5.8}$$

It can be shown that, when f(x) is analytic and mesh ratio sequence (5.8) is used all columns and diagonals converge, i.e.

$$\lim_{k \to \infty} T_{k,p} = If \quad , \quad \text{for all } p \; , \tag{5.9}$$

$$\lim_{p \to \infty} T_{k,p} = If \quad , \quad \text{for all } k \; . \tag{5.10}$$

For all other mesh ratio sequences, the columns converge. Whether or not the diagonals converge depends on the mesh ratio sequence.

The brief outline above follows a standard way of presenting Romberg integration. The impression left is that it has little if anything to do with the rest of numerical quadrature. This is a pity, since it is intimately related. So I am happy to take this opportunity to point out some of these connections.

One interesting and important property of the trapezoidal rule is that it is of trigonometric degree m−1 with respect to the integration interval; that is

$$R^{[m,1]}f = If \quad \text{when} \quad f(x) = \sum_{j=0}^{m-1} (a_j \cos 2\pi jx + b_j \sin 2\pi jx). \tag{5.11}$$

We note that in view of (5.7), the element $T_{k,p}$ is a linear sum of trapezoidal rule approximations and so a linear sum of function values. Thus it is a quadrature rule approximation. In fact, when the geometric mesh ratio sequence (5.8) is used, $T_{k,1}$ is a 2^k-copy of Simpson's rule and $T_{k,2}$ is a 2^k-copy of the four-panel Newton-Cotes rule. It is readily verified that $T_{k,p}$ = If when f is a polynomial of degree 2p+1. (In this case the remainder term in (5.2) is indeed zero and no approximation error is made in disregarding it.) Moreover $T_{k,p}$, being a linear sum of $R^{[m,1]}f$ with $m \geqslant m_k$, is of trigonometric degree m_k-1. In short, each element of the Romberg T-table represents a different quadrature rule approximation to If. Specifically, $T_{k,p}$ is of polynomial degree 2p+1 and trigonometric degree m_k-1.

To pursue the connections with standard quadrature theory, we have to release ourselves from the bondage of the trapezoidal rule. We need to define, in addition to (5.1), the offset trapezoidal rules

$$R^{[m,\nu]}f = \frac{1}{m}\sum_{j=0}^{m-1} f\left(\frac{j+t_\nu}{m}\right) \; ; \quad t_\nu = \frac{1+\nu}{2}, \quad -1 < \nu < 1. \tag{5.12}$$

Each of these m-point rules is also of trigonometric degree m-1 and each enjoys a minor variant of the Euler Maclaurin asymptotic expansion. Corresponding to (5.2) we have

$$R^{[m,\nu]}f = If + \sum_{q=1}^{2p+1} \frac{B_q(t_\nu)}{q!} \frac{f^{(q-1)}(1) - f^{(q-1)}(0)}{m^q}$$

$$+ (1/m^{2p+2}) \int f^{(2p+2)}(t)\, K_{2p+2}^{[m,\nu]}(t)\, dt \; . \tag{5.13}$$

Here $B_q(x)$ is the Bernoulli polynomial (see Abramowitz and Stegun (1964)). The most familiar of these offset rules is the mid-point rule, having $\nu = 0$ (and $t_\nu = \frac{1}{2}$). Since $B_q(\frac{1}{2}) = 0$ for q odd, expansion (5.13) becomes an even expansion in m^{-1} for the mid-point rule, as is (5.2) for the end-point rule.

We now take any quadrature rule Qf; for convenience we shall assume that the weights (if any) assigned to x = 0 and x = 1 are equal and that Q is of degree zero. Then

$$Qf = \sum_{k=1}^{N} w_k f(t_{\nu_k}) + \frac{w_0}{2}\,(f(0) + f(1)). \tag{5.14}$$

The m-copy version $Q^{(m)}$ of a rule Q is the rule obtained by subdividing the integration interval into m equal parts and applying a properly scaled version of Q to each. This gives

$$Q^{(m)}f = \sum_{k=1}^{N} \frac{w_k}{m} \sum_{j=0}^{m-1} f\left(\frac{j+t_{\nu_k}}{m}\right) + \frac{w_0}{m} \sum_{j=0}^{m}{}'' \; f\left(\frac{j}{m}\right) \tag{5.15}$$

which in view of (5.12) and (5.1) may be written in the form

$$Q^{(m)}f = \sum_{k=1}^{N} w_k\, R^{[m,\nu_k]} f + w_0\, R^{[m,1]}f \; . \tag{5.16}$$

That the m-copy version of a general rule can be expressed in terms of m-panel offset trapezoidal rules should be obvious geometrically after a moment's reflection. It also follows that one can apply the Euler Maclaurin asymptotic expansion (5.2) or (5.13) to each component. This gives an Euler Maclaurin expansion for $Q^{(m)}f$, namely

$$Q^{(m)}f = If + \sum_{s=1}^{2p+1} B_s(Q;\ f)/m^s + \frac{1}{m^{2p+2}} \int_0^1 f^{(2p+2)}(t)\, K_{2p+2}(Q,\ t)\, dt\ . \qquad (5.17)$$

Here

$$B_s(Q;\ f) = C_s(Q)\,(f^{(s-1)}(1) - f^{(s-1)}(0)) \qquad (5.18)$$

where

$$C_s(Q) = Q\,\varphi_s \qquad (5.19)$$

with

$$\varphi_s(x) = B_s(x)/s!\ . \qquad (5.20)$$

An aesthetically satisfying theory then assures us that when Q is a symmetric rule $C_s(Q) = 0$ when s is odd, and when Q has polynomial degree d, $C_s(Q) = 0$ for $s = 1,2,\ldots,d-1$.

A possible calculational procedure might be to construct a sequence of rules $G_\nu^{(m)}f$, these being m-copies of the ν-point Gauss-Legendre rule. Since this has polynomial degree $2\nu-1$, the extrapolation should be based on the asymptotic expansion

$$G_\nu^{(m)}f \sim If + \frac{B_{2\nu}}{m^{2\nu}} + \frac{B_{2\nu+2}}{m^{2\nu+2}} + \ldots\ . \qquad (5.21)$$

The rule $G_\nu^{(m)}f$, like all m-copy rules, is of trigonometric degree m-1. In this case the Neville algorithm as written should not be used. A simple modification exists. However, we take up this question in more generality in section 4 below.

Before proceeding to the N-dimensional generalization, we remark that the elementary theory of numerical quadrature can be developed through the Euler Maclaurin expansion (5.2) and its generalizations (5.13) and (5.17). The particular degree of symmetry of a rule Q can be determined by noting which terms drop out of (5.17). The Peano kernel and its variants are readily developed from the remainder terms in these expansions. One of the justifications for the use of higher degree rules is that the approach of a Riemann sum to its limit (5.1a) seems to be faster for (5.21) than for (5.3). The theory presented in this way connects up again shortly after this point with the standard theory.

We complete this section by stating the generalization of these results to the unit N-dimensional hypercube (Baker and Hodgson 1971)

$$H:\ 0 \leq x_i \leq 1, \qquad i = 1,2,\ldots,N\ . \tag{5.22}$$

(The user should remember that all the results apply with trivial modification to any region which is obtained from this by any affine transformation. For example, in two dimensions, to an arbitrarily located and oriented parallelogram.) Here Q is, as before, a degree zero quadrature rule. $Q^{(m)}f$ is the m^N-copy, obtained by subdividing H into m^N hypercubes each of side 1/m and applying a properly scaled version of Q to each. When $f(\underline{x})$ is analytic in each of its variables $x_1,x_2,\ldots,x_N$ over the hypercubes, one obtains without much difficulty an N-dimensional version of (5.13). This is

$$Q^{(m)}f = If + \sum_{s=1}^{p-1} \frac{B_s(Q;\ f)}{m^s} + R_p(Q^{(m)};\ f)\ , \qquad p \geq N\ . \tag{5.23}$$

Explicit expressions for the coefficients are known. These are

$$B_s(Q;\ f) = \sum_{\Sigma s_i = s} C_{s_1,s_2,\ldots,s_N}(Q)\ If^{(s_1,s_2,\ldots,s_N)} \tag{5.24}$$

where

$$C_{s_1,s_2,\ldots,s_N}(Q) = Q\ \varphi_{s_1,s_2,\ldots,s_N} \tag{5.25}$$

with

$$\varphi_{s_1,s_2,\ldots,s_N}(\underline{x}) = \frac{B_{s_1}(x_1)}{s_1!}\ \frac{B_{s_2}(x_2)}{s_2!}\ \cdots\ \frac{B_{s_N}(x_N)}{s_N!}\ . \tag{5.26}$$

Also

$$R_p(Q^{(m)};\ f) = O(m^{-p}) \ . \tag{5.27}$$

When Q is of polynomial degree d, or symmetric, or both, some of the coefficients $B_s(Q;\ f)$ vanish under precisely the same circumstances as in the one-dimensional case, described previously.

The derivation of this result is very straightforward. This is because the term $f^{(s-1)}(1) - f^{(s-1)}(0)$ may be written as the integral $If^{(s)}$, and the integral over a hypercube is a product of one-dimensional integrals <u>having constant limits</u>. Specific integral representations for the remainder term exist but are quite complicated (see Lyness and McHugh (1970)). It is worth noting that for general quadrature regions, the formula (5.23) is not valid. However, for simplectical regions (triangle, tetrahedron, ...) a formula corresponding to (5.23) exists. The proof for the simplex is far more difficult and the coefficients are given by expressions more complicated than (5.24).

When Q is an N-dimensional product of end-point or mid-point trapezoidal rules, one may use the Neville algorithm ((5.4) - (5.6)) for extrapolation. When Q is a rule of higher degree, the general extrapolation method described in section 4 may be used.

3. Some singular integrand functions

In the previous section we limited ourselves to functions $f(x)$ analytic in a region containing $[0,1]$. In fact the Euler Maclaurin expansion (5.2) as written is valid only for functions $f(x) \in C^{2p+3}[0,1]$. When $f(x)$ has a singularity in $[0,1]$, the expansion may not exist and cannot be used as a basis for extrapolation. Similar remarks are valid in an N-dimensional context.

However, for some singular integrands an alternative expansion of the same nature exists. In one dimension, integrand functions having algebraic and logarithmic singularities at one end or at both enjoy expansions like (5.29) given below. In the following, $g(x)$ is to be taken to be analytic and is a factor of $f(x)$, the integrand function. When $f(x) = g(x)x^{\alpha}$, $\alpha > -1$,

$$Q^{(m)}f \sim If + \sum_{q=0} \frac{A_{q+\alpha+1}}{m^{q+\alpha+1}} + \sum_{s=1} \frac{B_s}{m^s} \ . \tag{5.28}$$

Any function value at $x = 0$ may be ignored when $\alpha < 0$ (and also when $\alpha > 0$). If Q is symmetric $B_s = 0$ for odd s and if Q is of polynomial degree d

$$B_s = 0 , \quad 1 \leqslant s \leqslant d-1 .$$

When $f(x) = g(x)x^{\alpha}(1-x)^{\beta}$, $\alpha,\beta > -1$

$$Q^{(m)}f \sim If + \sum_{q=0} \frac{A_{q+\alpha+1}}{m^{q+\alpha+1}} + \sum_{q=0} \frac{C_{q+\beta+1}}{m^{q+\beta+1}} . \tag{5.29}$$

Expansions for some logarithmic singularities can be written down by differentiating these with respect to some incidental parameters, say α. Thus when $f(x) = g(x)\, x^{\alpha}\, \ln x\, (1-x)^{\beta}$,

$$Q^{(m)}f \sim If + \sum_{q=0} \frac{A'_{q+\alpha+1}}{m^{q+\alpha+1}} - \frac{A_{q+\alpha+1} \ln m}{m^{q+\alpha+1}} + \sum_{q=0} \frac{C'_{q+\beta+1}}{m^{q+\beta+1}} . \tag{5.30}$$

Successively simpler proofs of these expansions can be found in Navot (1961), in Lyness and Ninham (1967), and in Lyness (1971). Expressions for the coefficients are known. For example, when Q is the end-point trapezoidal rule $R^{[1,1]}$,

$$B_{2s}(Q;\, f) = \frac{B_{2s}}{(2s)!} f^{(2s-1)}(1) \tag{5.31}$$

and

$$A_{q+\alpha+1}(Q;\, f) = \frac{\theta^{(q)}(0)\zeta(-(q+\alpha))}{q!} , \tag{5.32}$$

where $\theta(x)$ is the regular part of $f(x)$ at $x = 0$, i.e., $\theta(x) = f(x)/x^{\alpha}$ and $\zeta(s)$ is the Riemann Zeta function. However, no simple integral representation of the remainder term is known to the author.

To complete the picture, we remark that expansions of the same general nature exist when $f(x)$ has algebraic or logarithmic singularities within the integration interval. However, the coefficients $A_{q+\alpha+1}$ etc. then depend on m in a somewhat complicated way, and extrapolation, while theoretically possible, is unduly complicated. In practice one should always arrange the integration interval so that the singularities fall at the end points of integration intervals.

It is just worthwhile to pause here to consider the use of rules of moderate polynomial degree for functions such as (5.28) and (5.29). We assume that α and β are not integers. The Gaussian rules are constructed to remove terms involving m^{-s}, $s = 1,2,\dots,d$ from the expansion. They do this by using more function values. In expansion (5.29) there is no point at all in doing this. In expansion (5.28), which does contain some terms of this type, there may be some point. If for example α is large, then removing the coefficients of m^{-s} with $s < \alpha$ is clearly advantageous. To remove any more in this way is probably not cost effective as the principal error term, $A_{\alpha+1}/m^{\alpha+1}$, remains.

We now turn to the possible generalizations of these expansions to the hypercube. The direct generalization in two dimensions obtained by taking a product would be to integrand functions which have singularities like $x^{\alpha}y^{\beta}$. The expansions for functions such as these are straightforward products of the corresponding one-dimensional expansions.

A much more challenging and probably more common situation is one in which there is a singularity of the form $f(\underline{x}) = r^{\alpha}g(\underline{x})$, $r = ||\underline{x}||$ being the radial distance of $\underline{x}$ from the origin (see Lyness (1976a, b)). The development of expansions for this followed the derivation of an expansion for homogeneous functions. A homogeneous function $f_{\gamma}(\underline{x})$ of degree γ about the origin is one satisfying

$$f(\lambda x_1,\lambda x_2,\dots,\lambda x_N) = \lambda^{\gamma}f(x_1,x_2,\dots,x_N), \text{ for all } \lambda > 0 . \quad (5.33)$$

Examples in two dimensions include

$$r^{\gamma}, \quad (\lambda x + \mu y)^{\gamma}, \quad (x \arctan y/x)^{\gamma} . \quad (5.34)$$

The product of homogeneous functions of degrees γ_1 and γ_2 respectively is one of degree $\gamma_1+\gamma_2$. The following theorem is basic.

Let $f_{\gamma}(\underline{x})$ be a homogeneous function of degree γ having no singularity within the closure of H except at the origin. Then

$$Q^{(m)}f_{\gamma} \sim If_{\gamma} + \frac{A_{N+\gamma}}{m^{N+\gamma}} + \frac{C_{N+\gamma}\ \ln m}{m^{N+\gamma}} + \sum_{s=1} \frac{B_{\gamma,s}}{m^{s}} . \quad (5.35)$$

However,

$$C_{N+\gamma} = 0 \quad \text{if} \quad \gamma \neq \text{integer} , \tag{5.36}$$

so in many cases the logarithmic term is absent. Moreover, when $Q^{(m)}$ is of degree d, $B_{\gamma,s}$ and C_s vanish for $s < d$ and when $Q^{(m)}$ is symmetric, $B_{\gamma,s}$ and C_s vanish for s odd.

The proof of (5.35) is relatively long. But once established, it opens the floodgate to a host of corollaries. These come about because a function like $f(\underline{x}) = r^{\alpha}g(\underline{x})$ may be expanded (using a multivariate Taylor expansion) into a form $f(\underline{x}) = f_{\alpha}(\underline{x}) + f_{\alpha+1}(\underline{x}) + \ldots$ with a final term $r(\underline{x})$. The functions $f_{\alpha+j}(\underline{x})$ are homogeneous functions of degree $\alpha+j$ and so enjoy expansion (5.35) with $\gamma = \alpha+j$. The final term $r(\underline{x})$ has a higher degree of continuity than $g(\underline{x})$ because the more significant part of the singularity has been subtracted. Thus one can use (5.23) with a moderate value of p for $r(\underline{x})$. After some manipulation (see Lyness (1976a)) one may establish an expansion valid for

$$f(\underline{x}) = r^{\alpha}\varphi(\underline{\theta})h(r)g(\underline{x}) , \qquad \alpha > -N , \tag{5.37}$$

where $(r, \underline{\theta})$ are the hyperspherical coordinates of $\underline{x}$ and the functions φ, $\underline{\theta}$ and f are analytic in their respective arguments. This expansion is

$$Q^{(m)}f - If \sim \left\{ \begin{array}{ll} \sum\limits_{t=0} \dfrac{A_{\alpha+N+t}}{m^{\alpha+N+t}} + \sum\limits_{s=1} \dfrac{B_s}{m^s} , & \alpha \neq \text{integer} , \\ \sim \sum\limits_{s=1} \dfrac{A_s+B_s}{m^s} + \sum\limits_{s=1} \dfrac{C_s \ \ln m}{m^s} , & \alpha = \text{integer} . \end{array} \right. \tag{5.38}$$

One may obtain expansions for functions of this form but with an additional $(\ln r)$ factor by differentiating with respect to α. (See Lyness (1976a).)

As is usual, some of these coefficients vanish; when Q is symmetric $B_s = C_s = 0$ for all s odd and when Q is of polynomial degree d, $B_s = C_s = 0$ for all s satisfying $0 < s < d$.

There are many other, sometimes unexpected, conditions in which terms in (5.38) vanish. One of them is quoted here.

If $h(r)g(\underline{x})$ is symmetric about the origin, i.e.

$$h(r)g(\underline{x}) = h(-r)g(-\underline{x})$$

then

$$A_{\alpha+s+N} = C_{\alpha+s+N} = 0 , \quad \text{for all } s \text{ odd.}$$

Since many terms may vanish under special circumstances it would be prudent for a serious user of these expansions to refer to references Lyness (1976a, b) before making extensive use of these techniques.

4. The mechanics of extrapolation (phase 3)

All the expansions mentioned in this article are of the form

$$Q^{(m)}f = If + a_1f_1(m) + a_2f_2(m) + \ldots + a_pf_p(m)$$
$$+ O(f_{p+1}(m)) \qquad (5.39)$$

where the functions $f_j(m)$ are functions like m^{-s}, $m^{-\alpha}$, $m^{-\alpha}\ell n\, m$ and satisfy

$$\lim_{m \to \infty} f_{s+1}(m)/f_s(m) = 0 . \qquad (5.40)$$

It is convenient to set $If = a_0f_0(m)$ with $f_0(m) = 1$. In section 2 we defined elements $T_{k,p}$ (here called $T_{k,p,0}$) of the Romberg T-table. These depend on the set of mesh ratios m_j where

$$1 \leqslant m_0 < m_1 < m_2 < \ldots \qquad (5.41)$$

and the initial column of the T-table is given by

$$T_{j,0,0} = Q^{(m_j)} f . \qquad (5.42)$$

The element $T_{k,p,0}$ could be obtained by omitting the remainder term in (5.39), considering the equations with m taking the p+1 values m_k , m_{k+1} ,..., m_{k+p} and solving for a_0. If we denote by $T_{k,p,s}$ the value of a_s obtained by this procedure, we find

$$T_{j,0,0} = \sum_{s=0}^{p} T_{k,p,s} f_s(m_j) \; , \quad j = k,k+1,\ldots,k+p \; . \tag{5.43}$$

We denote the matrix of these equations by $V^{[k,p]}$. Its elements are

$$V^{[k,p]}_{j,s} = f_s(m_j) \tag{5.44}$$

and so $T_{k,p,0}$ is the first element of the solution $\underline{x}$ of this set of equations, that is,

$$V^{[k,p]}\underline{x} = \underline{b} \; , \tag{5.45}$$

the elements of $\underline{b}$ being the quadrature rule approximations $T_{j,0,0}$.

There are many algorithms for extrapolation in special situations. The Neville algorithm (5.4) - (5.6) is simply the first of a long line. Recently efficient algorithms have been produced for the general situation (5.39) by Brezinski (1980) and Håvie (1979). The present author would like to emphasize that a good algorithm is a useful but not an essential part of this technique. So long as one has a linear equation solver (for example, LINPACK) available, one can go ahead and solve (5.45) to obtain extrapolated approximations. In multidimensional quadrature, the major computer time spent will be on the rule sum evaluations $Q^{(m_j)}f$. The time spent on calculating extrapolated values is much much shorter, whether one uses a well designed algorithm or uses a general linear equation solver.

5. Numerical stability

If our problem is one of those for which a suitable extrapolation expansion exists, we now have available at least one method for carrying out the extrapolation. All that remains is to decide on a rule Q and a mesh ratio sequence m_0 , m_1 , m_2 , ... Before doing this it is necessary to face what is really the only serious drawback to using extrapolation; this is the possibility of encountering gross numerical instability. In this section we describe this phenomenon briefly. With this background, we shall be able, in section 6, to return to the problem of choosing a rule and a mesh ratio sequence.

Numerical instability is the name given to a phenomenon in which individually very small errors in the early part of a calculation are amplified by the succeeding calculation causing a much larger error in the final result.

Experience has shown that in extrapolation, the only serious source of error is the condition error. To estimate this, we first assign a <u>noise level</u> ϵ_f to function evaluations. That is, we assume that the (rounded or truncated) function value $\tilde{f}(x_j)$ used in the calculation is given by

$$\tilde{f}(x_j) = f(x_j) + \varphi_j \epsilon_f , \qquad |\varphi_j| \leqslant 1 \tag{5.46}$$

for some φ_j. We then attempt to obtain a bound on $|Q\tilde{f} - Qf|$ in terms of ϵ_f.

In the context of N-dimensional quadrature, a convenient variant approach is as follows. We define a noise level for the quantity $Q^{(m_j)} f = T_{j,0,0}$ and denote this by ϵ_Q. We suppose

$$\tilde{T}_{j,0,0} = T_{j,0,0} + \theta_j \epsilon_Q , \quad |\theta_j| \leqslant 1 , \tag{5.47}$$

and that the algorithm used to calculate $T_{k,p,0}$ is based on the first column $\tilde{T}_{j,0,0}$. It can be shown that

$$|\tilde{T}_{k,p,0} - T_{k,p,0}| \leqslant K_{k,p,0} \, \epsilon_Q \tag{5.48}$$

where the condition number $K_{k,p,0}$ is well defined and easy to calculate. In fact, under two minor restrictions, $\pm K_{k,p,0}$ is the first element x_1 of the solution of

$$V^{[k,p]} \underline{x} = \underline{a} \tag{5.49}$$

where $V^{[k,p]}$ is given by (5.44) and

$$\underline{a} = [1,-1, 1,-1, 1,\ldots]^T . \tag{5.50}$$

The restrictions are that $m_0 < m_1 < m_2 < \ldots$ and that the final term $f_p(m)$ in (5.39) is of the form $m^{-\alpha}$ (and not $m^{-\alpha}\ln m$). In short, any algorithm used to calculate $T_{k,p,0}$ may be used again to calculate $K_{k,p,0}$ by simply replacing $T_{j,0,0}$ by $(-1)^j$.

Table 1

Values of $K_{0,p,0}$ for three ratio sequences and four error functional asymptotic expansions*

Negative powers of m in expansion	p	G	F	H
2, 4, 6, 8	4	1.95	6.2	6.2
10, 12, 14	7	1.96	7.4	55.8
16, 18, 20	10	1.96	9.2	552.8
0.5, 2, 2.5, 4	4	13.9	63	63
4.5, 6, 6.5	7	17.7	250	2,494
8, 8.5, 10	10	18.4	450	107,963
0.5, 1.5, 2, 2.5	4	20	95	95
3.5, 4, 4.5	7	39	883	8,931
5.5, 6, 6.5	10	46	2,720	904,363
0.5, 1.5, 2.5	3	12.2	22	22
3.5, 4.5	5	20.8	156	291
5.5, 6.5	7	23.8	379	3,783

* These have $f_j(m) = m^{-s_j}$, the values of s_j appearing in the first column.

The value of $K_{k,p,0}$ depends on the mesh ratio sequence and on the sequence $f_j(m)$ in (5.39). Table 1, taken from Lyness (1976b), gives some feeling for the possible size of these condition numbers $K_{0,p,0}$ for a few values of p. The mesh ratio sequences used here are

G: {m} = 1, 2, 4, 8, 16, 32, 64, ...

F: {m} = 1, 2, 3, 4, 6, 8, 12, ...

H: {m} = 1, 2, 3, 4, 5, 6, 7, ...

where G and H stand for Geometric and Harmonic. The F-sequence was suggested by Bulirsch (1964).

6. Remarks on choosing rule and mesh

The choice of rule and mesh sequence is an art rather than a science. The information on which such a choice could be made is contained in this article. Because of the possible numerical instability, the choice depends on the accuracy of the function values ϵ_f compared with the required accuracy ϵ_{req}.

If, for example, $\epsilon_f = 10^{-7}$ and we required a result for the unit hypercube $\epsilon_{req} = 10^{-5}$, then certainly no results $T_{k,p,0}$ for which $K_{k,p,0} > 100$ should be trusted. On the other hand if $\epsilon_f = 10^{-16}$ and $\epsilon_{req} = 10^{-5}$ then our limit is $K_{k,p,0} < 10^{11}$. Probably any realistic choice of rule and mesh ratios would satisfy this constraint.

In cases where numerical instability is a problem one should remember that using a rule Q of moderate degree d may remove some of the early terms in the expansion (5.39) leading to a smaller condition number.

One other point is worth making. When comparing the mid-point rule and the end-point rule in one dimension, one usually finds the end-point rule to be more economical in function values as function values may be re-used. This is a low-dimensional effect. In three dimensions and higher, the mid-point rule is almost invariably more economical. The 're-use' factor is $1/2^N$ which becomes less significant as the dimensionality increases.

With these various caveats in mind, one might start by considering the Harmonic sequence and the mid-point product trapezoidal rule. If this leads to too much numerical instability, reduce the condition number by switching to the sequence F or by using a quadrature rule Q of higher degree. The condition number can be evaluated without any function evaluation, and so is a 'cheap' calculation in this context.

At the present time, multidimensional quadrature is a slow and difficult problem. No fast automatic techniques for applying extrapolation efficiently and reliably are known to the author.

7. Other expansions

In this section, other expansions are briefly described or referenced.

Virtually all the expansions described for the hypercube in sections 2 and 3 have direct counterparts for the N-dimensional simplex

$$S_N : x_i \geqslant 0 \ , \quad i = 1,2,\ldots,N \ , \qquad \sum_{i=0}^{N} x_i \leqslant 1 \ ,$$

and by extension for any other simplex, since any simplex may be obtained as an affine transformation of this one. Even a statement of results may be complicated. The results for the triangle are given, at the same level as in this article, in Lyness (1983). The general theory is treated in Lyness (1978), Lyness and Monegato (1980), Lyness and Genz (1980), de Boor (1977), and de Doncker (1979). This latter paper introduces some expansions valid only when m is half an odd integer.

A family of expansions, which handles the one-dimensional integrand functions of section 3 using local interpolatory-type rules, has been constructed by de Hoog and Weiss (1973).

There are also expansions for closed contour integration of an analytic function in the complex plane, round a square, an equilateral triangle and a regular hexagon. When the m-panel trapezoidal rule is used on each edge, the expansion in all three cases is (like (5.3) above) in negative even powers of m. In the case of the square, the term in m^{-2s} is absent when 2s is a multiple of four. In the latter two cases the term in m^{-2s} is absent when 2s is a multiple of six. See Lyness and Delves (1967). Incidentally, when integrating a harmonic function f(x, y) over a square in the complex plane (i.e. f(x, y) = g(z) where z = x+iy) using the product m-panel trapezoidal rule, only terms in m^{-s} where s is a multiple of four remain. (Unlike the case above, here we restrict g(z) to being regular within the square.)

The situation with respect to the Cauchy Principal Value integral has been treated by Hunter (1972) and again by Lyness (1985). A two-dimensional integral having a Cauchy Principal Value type line singularity is treated in Monegato and Lyness (1979). In both these cases, expansions involving inverse powers of m are valid under certain delineated circumstances.

An excellent but dated source for expansions of this nature is a long detailed survey article by Joyce (1971). A continuation of this survey, up to 1983, appears in Kumar (1983). This covers all branches of Numerical

Analysis.

8. When to use extrapolation

Up to this point, the author has avoided discussing the somewhat subjective question of when to use extrapolation. He has contented himself, so far, with the questions of whether it is possible, how to do it, and which rule and mesh to use.

First, the limitations on <u>region</u>. The context is integration over a finite interval, a hypercube or a simplex. For other regions no suitable extrapolation expansion is known to the author. Second, the limitations on <u>integrand function</u>. Either there are no singularities and $g(\underline{x})$ is analytic in each variable, or the singularities are of the algebraic and logarithmic nature specified in (5.28), (5.29), (5.30), (5.35) and (5.37) and a few similar ones implied but not specifically mentioned. If the problem in hand falls into one of the categories mentioned above, extrapolation is a possibility.

Next we consider the cost effectiveness in terms of the number of function values required to attain a particular accuracy. Experience has shown that extrapolation is nearly always more expensive than using appropriate Gaussian rules by a factor of anything between 1.5 and 3. However, it is very much more efficient than simply using the trapezoidal rule or Simpson's rule blindly. In the absence of other constraining conditions, if Gaussian quadrature is available, it should be preferred to extrapolation.

However, for some problems (for example, the vertex singularity), it is not available at all. Even in straightforward cases, the weights and abscissae may not be easily available to a particular user. One advantage of extrapolation is that it does not need a long list of stored weights and abscissae. One requires only a subprogram to evaluate $Q^{(m)}f$ (with Q a product trapezoidal rule) and a linear equation solver.

Like many techniques in Numerical Analysis, extrapolation in numerical quadrature is very useful and even optimal for a small class of problems, is convenient but not optimal for a much wider class; but there is a huge class of problems for which it is not relevant or appropriate and should not be used. It is with the clear recognition of these respective classes that this article is principally concerned.

Acknowledgements

This work was supported by the Applied Mathematical Sciences subprogram of the Office of Energy Research, U.S. Department of Energy, under contract W-31-109-Eng-38.

Part B. Introduction to Algorithms for Automatic Quadrature

C.T.H. Baker

9. Basic quadrature rules

For the basic theory of numerical integration, one can cite the works of Davis and Rabinowitz (1984) and of Engels (1980), both of which contain reference to algorithms. The work of Rice (1983) includes an introduction to some computational aspects of quadrature. All these contain references to existing algorithms, but we should also mention the work of Piessens, de Doncker-Kapenga, Überhuber and Kahaner (1983) which is devoted entirely to a suite of algorithms. Surveys of quadrature algorithms are given by Lyness (1976c), de Doncker and Piessens (1976), Dixon (1974), Kahaner (1971), for example. Engels (1980) gives references for work prior to 1972.

The problem under consideration is the evaluation of an integral of a given function over a prescribed interval. For the sake of definiteness we consider the evaluation of

$$I(f) = \int_0^1 f(x)\,dx \quad , \tag{5.51}$$

where f is at least bounded Riemann-integrable, and preferably smooth. By affine transformations, (5.51) is seen to be equivalent to the general problem of evaluating

$$I[a, b; \varphi] = \int_a^b \varphi(t)\,dt \;. \tag{5.52}$$

Let us consider, briefly, some theoretical aspects. The common approximations assume the form

$$I(f) \simeq J_n(f) = \sum_{j=0}^{n} w_j^{(n)}\, f(x_j^{(n)}) \tag{5.53}$$

and

$$I[a,b;\varphi] \simeq J_n[a,b;\varphi] = (b-a) \sum_{j=0}^{n} w_j^{(n)} \varphi(a + (b-a)x_j^{(n)}) \quad (5.54)$$

respectively. It is possible to employ values of f' in the approximation to (5.51) in addition to f-values, but this is rarely done in practice. There exist, roughly speaking, two obvious ways of employing quadrature formulae of the type indicated. In the first, we take a sequence of integers $n_1 < n_2 < n_3 \ldots$ and a family of quadrature rules (Newton-Cotes, or Gauss, or Romberg, for example) so that

$$\lim_{r \to \infty} J_{n_r}(f) = \int_0^1 f(x)\,dx \quad (5.55)$$

and we evaluate $J_{n_1}(f)$, $J_{n_2}(f)$, $J_{n_3}(f)$, etc. . (In this context, it may be noted that many rules $J_n(f)$ correspond to Riemann sums (Baker 1968) approximating $I(f)$, in which case (5.55) holds if and only if $\lim_{n \to \infty} \sup_j w_j^{(n)} = 0$.) In the second, rather than employ a particular family of rules, we form partitions

$$P_m = \{0 = z_0^{(m)} < z_1^{(m)} < z_2^{(m)} < \ldots < z_m^{(m)} = 1\}, \quad (5.56)$$

in such a way that $\lim_{m \to \infty} \Delta(P_m) = 0$ where $\Delta \equiv \Delta(P_m) = \max_i (z_{i+1}^{(m)} - z_i^{(m)})$, and compute the approximations

$$J_m(f) = \sum_{i=0}^{m-1} J_n[z_i, z_{i+1}; f] \simeq I(f) \quad (5.57)$$

for $m = m_0, m_1, m_2, \ldots$, where $m_0 < m_1 < m_2 < \ldots$. Again we may observe that

$$\lim_{m \to \infty} J_m(f) = \int_0^1 f(x)\,dx \quad (5.58)$$

provided f is bounded, Riemann-integrable, and $\sum_j w_j^{(n)} = 1$. Note that in practice one may wish to replace $J_n[z_i, z_{i+1}; f]$ by $J_{n_i}[z_i, z_{i+1}; f]$, and a similar convergence result can be established. In terms of the theory, the distinction between the various approaches is somewhat artificial.

The existence of convergence results (5.55), (5.58) establishes that, in principle, one can obtain sufficiently accurate results by performing sufficient work. However, a practical technique requires a choice of formula(e) $J_n(f)$, a strategy for selecting $\{n_r\}$ or P_{m_r}, and most important, a criterion for judging that a given approximation based on a finite set of f-values satisfies a requested error tolerance. We shall return to the question of algorithms later.

Note that the integral (5.51) may be presented in a somewhat different form, as that of evaluating

$$I(f) = \int_0^1 \omega(x)\Psi(x)\, dx$$

where $\omega(x)$ is a standard function like $\sqrt{x}$ or $\cos r\pi x$, which may be badly behaved, or smooth and highly oscillatory. For Cauchy principal values, or weakly singular integrals, such a splitting $f(x) = \omega(x)\Psi(x)$ can be highly effective. We shall not dwell upon the differences which this formulation imposes. We should not pass by without mentioning, however, the usefulness of Filon's method (see Davis and Rabinowitz (1984)) for integrating $\cos \pi rx\, \Psi(x)$, $\sin \pi rx\, \Psi(x)$.

10. Some background theory

To choose a good strategy for a practical technique requires some background theory concerning (5.53). Generally the weights $w_j^{(n)}$ and abscissae $x_j^{(n)}$ of the rule (5.53) are such that it is exact for a certain class of problems. If the rule is exact for all polynomials of degree r, but not for $f(x) = x^{r+1}$, then r is the <u>degree of precision</u> of the rule. It is easily shown in such circumstances that $|I(f) - J(f)| \leq \{1 + \sum_j |w_j^{(n)}|\} E_r(f)$ where $E_r(f) = \inf \|f(x) - p_r(x)\|_\infty$, the infimum being taken over all polynomials $p_r(x)$ of degree at most r. (By Jackson's theorem,

$E_r(f) = O(r^{-k-1})$ if $f^{(k+1)}(x)$ is bounded.)

The last result provides a qualitative bound. If $f(x)$ is accurately approximated by a polynomial, quadrature will be successful. Indeed, let $f \in PC^{k+1}[0,1]$; then the Taylor expansion with integral remainder shows that

$$f(x) = f(0) + xf'(0) + \ldots + \frac{x^k}{k!} f^{(k)}(0) + R(x) \tag{5.59}$$

where

$$R(x) \equiv R_k(f;\, x) = \frac{1}{k!} \int_0^1 (x-t)_+^k\, f^{(k+1)}(t)\, dt, \quad 0 \leqslant x \leqslant 1. \tag{5.60}$$

It follows that, if $r \geqslant k$,

$$I(f) - J_n(f) = I(R) - J_n(R)\ . \tag{5.61}$$

In consequence we can show that

$$I(f) - J_n(f) = \int_0^1 K_{k+1}(t)\, f^{(k+1)}(t)\, dt \tag{5.62}$$

where

$$K_{k+1}(t) = \frac{1}{k!} \left\{ \int_0^1 (x - t)_+^k\, dx - \sum_{j=0}^{n} w_j (x_j - t)_+^k \right\}. \tag{5.63}$$

The equation (5.62) is an exact result from which we obtain bounds of the form $||K_{k+1}||_p||f^{(k+1)}||_q$ where $p^{-1} + q^{-1} = 1$; in particular, for $k \leqslant r$,

$$\left.\begin{aligned} |I_n(f) - J_n(f)| &\leqslant c_{k+1}(J_n)||f^{(k+1)}||_\infty, \\ c_{k+1} &= \int_0^1 |K_{k+1}(x)|\, dx\ . \end{aligned}\right\} \tag{5.64}$$

The constants $c_k \equiv c_k(J)$, $k = 1,2,\ldots,r$, are known as Peano constants. For (5.54) the bound corresponding to (5.64) is

$$\left|I_n[a, b; \varphi] - J_n[a, b; \varphi]\right| \leqslant (b-a)^{k+2} c_{k+1}(J_n) \sup_{[a,b]} \left|\varphi^{(k+1)}(.)\right|,$$
$$k \leqslant r. \qquad (5.65)$$

In (5.64), $f^{(k+1)}$ can be replaced by $(d/dx)^{k+1}F(x)$, where $F(x) = f(x) - p_r(x)$ for any polynomial $p_r(x)$ of degree r. The bounds (5.64) and (5.65) are attainable. From (5.65) there follows the result that the error in (5.57) is $O(\Delta^{k+1})$, $k \leqslant r$.

If one examines the bounds (5.64), (5.65) one may favour the use of rules of high degree of precision (say the Gauss-Legendre, or Radau, or Lobatto rules) with certain strategies. Some authors have considered the choice of $\{x_j^{(n)}, w_j^{(n)}\}$ which minimises a given Peano constant, but the resulting rules do not seem to be employed in practical algorithms. Nor are the bounds (5.64), (5.65) necessarily of direct help: there appear to be no algorithms which employ estimates of such bounds to return error parameters, and indeed they do not explain such features as the high order of accuracy obtained when the repeated trapezium rule is applied to a periodic integrand.

The latter result is explained by considering the expansion, for $f \in PC^{2N+1}[0,1]$,

$$\int_0^1 f(x)\, dx - h \sum_{i=0}^{m}{}'' f(ih)$$
$$= \sum_{k=1}^{N} h^{2k} \gamma_k [f^{(2k-1)}(1) - f^{(2k-1)}(0)] + O(h^{2N+1}), \qquad (5.66)$$

the Euler Maclaurin formula. The sum on the left-hand side is an expression of the type (5.57) where $z_i - z_{i-1} = h$ and $n = 1$, $J_n(f) = \frac{1}{2}\{f(0) + f(1)\}$, the trapezium rule. (The results and ideas extend to more general rules used with a uniform partition P_m; see Baker and Hodgson (1971).) According to (5.66), if $f(\ )$ is a smooth periodic function ($f^{(2k-1)}(1) = f^{(2k-1)}(0)$, $k = 1,2,3, \ldots$) the order of the error in the repeated trapezium rule is $O(h^{2N+1})$ for $f \in PC^{2N+1}[0,1]$, instead of $O(h^2)$ as indicated by the use of (5.65). Extrapolation methods based on (5.66) are discussed in section 2 above.

Our results so far refer to discretisation error; for the effect of rounding and its control see Piessens et al. (1983). All carefully written routines require machine constants associated with the particular machine arithmetic to be supplied.

Algorithms of the Patterson-Kronrod type approach the question of controlling the number of function evaluations in a different way. It would be consistent with the remarks following (5.55) to take for $J_{n_r}(f)$ the Gauss-Legendre rule with n_r+1 points appropriate to integration on [0,1]. Then the calculation of $J_{n_{r+1}}(f)$ is totally independent of the calculation of $J_{n_r}(f)$ because the two rules have different abscissae and weights. In schemes of the Patterson-Kronrod type (of which there is an example in the NAG Library) one starts with a Gauss-Legendre rule $J_{n_0}(f)$, but $J_{n_1}(f)$ is then constructed by the addition of abscissae to the set $\{x_j^{(n_0)}\}$ used in $J_{n_0}(f)$ in such a way as to increase the degree of precision, and so on. The weights and abscissae of such schemes (e.g. the 10-point Gauss, and 21-, 43- and 87- point Patterson rules) are illustrated in Piessens et al. (1983); see also Patterson (1968), Kronrod (1965). Evans, Forbes and Hyslop (1983) use transformations of singular integrals to permit the application of Patterson schemes to such problems.

So far, no mention has been made of tailoring the approximation to the varying behaviour of f(x) as x varies between 0 and 1. It is implicit in the description of (5.57) that the problem of determining $\int_0^1 f(x)\,dx$ is broken down to the problem of determining $\int_{z_i}^{z_{i+1}} f(x)\,dx$, and one might wish to determine sub-intervals where f(x) does not vary smoothly and give them special attention. (This process may result in intervals $[z_i, z_{i+1}]$ which are too small to be meaningful in the computer, and warnings should then be given.) Schemes for organising sub-interval treatment in this way are mentioned by Rice (1983).

11. Types of algorithm

With the indications above in mind, it is now appropriate to consider practical algorithms: what type of algorithms are and should be provided, and how may they be used?

We can isolate various cases. The user may wish to compute a single integral, a number of different integrals, or a number of similar integrals in a family depending upon one or more parameters. He may require rough estimates only (low accuracy) or he may need to achieve a certain relative or absolute error. Efficiency of the algorithm may or may not be a consideration, depending on the number of cases to be treated.

Rough estimates of an integral can frequently be computed by application of a basic quadrature rule (5.53), and in any event a program for generating the weights and abscissae of a basic rule may be needed as a subroutine for more general tasks.

An <u>automatic routine</u> is one which applies a strategy to generate approximations to I(f); the strategy is determined in advance and depends upon f only to the extent that an attempt is made to estimate the error at each stage, in terms of f; if the estimated error is too large, the strategy is continued. An example of this approach is given by Deuflhard and Bauer (1982). By contrast, an <u>adaptive routine</u> is one in which the strategy is adapted in accordance with the varying behaviour of f over the interval of integration. As example we have the IMSL program DCADRE, a pioneer of its time for which a simplified version is given in Davis and Rabinowitz (1984, Appendix 2); see also de Boor (1971a).

The organisation of an adaptive routine (de Boor 1971b) is complex; where a sequence of similar integrands with varying parameters occurs it may be beneficial to record the weights and abscissae associated with the final strategy in the first call of the routine, and to employ this strategy for other members of the family. However, the changing parameter may change the nature of the integrand, for example $f_\alpha(x) = x^2(1.2 - x)(1 - \exp \alpha(x - 1))$ is very different with $\alpha = 0.2$ and with $\alpha = 20$ (see Rice (1983)).

In general it seems that adaptive routines are more expensive than necessary but give greater confidence in the results. Rice (1983) lists examples of automatic and adaptive routines in the ACM Library.

All good suites of library routines provide decision trees by which the user can decide upon a suitable choice of routine for the problem he has to hand. Notwithstanding such provision, the user who has an extended set of problems may wish to test the available routines in order to select the most appropriate. The practice of battery testing of routines on an artificial test set has largely been replaced by a process of performance evaluation due to Lyness and Kaganove (1976); a program for the evaluation of routines is provided by Rice (1983, p.204 et seq.).

When one examines the strategies used in routines for accepting an approximate value for I[a, b; f] one frequently finds them to be based upon nothing more elaborate than a comparison of $J_n^*[a, b; f]$ with $J_N^{**}[a, b; f]$ for the nth and Nth member of families of rules J_m^* and J_m^{**} respectively. It should not go unremarked that although the extrapolation process associated with classical Romberg quadrature has not found great support with the authors of library programs, the ϵ-algorithm (de Doncker 1978, Piessens et al. 1983) has been given a practical rôle. Further, there are many ad-hoc strategies which have been developed by experience of the algorithms in a test set (one suspects these may not always represent the features presented by the user). Davis and Rabinowitz have stated that "the aim of an automatic integration scheme is to relieve the user of any need to think"; whether this aim is attainable is another matter. Piessens et al. (1983) point out that there is nothing to prevent the misuse of an algorithm. However, this author's view (which appears to be shared by Piessens et al.) is that there also remains scope for further research and development in the construction of reliable algorithms.

The foregoing relates to one-dimensional quadrature, and the difficulties apply a fortiori in the problem of multiple integration. Whilst research into numerical integration formulae over regular regions (spheres, cubes, pyramids, etc.) is fruitful and rewarding the possibility of irregularly shaped domains of integration, and the occurrence of a large number of dimensions, means that Monte Carlo methods and number-theoretic approximations find a rôle in many library programs for multiple integration (Haber 1970). Such methods do not appear to come with deterministic error bounds, but we have seen that such rigorous error bounds are rarely computed even in the one-dimensional cases where they are available.

12. Available software

To the reader who does not have ready access to a library of programs, the publication QUADPACK by Piessens et al. (1983) may prove the most useful for one-dimensional integration. It incorporates the listings of a suite of routines written in standard Fortran for the approximation of one-dimensional integrals, and the different routines follow a consistent naming convention for parameters. Perhaps of great importance if one is to avoid misuse of the algorithms is the provision of guidelines and decision lines to assist the choice of algorithm.

Versions of QUADPACK have been incorporated in the NAG Library and in the SLATEC Library.

The NAG Library chapter D01 on quadrature and cubature provides a wide selection of algorithms for a variety of types of problem. The suite is not completely infallible but one can hope that the failure parameter will give a record of difficulty whenever the results are unreliable.

Basic quadrature routines are also provided in IMSL.

CHAPTER 6

Numerical Treatment of Integral Equations

Part A. Abel and Volterra Equations

C.T.H. Baker

1. Analytical background

An integral equation is an equation in which the unknown function appears as part of an integrand. Volterra and Abel integral equations arise in a number of practical situations, for example renewal theory, the theory of competing populations, the theory of superfluidity, that of nuclear reactors, and in the indirect treatment of parabolic equations and of moving-boundary problems. Some general references are Burton (1983), Miller (1971), Corduneanu (1971). As an illustration of the wealth of recent literature existing in the study of Volterra integral equations we may refer to Tsalyuk (1979), who, in a survey covering 1966-1976, reviewed as many as 515 papers.

The 'kernel' $H(x, y, v)$, to which we refer below, denotes a prescribed function which is assumed to be continuous for $-\delta \leqslant y \leqslant x+\delta \leqslant X+\delta$ and $|v| < \infty$ for some $\delta \geqslant 0$, $X \leqslant \infty$. Furthermore, $g(x)$ denotes a prescribed function on $[-\delta, X+\delta]$.

<u>The classical Abel equation</u> is the equation

$$\int_0^x \frac{f(y)}{(x-y)^\alpha}\,dy = g(x), \qquad 0 < \alpha < 1\ , \qquad x \geqslant 0\ , \tag{6.1}$$

where α is given. This equation may be called ill-posed. Under appropriate restrictions on $g(x)$, it has the solution

$$f(x) = \frac{\sin \alpha\pi}{\pi}\ \frac{d}{dx}\int_0^x \frac{g(y)}{(x-y)^{1-\alpha}}\,dy\ .$$

The generalised Abel equation of the first kind, in which the solution $f(x)$ is sought, has the form

$$\int_0^x \frac{H(x, y, f(y))}{(x^p - y^p)^\alpha}\, dy = g(x) , \quad 0 < \alpha < 1 , \quad p = 1 \text{ or } 2 , \quad x \geqslant 0 . \tag{6.2}$$

The linear case of equation (6.2)

$$\int_0^x \frac{K(x, y)\, f(y)}{(x^p - y^p)^\alpha}\, dy = g(x) , \quad 0 < \alpha < 1 , \quad p = 1 \text{ or } 2 , \quad x \in [0,X] , \tag{6.3}$$

where $K(x, y)$ is continuous at least for $0 \leqslant y \leqslant x$, is also of interest. We consider only $p = 1$, here.

If, in equation (6.2), $\alpha = 0$ and $H(x, y, f)$ is a smooth kernel we obtain a Volterra equation of the first kind

$$\int_0^x H(x, y, f(y))\, dy = g(x) , \qquad x \in [0,X] , \tag{6.4}$$

where X may be taken arbitrarily large. A basic example of (6.4) is the equation

$$\int_0^x f(y)\, dy = g(x) .$$

On differentiating, we obtain the solution $f(x) = g'(x)$ provided $g'(x)$ exists and $g(0) = 0$. Thus we see that the equation has a solution only if $g(x)$ has special properties. This suggests that care must be taken lest the problem of solving (6.4) is 'improperly' posed.

The corresponding Abel and Volterra equations of the second kind are, respectively,

$$f(x) - \int_0^x \frac{H(x,\ y,\ f(y))}{(x^p - y^p)^{\alpha}}\, dy = g(x)\ ,\ 0 < \alpha < 1,\ p = 1 \text{ or } 2\ ,$$

$$x \in [0,X] \qquad (6.5)$$

(in the classical case, which we consider here, $p = 1$) and

$$f(x) - \int_0^x H(x,\ y,\ f(y))\, dy = g(x)\ ,\quad x \in [0,X]\ . \qquad (6.6)$$

For the Volterra equation (6.6), it may be noted that, provided the kernel is sufficiently smooth, the smoothness of the solution f depends on the smoothness of g. It will be observed that the Abel equations (6.2), (6.3) and (6.5) exhibit weak singularities in the integrand. In consequence, when $g(x)$ is <u>well behaved</u> at $x = 0$ the solution $f(x)$ of the classical Abel equation of the second kind may be expected to be <u>badly behaved</u>.

If $f(x)$ has been found for $x \in [0,A]$, $A < X$, then

$$f(x) - \int_A^x H(x,\ y,\ f(y))\, dy = g(x) + \int_0^A H(x,\ y,\ f(y))\, dy\ ,$$

where the right-hand term $g_A(x)$, say, is now known for $x \geqslant A$. The new equation is of the form

$$f(x) - \int_A^x H(x,\ y,\ f(y))\, dy = g_A(x)\ ,\quad x \in [A,X]\ , \qquad (6.7)$$

which, taking $\varphi(x) = f(x+A)$, can be recast in the original standard form as

$$\varphi(x) - \int_0^x H(x+A,\ y+A,\ \varphi(y))\, dy = g_A(x+A)\ ,\quad 0 \leqslant x \leqslant X-A. \qquad (6.8)$$

Thus a method for advancing the solution of (6.6) from $x = 0$ yields, when applied to (6.8), a technique for advancing the solution $f(x)$ from $x = A$.

Special subclasses of integral equations occur frequently. The kernel $H(x,\ y,\ f)$ is known as a convolution kernel if $H(x,\ y,\ f) = h(x - y;\ f)$ and <u>nonlinear Volterra integral equations of convolution type</u> of the first and second kinds, are, respectively,

$$\int_0^x h(x - y; f(y))\,dy = g(x) \tag{6.9}$$

and

$$f(x) - \int_0^x h(x - y; f(y))\,dy = g(x) . \tag{6.10}$$

Equations (6.4) and (6.6) with $H(x, y, f) = K(x, y)f$ give rise to linear Volterra integral equations of the form

$$\int_0^x K(x, y)\, f(y)\,dy = g(x) \tag{6.11}$$

and

$$f(x) - \int_0^x K(x, y)\, f(y)\,dy = g(x) \tag{6.12}$$

where $g(x)$ is termed the forcing function.

We often encounter a special type of nonlinear convolution equation in which the kernel H is such that

$$H(x, y, f) = k(x - y)\,\varphi(y, f) . \tag{6.13}$$

Of interest also is the case where $H(x, y, f)$ is finitely decomposable:

$$H(x, y, f) = \sum_{r=1}^{R} X_r(x)\, Y_r(y, f) . \tag{6.14}$$

Whereas all the above cases are classical, integral equations also arise quite frequently in non-standard forms, for example as

$$f(x) - \int_{\alpha(x)}^{\beta(x)} H(x, y, f(y))\,dy = g(x) , \quad x \in [0,X] , \tag{6.15}$$

or with nonlinear free terms, and these require further study.

In general, solution of equations of the first kind is more difficult than solution of equations of the second kind. However, under certain conditions Volterra equations of the first kind may be transformed into equations of the second kind (Baker 1977). We shall illustrate by considering linear equations. Consider the linear Volterra equation of the first kind

$$\int_0^x K(x, y)\, f(y)\, dy = g(x) \tag{6.16}$$

where $K(x, y)$, $g(x)$ are continuously differentiable. We seek a continuous solution and require, therefore, that $g(0) = 0$. Given that $\frac{\partial}{\partial x} K(x, y) = K_x(x, y)$ is continuous for $y \leqslant x$ and that $g'(x)$ is continuous, we obtain, on differentiating (6.16),

$$K(x, x)\, f(x) + \int_0^x K_x(x, y)\, f(y)\, dy = g'(x)\ .$$

If $K(x, x) \neq 0$, then

$$f(x) + \int_0^x K_1(x, y)\, f(y)\, dy = \frac{g'(x)}{K(x, x)}\ ,$$

where $K_1(x, y) = \dfrac{K_x(x, y)}{K(x, x)}$. Thus we obtain a Volterra equation of the second kind. Determination of $g'(x)$ indicates the ill-posed nature of the problem. If, on the other hand, $\frac{\partial}{\partial y} K(x, y) = K_y(x, y)$ is continuous for $y \leqslant x$, we may write

$$F(x) = \int_0^x f(y)\, dy$$

which is a Volterra equation of the first kind for $f(x)$, with solution $f(x) = F'(x)$. A Volterra equation of the second kind may be derived for $F(x)$; we find:

$$K(x, x)\, F(x) - \int_0^x K_y(x, y)\, F(y)\, dy = g(x)\ .$$

Provided $K(x, x) \neq 0$ this results in

$$F(x) - \int_0^x K_2(x, y) F(y)\, dy = \frac{g(x)}{K(x, x)}$$

where $K_2(x, y) = (K_y(x, y) / K(x, x))$. The function $f(x)$ is then obtained by differentiating the solution $F(x)$ of this equation (Baker 1977, p. 9).

Consider the linear Abel equation of the first kind

$$\int_0^x \frac{K(x, y) f(y)}{(x - y)^{\alpha}}\, dy = g(x) . \tag{6.17}$$

It may be shown that, provided $K(x, x)$ and $g(x)$ satisfy certain conditions, equation (6.17) yields a Volterra equation of the second kind

$$f(x) + \int_0^x K^{\#}(x, y) f(y)\, dy = g^{\#}(x) ,$$

where

$$K^{\#}(x, y) = \frac{\sin \alpha\pi}{\pi} \int_0^1 \left\{\frac{z}{1 - z}\right\}^{1-\alpha} K_x(z(x - y) + y, y)\, dz$$

and

$$g^{\#}(x) = \frac{\sin \alpha\pi}{\pi} \frac{d}{dx} \int_0^x g(y) (x - y)^{\alpha-1}\, dy$$

(Baker 1977).

Consider the integro-differential equation

$$f'(x) = G(x, f(x), \int_0^x H(x, y, f(y))\, dy) , \quad x \geqslant 0 \tag{6.18a}$$

with prescribed initial condition

$$f(0) = f_0 . \tag{6.18b}$$

Equations of this type have practical applications.

We may rewrite equation (6.18) to obtain the system

$$\left.\begin{aligned} f_1(x) &= \int_0^x G(y,\ f_1(y),\ f_2(y))\,dy + f_0\ , \\ f_2(x) &= \int_0^x H(x,\ y,\ f_1(y))\,dy\ , \end{aligned}\right\} \tag{6.19}$$

which is a coupled pair of integral equations. Using vector notation $\underline{\varphi}(x) = [f_1(x)\ \ f_2(x)]^T$ we obtain the form

$$\underline{\varphi}(x) = \int_0^x \underline{\Psi}(s,\ \underline{\varphi}(s))\,ds + \underline{\varphi}(0)\ . \tag{6.20}$$

Thus, the integro-differential equation is written as a special case of a Volterra equation of the second kind and hence techniques for the numerical solution of Volterra integral equations may be adapted to treat the system (6.20). However, this may not always be the most practical arrangement, as a wider range of techniques may be available if each equation in (6.16) is treated separately. While the preceding approach (conversion of an integro-differential equation to a pair of integral equations) may provide theoretical insight, we suspect that the converse process (conversion of an integral equation to an integro-differential equation) may be more practically useful. Thus (6.6) yields

$$f'(x) = H(x,\ x,\ f(x)) + g'(x) + \int_0^x H_x(x,\ y,\ f(y))\,dy\ ,$$

which is of the form (6.18a).

A tool in the study of numerical methods for second-kind Volterra equations, due to Pouzet (1962), is the imbedding of the integral equation in a differential equation containing a parameter. See Wolkenfelt, van der Houwen and Baker (1981). The definition of the kernel function $H(x,\ y,\ f)$ of (6.6) is extended for $y > x$ and we define

$$\Psi(t,\ x) = g(x) + \int_0^t H(x,\ y,\ f(y))\,dy\ , \qquad 0 \leqslant t \leqslant X\ , \tag{6.21}$$

where $f(x)$ satisfies the integral equation (6.6). Clearly

$$f(x) = \Psi(x, x) ,$$

so that we may write (6.21) as

$$\Psi(t, x) = g(x) + \int_0^t H(x, y, \Psi(y, y))\, dy . \tag{6.22}$$

On differentiating (6.22) with respect to t we obtain the initial-value problem (Pouzet 1962)

$$\left.\begin{aligned} &\frac{\partial}{\partial t} \Psi(t, x) = H(x, t, \Psi(t, t)) , \quad 0 \leqslant t, x \leqslant X , \\ &\Psi(0, x) = g(x) . \end{aligned}\right\} \tag{6.23}$$

This latter equation (6.23) may be regarded as a partial differential equation, although in the study of numerical methods it has proved to be more convenient to treat it as an ordinary differential equation with x as parameter and t the independent variable. Differential equation solvers can now be employed in the t-variable to generate numerical methods for determining the solution of the integral equation: see the thesis of Wolkenfelt (1981), for example.

It should be apparent from what we have said here that the theory is of some importance, and those with Abel or Volterra equations to solve should first decide which of a number of equivalent formulations is most appropriate for treatment. In particular, it will be noted that if the differentiations can be performed, it may be possible to transform a first-kind equation into a second-kind equation or an integro-differential equation.

2. Methods for numerical solution

Numerical methods can be approached from various viewpoints. The surveys of Baker (1982) and of Brunner (1982) give a starting point; we particularly recommend the latter for its comprehensive introduction to the topic.

We commence with the easiest methods, the quadrature and Runge-Kutta methods for Volterra second-kind equations. To develop these simply, we shall 'borrow' some formulae from methods for the ordinary differential equation

$$f'(x) = F(x, f(x)) , \quad x \geqslant 0 , \quad f(0) = f_0 . \tag{6.24}$$

We shall assume some familiarity with numerical methods for (6.24), in particular the linear multistep formula with characteristic polynomials

$$\left.\begin{aligned} \rho(\mu) &= \alpha_0\mu^k + \alpha_1\mu^{k-1} + \ldots + \alpha_k , \\ \sigma(\mu) &= \beta_0\mu^k + \beta_1\mu^{k-1} + \ldots + \beta_k \end{aligned}\right\} \tag{6.25}$$

(see Chapter 3, section 3). Denoting by E the forward shift operator corresponding to a step $h > 0$: $E f_n = f_{n+1}$, the formula for (6.24) can be written

$$\rho(E) f_n = h\sigma(E) F_n ; \quad F_n \equiv F(nh, f_n) . \tag{6.26}$$

Cyclic linear multistep methods involve a set of q such formulae, applied cyclically,

$$\rho_r(E) f_n = h\sigma_r(E) F_n , \quad r = 0,1,\ldots,q-1 , \; n \equiv n_0 + r \bmod q . \tag{6.27}$$

The methods are assumed consistent and zero stable (hence convergent). Finally a Runge-Kutta method for (6.24) can be defined in terms of the elements a_{rs} $(r, s = 0,1,\ldots,p)$ of a square generating matrix A by the formulae

$$\left.\begin{aligned} f_{n+1} &= f_{n,p} , \\ x_{n,r} &= nh + \theta_r h , \\ f_{n,r} &= f_n + h \sum_{s=0}^{p} a_{rs} F(x_{n,s} , f_{n,s}) \end{aligned}\right\} \begin{aligned} & r = 0,1,\ldots,p , \\ & n \geqslant 0 , \end{aligned} \tag{6.28}$$

where

$$f_{n,r} \simeq E^{\theta_r} f(nh) , \quad \theta_r = \sum_{s=0}^{p} a_{rs} , \quad \theta_p = 1 . \tag{6.29}$$

(Thus E^{θ_r} denotes the shift operator for a step $\theta_r h$, and $E^{\theta_p} \equiv E$.) For (6.24) it is conventional to choose A so that $a_{rp} = 0$, $r = 0,1,\ldots,p$.

If we apply our formulae to the problem

$$f'(x) = \varphi(x)$$

we obtain methods for finding approximations $f_n \simeq f(nh) = \int_0^{nh} \varphi(s)ds$ derived from multistep formulae and their cyclic versions corresponding to formulae

$$Q : \int_0^{nh} \varphi(s)\, ds \simeq h \sum_{j=0}^{n} w_{nj}\, \varphi(jh) , \qquad n \geqslant k , \tag{6.30}$$

and the Runge-Kutta method yields approximations

$$\int_0^{nh+\theta_r h} \varphi(s)\, ds \simeq h \sum_{j=0}^{n-1} \sum_{s=0}^{p} a_{ps}\, \varphi(jh + \theta_s h) + h \sum_{s=0}^{p} a_{rs}\, \varphi(nh + \theta_s h) . \tag{6.31}$$

Formulae such as (6.30) can occur quite naturally in their own right. They can be derived by the application of primitive rules such as the trapezium rule and Simpson's rule, as illustrated in Baker (1977) and in the paper of Baker and Keech (1978) to which we refer for details. It is often not necessary to employ ρ, σ in the construction. Sometimes, one is tempted to modify (6.31) by the use of

$$\int_0^{nh+\theta_r h} \varphi(s)\, ds \simeq h \sum_{j=0}^{n} w_{nj}\, \varphi(jh) + h \sum_{s=0}^{p} a_{rs}\, \varphi(nh + \theta_s h), \tag{6.32}$$

where the weights w_{nj} are those of Q in (6.30). The quadrature rules (6.31), (6.32) are called respectively, the extended Runge-Kutta rules, and the mixed-quadrature Runge-Kutta rules.

With the aid of the integration rules above, it is simple to discretise

(6.6). Putting $x = nh$ in (6.6), the rule Q of (6.30) yields the equations

$$f_n - h \sum_{j=0}^{n} w_{nj} H(nh, jh, f_j) = g(nh) . \qquad (6.33)$$

Setting $x = nh + \theta_r h$ and using one of (6.31) or (6.32) yields equations of the form

$$f_{n,r} - \sum_{j=-1}^{n} \sum_{s=0}^{p} \Omega_{n(p+1)+r+1,j(p+1)+s+1} H(nh+\theta_r h, jh+\theta_s h, f_{j,s}) = g(nh + \theta_r h) . \qquad (6.34)$$

For $j = n(p+1)+r+1$, we have, respectively, the weights $\Omega_{jk}(A)$ of the extended RK rule (6.31), or the weights $\Omega_{jk}[Q, A]$ defined by (6.32); (6.34) also admits (6.33) as a special case. In all cases $\Omega_{jk} = 0$ if $k < 0$, or if $k > (p+1)([j/(p+1)]+1)$, where $[z]$ is the integer part of z. This formulation has the charm of unifying a number of methods, but it does not include the methods of Beltyukov (1965) extended recently by Brunner, Hairer and Nørsett (1982), nor the modified methods of van der Houwen (1980), Wolkenfelt (1981).

We note the economy of effort in using (6.32) as opposed to (6.31) for large n. In comparing methods, the work due to implicitness arising from a choice A which is full may be of less significance than the amount of work in approximating the 'lag' $\int_0^{nh} H(x, y, f(y))\, dy$, $x = nh+\theta_r h$, which is performed more economically (per step) using the mixed-quadrature RK method rather than the extended method. When the equation is of convolution type this can be exploited (i) to avoid unnecessary evaluations of the kernel, and (ii) to use FFT techniques (see Chapter 8, section 5) in the efficient evaluation of the lag, with appropriately chosen RK, mixed RK, or quadrature formulae.

The methods derived here and associated with methods for (6.24) can be constructed by applying ODE techniques to the imbedded problem (6.23), or by examining the degenerate kernel case (6.14) and taking the limit as R is allowed to grow larger.

We observe that not all the formulae presented by our discussion are equally satisfactory. The convergence and order are readily established but the stability properties also need to be considered in making a choice. Observe also that some methods require the kernel to be defined for $y \geqslant x$ as well as $y \leqslant x$, and a smooth extension may then be required. The work involved and storage may become paramount.

Turning to other equations, we note that methods for Volterra equations of the first kind can frequently be constructed as analogues of the methods for equations of the second kind, but this process is dangerous. Whilst under the mildest conditions the above methods converge as h tends to zero for second-kind equations, similar guarantees for the first-kind Volterra equation are harder to secure. Even when convergence results can be established the results may be unsatisfactory, for example when using the trapezium rule to discretise certain types of equation (6.16): use of the mid-point rule is better. However, product integration methods and collocation methods, which are especially suited to Abel equations, can be used for a variety of problems, with some care over the details.

We shall outline the product integration method for Abel equations, and it should be clear how the techniques might be adapted for Volterra kernels where $H(x, y, f)$ admits a suitable decomposition as a product, say $k(x - y)\, H^*(x, y, f)$; in the case of Abel equations (6.2) with $p = 1$ we have such a decomposition with $k(s) = s^{-\alpha}$. Now the analogue of certain quadrature methods involves the construction of product integration rules or weighted integration rules with weight function $k(x - y)$:

$$\int_0^x k(x - y)\, H^*(x, y, f(y))\, dy \simeq \sum_j \nu_j(x)\, H^*(x, jh, f(jh)) ,$$

$$x \in \{ih,\ i = 1,2,\ldots .\}. \qquad (6.35)$$

The use of such discretisations to obtain a system of equations for the approximate solution of (6.3) or (6.5) with $p = 1$ requires no elaboration. In the case of the Abel equation, the expressions for the weights $\nu_j(x)$ can be obtained analytically. (The technique can be varied to give formulae which are modifications of extended Runge-Kutta formulae used for Volterra equations, on taking $x \in \{ih + \theta_r h;\ i = 0,1,2,\ldots\ ;\ r = 0,1,\ldots,p\}$ and summing over the appropriate set of integrand values.) The valuable feature is the appearance of a convolution structure in the pattern of the integration weights. Space does not allow the presentation of all the details here, but the reader can readily derive the weighted trapezoidal rule appropriate for the classical Abel equation of the second kind.

Finally, we come to the collocation methods: here the space of approximating functions can be taken to be piecewise polynomials or splines. The formal aspects of collocation are the same as for Fredholm equations (see Part B), but because the equation is of Volterra type the resulting equations can be solved in block-by-block fashion (compare extended implicit Runge-Kutta methods in the Volterra case). Since the solution of an Abel equation may be badly behaved, special steps have to be taken in this case to obtain accuracy. The reader may refer in particular to te Riele (1982), Brunner (1982) and Brunner and te Riele (1982). In particular, for smooth solutions, the superconvergence results associated with Brunner apply.

Collocation techniques readily extend to Volterra integro-differential equations: so also do the quadrature and weighted integration techniques if associated with a linear multistep formula for the derivative part, whilst Runge-Kutta methods have as a natural development the extended or the mixed-quadrature RK integration techniques, with counterparts in weighted integration for weakly singular integrands.

PART B. FREDHOLM EQUATIONS

L.M. Delves

3. Introduction

In this part we consider the solution of equations involving Fredholm operators. It is necessary to consider these types of operators separately because an equation involving a Volterra operator is akin to an initial-value differential equation (and can often be converted to that form). A Fredholm operator corresponds (roughly) to a boundary-value differential equation; more precisely, a two-point boundary-value second-order differential equation can always be converted to a Fredholm integral equation. The result of this rough correspondence is that techniques for solving Volterra equations are reminiscent of those for initial-value ordinary differential equations; those for Fredholm equations are not.

The standard classification of Fredholm equations recognises three distinct types of equation:

(a) Fredholm equation of the first kind:

$$\int_a^b K(x, y)\, f(y)\, dy = g(x)\ . \tag{6.36}$$

(b) Fredholm equation of the second kind:

$$f(x) - \int_a^b K(x, y)\, f(y)\, dy = g(x)\ . \tag{6.37}$$

(c) Fredholm equation of the third kind (Eigenvalue):

$$f(x) - \lambda \int_a^b K(x, y)\, f(y)\, dy = 0\ . \tag{6.38}$$

In these equations, f(x) is assumed unknown; g(x) is a known function, referred to as the driving term; $K(x, y)$, referred to as the kernel function, is also known. A solution is sought for $a < x < b$. The three types of equations require rather different treatments. A first-kind equation arises typically from some sort of measuring process in which the instrument used has a finite resolution, represented by the kernel function; $g(x)$ represents the raw measurements, and the unknown $f(x)$ represents the measured values after correction for the instrument's finite resolving power. Such a correction process is typically ill-conditioned, and the solution process must take account of this feature; these equations are both common and relatively difficult to handle.

Second- and third-kind equations, on the other hand, are relatively straightforward to treat, provided g and K are smooth and the interval $[a,b]$ is finite. In these equations, the term $f(x)$ stabilises the problem, with the result that the algebraic equations produced by approximate methods are usually reasonably well conditioned. Eigenvalue problems seem to occur fairly often in practice; second-kind equations (with smooth g, K) not very often. However, we shall treat second-kind equations fairly fully, because most numerical methods for integral equations have three recognisably distinct components, which may be illustrated for this case:

(i) The integral operator is discretised in some way;

(ii) The discretisation is inserted into the integral equation and a set of algebraic equations, which approximates the original integral equation, is set up and solved;

(iii) An estimate of the error in the discretised solution is made.

Techniques for discretising the integral operator are common to all three types of equation, so we need consider them in detail only once. However, the details of the algebraic equations, and the production of error estimates, depend strongly on the type of equation.

4. Fredholm second-kind equations

4.1 Basic Nystrom method

We begin with equation (6.37); and assume for simplicity that g, K are smooth and easy to evaluate, and that the interval $[a,b]$ is finite. We warn the reader that these assumptions are rarely satisfied in practice. If they are, then for fixed x we may replace the integral in (6.37) by a numerical quadrature. We choose a rule with N points y_i and corresponding weights w_i ,

$i = 1,2,\ldots,N$, such that

$$\int_a^b h(y)\,dy = \sum_{i=1}^{N} w_i\, h(y_i) + E_N(h)\,, \qquad (6.39)$$

where $E_N(h)$ is the error in the rule. Applying this rule to (6.37), we obtain the identity

$$f(x) = g(x) + \sum_{j=1}^{N} w_j\, K(x, y_j)\, f(y_j) + E_N(K(x,\ \ )\, f(\ \ ))\,. \qquad (6.40)$$

The Nystrom or collocation method consists in ignoring the error term E_N and satisfying the resulting approximate identity at the N points $x = y_i$, $i = 1,\ldots,N$. Writing f_i for the approximate value of $f(y_i)$, and $\underline{f} = [f_1\ f_2\ \ldots\ f_N]^T$, this yields the $N \times N$ set of linear defining equations

$$(I - K)\,\underline{f} = \underline{g}\,, \qquad (6.41)$$

where I is the $N \times N$ identity matrix and K and $\underline{g}$ have elements

$$K_{i,j} = w_j\, K(y_i\,,\, y_j)\,, \qquad g_i = g(y_i)\,.$$

These equations have a unique solution provided $(I - K)$ is nonsingular; this solution defines an approximation to $f(x)$ at the quadrature points y_i. It is a simple matter to interpolate between these points if necessary, and traditional to use (6.40) (excluding the error term E_N) as defining a 'natural interpolant' $f_N(x)$. Sufficient conditions under which the solution of (6.41) will exist, and under which $f_N(x)$ will converge to $f(x)$, are discussed in Baker (1977), Baker and Miller (1982), Delves and Mohamed (1985). Provided (6.37) has a unique solution, most 'sensible' sequences of N-point rules will yield convergence for continuous K, g; this includes any Newton-Cotes or Gauss-Legendre repeated rule of fixed degree, and the sequence of Gauss-Legendre or Gauss-Patterson N-point rules. However, we can make the following observations when trying to choose an appropriate rule:

(a) Most implementations of the method solve (6.41) directly, using some form of Gauss elimination routine. The cost of this is $O(N^3)$ operations, an often overlooked feature of Fredholm equations which makes 'difficult' problems (those requiring large N) surprisingly expensive to solve compared with a differential equation.

(b) Iterative methods exist which reduce this cost to $O(N^2)$; see for example Atkinson (1976) or Delves and Mohamed (1985, section 4.6). However, these methods are not simple to implement effectively. Multigrid methods are also applicable, and are very similar conceptually to the schemes of Atkinson; they are also not simple to implement effectively (Hemker and Schippers 1981).

(c) It therefore pays to try to keep N small. This suggests the use of a high-order quadrature rule: a sequence of Gauss N-point rules, for example, or at least a repeated high-order rule, rather than the (simple to code) repeated trapezium or Simpson's rules.

4.2 Error estimates

A numerical method without some sort of error estimate is useless; we have no way of rejecting or accepting the results. Unfortunately, error estimates are hard to provide within the Nystrom framework. It is always possible to repeat the calculation with larger N and compare results, but this is an expensive procedure. An alternative approach is to try to estimate the error term E_N in (6.40) directly. This is done for the repeated Simpson's rule in Thomas (1975), but again it costs as much to produce the error estimate as to compute the numerical solution. Alternative approaches, based on a difference correction or deferred approach to the limit, are discussed briefly in Delves and Mohamed (1985, section 4.3). These can be effective, but their effectiveness depends on the fact that they are based on a low-order rule, which is itself not cost-effective.

4.3 Treatment of singularities

If the kernel K or the function g is not smooth, the straightforward Nystrom method is likely to converge only slowly (though there are cases when this is not true; see Delves and Mohamed (1985, example 4.2.5a)). Singular kernels arise quite often in practice. Often the singularity can be factored out:

$$K(x, y) = K_0(x, y)\, K_1(x, y)$$

where K_0 is some 'standard' singular function and K_1 is well behaved. It is not necessary for K_0 to be unbounded for it to cause trouble; for example, if $K_0(x, y) = (x - y)^{1/2}$, a standard Nystrom calculation will converge only slowly.

The obvious way of taking account of the nature of K_0 is via the use of product integration. With $x = y_i$, the integral operator in (6.37) takes the form:

$$\int_a^b K_0(y_i, y)\, K_1(y_i, y)\, f(y)\, dy = \sum_{j=1}^{N} w_{ij}\, K_1(y_i, y_j)\, f(y_j) + E(y_i, N), \qquad (6.42)$$

where $E(y, N)$ represents the error in the quadrature rule. We now construct the weights w_{ij} by insisting that E be zero when $K_1(y_i, y)$ is a polynomial of degree $N-1$ for each y_i, $i = 1,2,\ldots,N$. Clearly the weights will depend on K_0; in general they will depend also on i, N, and this is a drawback of the method. However, it is straightforward to compute appropriate values for the w_{ij}, for any given K_0, and the method then yields an approximating set of linear equations for the values $f(y_i)$ which have already taken account of the singular nature of K. Examples of the construction of the weights are given in Atkinson (1976), Sloan (1979). The convergence of these methods is studied in Chandler (1980). The case $K_0 = \ln(x - y)$ is described in Delves and Mohamed (1985, section 4.4). The paper of Sloan is of special interest since he uses the standard Chebyshev points, and gives explicit results for the weights, for algebraic and logarithmic singularities. Product formulae for Cauchy singular-value integral equations are discussed in Krenk (1978).

It is not always possible to factor out the singularity completely. However, the technique generalises easily to the case when

$$K(x, y) = \sum_{\ell=1}^{M} k_\ell(x, y)\, K_\ell(x, y), \qquad (6.43)$$

where the k_ℓ are singular and the K_ℓ smooth. For example, we may write the kernel $K(x, y) = \ln[\cos(x) - \cos(y)]$ in the form (6.43) with

$$k_1 = 1, \quad K_1 = \ln\left[\frac{\sin((x - y)/2)\,\sin((x + y)/2)}{((x - y)/2)(x + y)(2\pi - x - y)}\right],$$

$$k_2 = \ln(x - y) , \qquad K_2 = 1 ,$$

$$k_3 = \ln(2\pi - x - y) , \quad K_3 = 1 ,$$

$$k_4 = \ln(x + y) , \qquad K_4 = 1 ;$$

an example using this decomposition is given in Atkinson (1976). It is clear that the approach may in practice be messy to apply, but it seems that any method which aims at treating singularities requires to do so in detail if it is to be successful; we can either take the necessary analytic care, or apply brute force (large N). Which is more cost-effective depends on circumstances; common forms of singularity are well worth the effort expended.

An alternative approach is to subtract out the singularity; we illustrate via an example. Consider the singular problem

$$f(x) - \int_a^b \frac{f(y)}{|y - x|^{\alpha}}\, dy = g(x) .$$

For any point y_i ,

$$f(x) - \int_a^b \frac{f(y_i)}{|y - x|^{\alpha}}\, dy - \int_a^b \frac{f(y) - f(y_i)}{|y - x|^{\alpha}}\, dy = g(x) .$$

Now if we use the Nystrom scheme on this subtracted equation, we obtain the defining equations:

$$f(y_i)\left[1 - \int_a^b \frac{dy}{|y - x|^{\alpha}}\right] - \sum_{\substack{j=1 \\ j\neq i}}^{N} w_j \frac{[f(y_j) - f(y_i)]}{|y_j - y_i|^{\alpha}} = g(y_i) .$$

The remaining integral must be evaluated exactly if the technique is to be of value; we then see that the subtraction has 'smoothed' the singularity, but has not entirely removed its effect; the integrand even after subtraction is still not very smooth.

It is possible however to perform systematic repeated subtractions to make the residual integrals as smooth as desired; see Abd-Elal and Delves (1976), Graham (1980), Ioakimidis and Theocaris (1980), and also Anselone

(1981). These techniques are quite messy in practice.

4.4 Infinite regions

Problems for which either or both of the limits (a,b) in the integral operator are infinite, are formally singular, and special action of some sort is needed. The available approaches are independent of the numerical technique to be used. If we consider for definiteness a semi-infinite region, and a second-kind problem:

$$f(x) - \int_a^\infty K(x, y)\, f(y)\, dy = g(x)$$

with interval $(a,b) = (a,\infty)$, we may try any of the following:

(i) Use either a Nystrom or an expansion method, with expansion and/or quadrature rule directly handling the infinite interval. This direct approach is normally very effective.

(ii) Truncate the region: we replace (a,∞) by (a,R) and compute a solution $f_R(x)$. We should then check convergence as R increases; the rate of convergence depends on the behaviour of the kernel $K(x, y)$ as a function of y, and may be slow. This technique is popular for its simplicity, but has little to recommend it otherwise.

(iii) We may map the infinite region onto a finite region. Let

$$z = z(x): (a,\infty) \to (-1,1) \text{ (say)}$$

$$f(x) = F(z(x)); \; K(x, y) = K_M(z(x), z(y)); \; g(x) = G(z(x)) .$$

Then

$$F(z) - \int_{-1}^{1} K_M(z, t)\, J(t)\, F(t)\, dt = G(z) ,$$

where $J(t)$ is the Jacobian of the mapping z. We may now solve on the interval $(-1,1)$. This approach can also be very effective; but the Jacobian $J(t)$ is likely to be singular near $t = 1$, and the nature of this singularity should be taken into account.

(iv) Sometimes neither of the approaches (i) or (iii) is easy to apply. We may however know something about the asymptotic form of the solution

$f(x)$. For example, in scattering problems we may know that for large x, $f(x)$ takes the form

$$f(x) = Af_1(x) + Bf_2(x) ,$$

where f_1 , f_2 are known but A, B are not. Then for sufficiently large R we may write

$$f(x) - \int_a^R K(x, y)\, f(y)\, dy - \left[\int_R^\infty K(x, y)\, f_1(y)\, dy\right] A - \left[\int_R^\infty K(x, y)\, f_2(y)\, dy\right] B = g(x)$$

and the accuracy of this representation will probably be much better than that of approach (ii), for a given R.

4.5 Expansion methods

An expansion method is one in which we approximate the unknown $f(x)$ by a truncated expansion of the form

$$f(x) \simeq f_N(x) = \sum_{i=1}^{N} a_i\, h_i(x) . \qquad (6.44)$$

The h_i are a set of basis functions; the expansion coefficients a_i are to be determined. The h_i may depend explicitly on N, but we suppress any such dependence here. Expansion methods may be applied to a wide class of problems, including integral equations. They tend to be more complicated to implement than the simple Nystrom method, but have compensating advantages.

An expansion method is defined by:

(i) The choice of the functions h_i ,

(ii) The method used to derive and set up defining equations for the coefficients a_i ,

(iii) The method used to solve these equations,

(iv) The method used to derive an error estimate.

The first two aspects are the most important; the second, the choice of defining equations, is usually taken as categorising the method. We describe briefly the Galerkin method as an example.

Let f_N be an approximation to f; then we introduce the residual function r_N and error function e_N :

$$r_N = g - Lf_N , \qquad e_N = f - f_N . \tag{6.45}$$

The Galerkin method sets N moments of the residual to zero. We introduce a set of 'test functions' $n_i(x)$, i = 1,...,N, and impose the conditions

$$(n_i , r_N) = \int_a^b n_i(x)\, r_N(x)\, dx = 0 , \quad i = 1,\ldots,N . \tag{6.46}$$

The integrals in (6.46) must be evaluated numerically, and we assume for efficiency that a single Q-point rule is used for each i. Then inserting (6.44), (6.46) takes the form

$$P^T WM \underline{a}_N = P^T W \underline{g} , \tag{6.47}$$

where

$$P_{ij} = n_j(x_i) \quad i = 1,\ldots,Q , \quad j = 1,\ldots,N , \quad W = \mathrm{diag}[w_i] .$$

A common choice is to take $n_i = h_i$, yielding the 'symmetric Galerkin' method. Alternatively, a weighted Galerkin method results from choosing

$$n_i = w(x)\, m_i . \tag{6.48}$$

Galerkin methods are discussed in detail in Delves and Mohamed (1985, Ch. 6-8).

4.6 The Fast Galerkin method

This method (Delves 1977; Delves, Abd-Elal and Hendry 1979) also maps the interval [a,b] to [-1,1] and uses a Chebyshev expansion for f_N, taking the test functions to be

$$n_i(x) = T_{i-1}(x)\,(1 - x^2)^{-1/2} \; . \tag{6.49}$$

This choice of weighted test function simplifies the algebra involved, but is not essential to the achievement of the major aim of the method. This aim is to reduce the cost of both the matrix set-up and solution phases of the method. The details are given in the references cited above, or in Delves and Mohamed (1985, section 8.5). The basic idea can be seen by noting that, with this choice of test function, the right-hand side of the Galerkin equations

$$L_G\, \underline{a}_N = \underline{g}_G \tag{6.50}$$

has ith component

$$g_{Gi} = \int_a^b g(x)\, T_{i-1}(x) \,/\, (1 - x^2)^{1/2}\, dx \; ,$$

which, apart from a constant factor, identifies g_{Gi} as the ith coefficient in the Chebyshev expansion of g(x). Provided g(x) is smooth, these coefficients can be computed accurately using FFT techniques (see Chapter 8, section 5); it is shown in Delves (1977) that an N-term FFT retains sufficient accuracy. Similar use can be made of FFT techniques to evaluate the left-hand side matrix, although due account has to be taken of the <u>lack</u> of a factor $(1 - y^2)^{-1/2}$ under the integral in equation (6.37); for details, see Delves (1977). The resulting algorithm has set-up costs of $O(N^2 \ell n\, N)$, reasonably comparable with the $O(N^2)$ cost of the Nystrom method. Moreover, because the algorithm is based on an orthogonal expansion, the structure of the matrix produced allows a straightforward and rapidly convergent iterative solution scheme, with cost $O(N^2)$.

The $O(N^2 \ell n\, N)$ set-up cost can be reduced still further to $O(N^2)$ in the case of convolution kernels, where

$$K(x, y) = k(x - y)$$

(see Abd-Elal and Delves (1984)).

5. Fredholm first-kind equations

Although second-kind Fredholm equations with singular kernels occur in a number of applications (boundary integral treatment of partial differential equations, for example), first-kind problems (6.36) are much more widely encountered; most problems in which a measuring instrument of finite resolution is involved, will yield a first-kind equation when modelled mathematically. The numerical treatment of (6.36) must take into account the following features:

(a) The driving term $g(x)$, and the kernel $K(x, y)$ are often not known exactly but are subject to errors of measurement.

(b) The problem (6.36) is inherently <u>ill-posed</u>. If we perturb the driving term g and measure the resulting change in the solution f:

$$g \to g + \delta g ,$$

$$f \to f + \delta f ,$$

then the ratio $||\delta f|| \ / \ ||\delta g||$ may be arbitrarily large.

(c) For a given kernel K, the existence of a solution of (6.36) may depend on the form of the driving term g.

These difficulties are discussed further in Delves and Mohamed (1985). We note here that they make a general treatment of the numerical problem rather difficult; there is <u>no single best method</u> for first-kind problems. The attainable accuracy depends crucially on the form of the kernel; a smooth kernel yields a very ill-posed numerical problem, and a singular kernel in general one which is less ill-posed numerically. To achieve the maximum accuracy for a given problem, any available information about the solution (such as positivity or monotonicity) should be fed into the solution process; moreover, because the potentially available accuracy is likely to be limited, it is helpful to know what properties are required of the solution (maximum amplitude ? position of peak ? smoothness ?) before choosing the numerical method to be used.

Many special methods have been proposed for particular problem areas.

We consider here only the most common of these; in particular, methods which are inherently capable of being automated in a straightforward manner. This last requirement rules out methods which rely on finding a sequence of approximate solutions and then selecting the one which 'looks best'; however no criticism of such methods is intended - they may yield the best results for a particular problem if suitable experience is available to guide the selection.

5.1 Tikhonov regularisation

We write (6.36) in the operator form

$$Kf = g \,. \tag{6.51}$$

The integral operator K may be discretised effectively using any of the techniques described for second-kind equations; the difficulty in solving (6.51) stems not from the operator itself but from the ill-conditioning of the equation and hence of the resultant equations for the approximate solution. The standard approach for quenching the ill-conditioning is to solve, not (6.51) but the regularised problem:

$$\text{minimise } \{ ||Kf - g||^2 + \alpha ||Lf||^2 \} \,, \tag{6.52}$$

where α is called a regularisation parameter, and L is some linear operator chosen for convenience or after some plausibility analysis. The most common choice is to set

$$L = \frac{d^n}{dx^n} \tag{6.53}$$

yielding 'nth order' regularisation. The purpose of the term involving L in (6.52) is to ensure that the solution f remains smooth in a sense determined by the choice of L; with $n = 2$, for example, we control the size of the second derivative of the solution. The way in which (6.52) reduces the ill-conditioning can be seen by setting $n = 0$, yielding $L = I$, and using the L_2 norm in (6.51). Then the optimal f satisfies the equation

$$[K^+K + \alpha I]f = K^+g , \tag{6.54}$$

that is,

$$\int_a^b K_R(x, y)\, f(y)\, dy + \alpha f(x) = \int_a^b K(y, x)\, g(y)\, dy$$

where

$$K_R(x, y) = \int_a^b K(z, x)\, K(z, y)\, dz .$$

Equation (6.54) is a second-kind Fredholm equation; for non-zero α it is therefore well posed. The larger we choose α, the better posed the numerical problem, and the further (6.54) is from the original problem (6.51). It is then necessary to give a prescription for choosing α. We treat only the most successful of such methods (Wahba 1977, 1980; Wahba and Craven 1979; O'Brien and Holt 1981; Anderssen and Prenter 1981), that of cross-validation. The idea involved is very general; we outline it in the context of a zero-order regularisation as described by (6.52) with $L = I$. We choose a discrete L_2 norm over a point set x_i , $i = 1,2,\ldots,P$; then (6.52) becomes

$$\text{minimise} \sum_{i=1}^{P} [(Kf(x_i) - g(x_i))^2 + \alpha f(x_i)^2] . \tag{6.55}$$

We now remove the term involving point x_j from the sum, and solve for the minimising function f. This solution, f_j , will depend on j; if α has been chosen well, it will yield a small residual at each of the points x_i, including in particular the point x_j which was not used in defining f_j . A measure of the overall success of the value of α chosen, is

$$V(\alpha) = \sum_{j=1}^{P} [(Kf_j)(x_j) - g(x_j)]^2 ,$$

and we therefore adjust α to minimise $V(\alpha)$.

This procedure requires no prior knowledge of the problem or of a suitable α; in practice it seems to be very effective. Although we need to solve P problems rather than the original single problem, these are closely related, and it is possible to arrange the details of the calculations to make use of this fact. Note that we have not specified how the operator K is to be handled; we can use cross-validation in either a Nystrom or an expansion framework.

It is also possible to impose other constraints on the solution, during the solution process. Typical constraints, based (presumably) on prior knowledge, take one of the forms

$$f(x) > 0\ , \qquad f'(x) > 0\ , \qquad 0 < f(x) < A\ .$$

Constraints such as these, expressed on the point set $\{x_i\}$, can be imposed during the minimisation in (6.55); note however that the resulting minimisation problem becomes significantly more difficult.

5.2 Regularisation in the Fast Galerkin framework

An alternative form of regularisation is attractive within methods based on orthogonal expansions of the solution (Babolian and Delves 1979). Suppose that we represent the solution f(x) by an expansion of the form

$$f(x) = \sum_{i=1}^{N} a_i\, h_i(x)\ .$$

If the set h_i is orthonormal, then the very weak assumption that f is square integrable means that the coefficients a_i must tend to zero for large i. We can model this behaviour by assuming that for some C, r,

$$|a_i| < Ci^{-r}\ . \tag{6.56}$$

The parameter r depends on the smoothness properties of the solution f; the smoother f(x) over the region (a,b), the larger the value of r which can be used in (6.56).

If we now construct the Galerkin equations for the problem (6.36):

$$\sum_{j=1}^{N} B_{ij} a_j = g_i\ , \qquad i = 1,2,\ldots,M\ , \tag{6.57}$$

we may 'solve' (6.57) in some suitable norm, subject to the constraints (6.56). The number M of Galerkin equations used is usually taken to be N, but this is not essential to the method. Using the L_1 norm, this procedure leads to a linear programming problem for which efficient solution techniques are available (see Chapter 11, section 2).

This approach has the advantage that further constraints on the solution can be added without changing the nature of the minimisation problem. Thus, positivity constraints (on a point set) take the form

$$\sum_{j=1}^{N} a_j h_j(x_i) > 0 , \qquad i = 1,2,\ldots,P ,$$

and these constraints merely form extra rows in the LP problem.

The success of such an algorithm depends on the choice of the parameters r, C, which regularise the problem. Methods for choosing r, C were discussed in Babolian and Delves (1979); we may alternatively adapt the cross-validation process to determine suitable values (Essah, Delves and Belward 1986). Let $\underline{a}^{(j)}$ be the solution of the constrained problem defined by (6.56), (6.57) with the jth equation of (6.57) dropped. Then the residual for this jth equation is

$$V_j = \sum_{k=1}^{N} B_{jk}(\underline{a}^{(j)})_k - g_j .$$

We compute

$$V(C, r) = \sum_{j=1}^{M} |V_j|$$

and choose C, r to minimise V(C, r). Again, this yields a relatively expensive but effective procedure; the sequence of LP problems which are solved are highly related, and advantage can be taken of this to reduce the cost of the method.

As noted above, the (maximum) value of r which can be inserted in (6.56) depends on the smoothness properties of the solution f(x), and in particular on the number of continuous derivatives which it possesses on the interval [a,b]. If this is known in advance, then r can be pre-assigned; in any case, the computed value of r forms a useful check on the 'numerical smoothness' of the solution.

6. Software

Implementations of these methods can be of two types. A 'non-automatic' routine is one which requires that the user specify the parameter N; it returns an approximate solution and (perhaps) an error estimate. An 'automatic' procedure asks the user to provide a required accuracy, and chooses N to satisfy this accuracy.

6.1 Non-automatic routines

The Nystrom technique is so simple to implement for second-kind linear equations that most users write their own routine. However, they are then tempted to use a simple rule, and obtain slow convergence as a result. The NAG (Algol 68) Library contains a non-automatic Nystrom routine, D05CAB, which provides for the use of any of the available quadrature rules, including Gauss-Legendre and Newton-Cotes rules. By using the Gauss-Rational, or the modified Gauss-Laguerre or Gauss-Hermite rules in the Library, problems on semi-infinite or infinite regions can be solved. The El-gendi algorithm, based on Chebyshev expansion (El-gendi 1969), is provided in the Fortran Library via routines D05AAF (for smooth kernels) and D05ABF (for kernels K(x, y) which have a discontinuous derivative across the line x = y; such kernels are no problem for a Nystrom technique, but need special attention with the El-gendi or Fast Galerkin techniques). None of the three NAG routines returns an error estimate, although the El-gendi routines could be made to do so with little effort.

Routine FE2SR (Thomas 1976) implements the Nystrom technique with repeated Simpson's rule, but has the advantage of returning an error estimate. An implementation (FFTNA) of the Fast Galerkin scheme is given in Delves and Abd-Elal (1977); improved versions are given in Riddell (1981) (PMODNA) and in the package FAG1 (Delves, Abd-Elal and Hendry 1981); all of these routines provide error estimates.

It is difficult to provide a brief account of the strengths and weaknesses of these routines; they behave differently on different classes of problem. An attempt at a comparison is given in Riddell and Delves (1980). There the routines were run on a large number of problems in four problem classes, and on the basis of their average performance ranked for accuracy, speed, and reliability, both without and with error estimates. In the latter case, those routines not providing error estimates were run twice with different N to provide a 'brute force' estimate. The final rankings show that overall best performance was given by the Nystrom routine D05CAB using a sequence of Gauss-Legendre rules, and by the Fast Galerkin routine PMODNA. The low-order routine FE2SR behaved relatively badly on most problem families, as predicted on qualitative grounds above.

6.2 Automatic routines

These require much more development effort to yield a satisfactory performance over a wide range of problems, and hence the potential user is much more dependent on specialist software. Three Nystrom routines are available from Atkinson (1976, and more recent versions):

SIMP: A routine using a sequence of Simpson rules;

BOOLE: A version of SIMP using the higher order Boole's rule;

GAUSS: A re-written version using a sequence of Gauss rules, following the numerical evidence from the comparison of non-automatic routines.

These routines use the two-level iterative scheme outlined in Atkinson (1976), to solve the defining equations. A later routine using a multigrid technique is given in Schippers (1980).

Schippers tested his routine against the Simpson routine of Atkinson; he concluded that the multigrid approach saved about 30% in time for an average problem. The Atkinson routines were also tested by Riddell (1981), along with two automated versions of the Fast Galerkin algorithm. Again, the high-order methods showed signs of superiority, but none of the routines was uniformly better than the others, all having trouble in stopping soon enough to give the required accuracy at minimum cost; this is in fact more difficult for a high-order routine because the accuracy increases more rapidly with N.

6.3 Singular problems

Of the above routines, only FAG1 makes any attempt to deal efficiently with singular problems. This package contains a library of singular functions and kernels which it can deal with. The library includes:

Algebraic singularities: $(1 - x)^a$, $a > -1/2$

Logarithmic singularities: $\ell n(x - y)$, $\ell n(1 - y)$, $\ell n(1 + y)$

Green's Function and Volterra kernels

Cauchy Singular-Value equations

FAG1 provides a problem-description language which allows the user to describe his own, possibly non-standard problem; thus it can handle non-standard forms of second-kind equation.

6.4 First-kind and eigenvalue problems

Despite the importance of the problem, there is a dearth of software for first-kind equations. The package FAG1 contains a facility based on the Galerkin technique of section 5.2, for first-kind (or other ill-posed) problems; it is the only general-purpose solver known to the author. This situation presumably reflects the remark made earlier, that special methods for particular problems seem endemic; it is not, however, very satisfactory for the casual user.

There is little software directly available for eigenvalue integral equations. The Fast Galerkin package FAG1 handles eigenvalue problems with either smooth or singular kernels, using inverse iteration to solve the approximate equations; it assumes that the eigenvalues sought are real, and will fail if this is not so. FAG1 also produces an error estimate for the eigenvector. The NAG Library contains a number of routines for the algebraic problem. Producing the defining equations for the Nystrom method is straightforward for smooth problems, but less so for singular problems, and the lack of readily available software is a little surprising.

CHAPTER 7

Finite-Element Methods

R.W. Thatcher

1. Introduction

The finite-element method is a way of solving large and complicated physical problems. It was developed by engineers in the late 1950's and the early 1960's, e.g. Turner et al. (1956), and simple applications were analysed by mathematicians in the late 1960's and early 1970's, e.g. Zlamal (1968), Strang and Fix (1973). It was viewed by the early engineering workers as an energy method, in particular as a method for splitting a continuum problem into a discrete problem by considering it as a collection of 'structural elements' and minimising the potential energy; such a simple analogy is not apt for most applications. The early success of the method was dependent on the insight and ability of the engineer to choose a suitable collection of elements to model the problem appropriately. The popularity of the method increased as the power of the computer increased for it leads to a highly repetitive and computationally efficient algorithm. It is this aspect of the method together with its ability to handle problems in complex geometries and also its rigorous mathematical foundation for many applications, that has brought the finite-element method to its prominent position. The history of the method goes back earlier than the 1950's, for example a similar method was proposed by Courant (1943), but computing power was not sufficiently developed at that stage for its full potential to be realised and exploited.

There are now many useful books on the finite-element method. A good idea of the scope of the method and the range of problems that have been tackled can be obtained from Zienkiewicz (1977), Connor and Brebbia (1978). Some good general presentations of the finite-element method are given by Davies (1980), Mitchell and Wait (1977), Akin (1982), Huebner (1975), Carey and Oden (1983), Ciarlet (1976). Finally two books of more specialised interest respectively in applied mechanics and fluid mechanics are Oden (1972), Thomasset (1981).

2. Splitting up a problem into elements

The finite-element method can be viewed as a computationally efficient method for defining test and trial functions. The domain of the problem is split up into a number of sub-domains called elements. Although it is not essential, it is usual to cover the domain with topologically similar elements; for example either all quadrilateral elements or all triangular elements for problems in two dimensions.

Within each element a local trial function is assumed; these functions are usually polynomial and therefore easy to handle by computer, in particular to integrate and differentiate. Again, it is usual to have the same degree of polynomial in all elements for a given problem.

These simple rules are quite often broken for three-dimensional problems when 'brick' and 'wedge' shaped elements may be used together to fill a region in a simple fashion. They may also be broken if there is a singularity present. Without going into details, by a singularity it is understood that the solution of the problem or one of its low-order derivatives is either discontinuous at some point or very large in a small region, for example at the tip of a crack or along a wall for a fluid at high Reynolds number. In such situations special elements of one form or another are often chosen in such a way that the true analytic behaviour of the solution is properly approximated; see for example Fix, Gulati and Wakoff (1973), Hutton (1979), Akin (1976). By this device the solution can often be represented more accurately than can be achieved by simple polynomials unless a very high degree of grid refinement is adopted.

When the trial function in an element is a polynomial, it may be written as a linear combination of 'nodal coordinates' multiplying 'nodal interpolating functions (polynomials)'. The nodal coordinates are usually values of the local function and/or some of its partial derivatives at specified points called 'nodes', most of which are placed around the periphery of the element. By equating nodal coordinates in neighbouring elements the required inter-element continuity conditions are obtained.

As an example we consider the cubic Hermite triangular element illustrated in Fig. 1 in which the local function φ is the cubic polynomial

$$\varphi = A_1 + A_2x + A_3y + A_4x^2 + A_5xy + A_6y^2 + \dots + A_{10}y^3 . \quad (7.1)$$

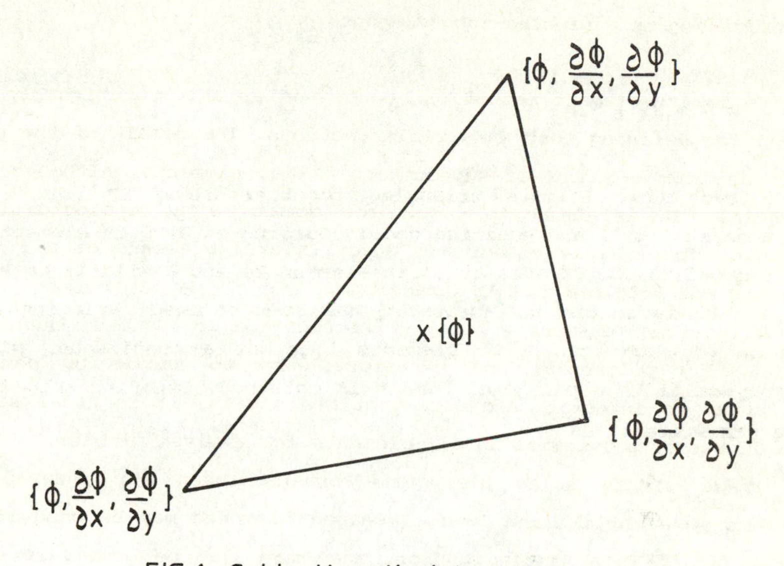

FIG.1 Cubic Hermite triangular element

This element has four nodes, the three vertices and the centroid, and ten nodal coordinates, $\{\varphi, \frac{\partial\varphi}{\partial x}, \frac{\partial\varphi}{\partial y}\}$ at each vertex and $\{\varphi\}$ at the centroid. The nodal coordinates are denoted by $\{\gamma_i\}_{i=1}^{10}$ where

$$\left.\begin{aligned} \gamma_{3i-2} &= \varphi \quad \text{at the ith vertex,} \\ \gamma_{3i-1} &= \frac{\partial\varphi}{\partial x} \quad \text{at the ith vertex,} \\ \gamma_{3i} &= \frac{\partial\varphi}{\partial y} \quad \text{at the ith vertex,} \\ \gamma_{10} &= \varphi \quad \text{at the centroid.} \end{aligned}\right\} \tag{7.2}$$

There is a unique correspondence between the full cubic (7.1) and the ten nodal coordinates (7.2). Denoting by N_i the cubic polynomial corresponding to the set of $\{\gamma_i\}$,

$$\left.\begin{aligned} \gamma_j &= 0 \qquad \text{for } j = 1,2,3,\ldots,10, \quad j \neq i, \\ \gamma_i &= 1, \end{aligned}\right\} \tag{7.3}$$

then N_i is called the 'nodal interpolating polynomial' for the ith nodal coordinate and

$$\varphi = \gamma_1 N_1 + \gamma_2 N_2 + \ldots + \gamma_{10} N_{10} \ . \qquad (7.4)$$

Bringing two cubic Hermite triangles together along PQ (or P'Q') as illustrated in Fig. 2 and equating nodal coordinates at Q in element 1 with the corresponding coordinates at Q' in element 2, and similarly at P, P', a function that is continuous across PQ has been defined. Bringing a whole collection of such triangular elements together a continuous, piecewise cubic polynomial in a polygonal domain is obtained. Denoting this function by Φ we can write it as

$$\Phi = \sum_{i=1}^{N} r_i \varphi_i \qquad (7.5)$$

where the r_i will be called 'global coordinates', representing values of Φ or one of its partial derivatives at the nodes of the triangulation which for the element above will be the vertices and the centroids.

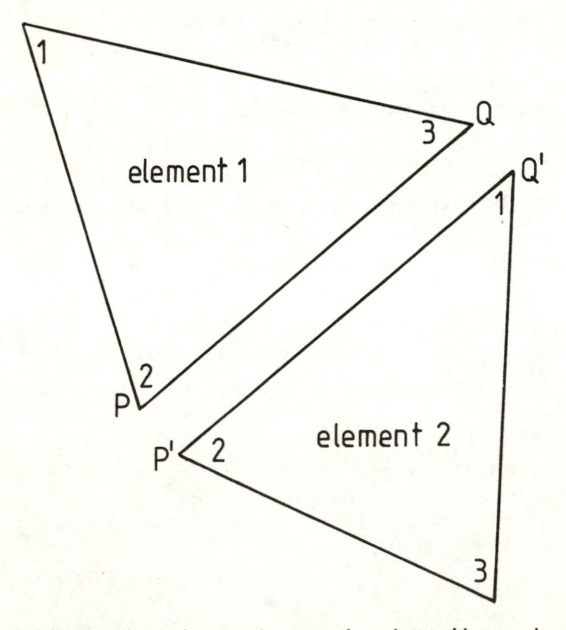

FIG. 2 Bringing two elements together along PQ

The functions $\{\varphi_i\}$ are said to have 'local support', that is to say the φ_i are zero over most of the region. In particular, if r_i is a global coordinate at node I then φ_i is non-zero only in those elements which include the node I. It is this feature of the functions used in the finite-element process that contributes to its great efficiency.

3. Using finite-element functions

The finite-element method can be used to define the functions used in a variational method, or both test and trial functions for a weighted-residual method or a variational-inequality method. Some typical examples will be outlined below.

As an example of a variational method, the function that minimises the functional

$$I_1[\varphi] = \int_\Omega \{k\,|\nabla\varphi|^2 - 2f\varphi\}\,d\Omega\,, \qquad k(x, y) > 0\,, \tag{7.6}$$

is sought over all φ belonging to some set D_1 , for example the set of functions for which

$$\int_\Omega k\,|\nabla\varphi|^2\,d\Omega \quad \text{exists} \tag{7.7}$$

and $\varphi = 0$ on $\partial\Omega$ the boundary of Ω. A second and similar variational method is to find the minimum of

$$I_2[\varphi] = \int_\Omega \{W(\varphi, \varphi) - 2f\varphi\}\,d\Omega \tag{7.8}$$

where

$$W(u, v) = \nabla^2 u\,\nabla^2 v + c\left\{2\frac{\partial^2 u}{\partial x\partial y}\,\frac{\partial^2 v}{\partial x\partial y} - \frac{\partial^2 u}{\partial x^2}\,\frac{\partial^2 v}{\partial y^2} - \frac{\partial^2 u}{\partial y^2}\,\frac{\partial^2 v}{\partial x^2}\right\}\,, \tag{7.9}$$

$$c = \text{const}\,,$$

over some set D_2 ; a function φ belonging to D_2 would satisfy

$$\int_{\Omega} W(\varphi, \varphi)\, d\Omega \quad \text{exists} \tag{7.10}$$

together with certain boundary conditions. A third variational method is one in which a part γ of the boundary $d\Omega_{\gamma}$ of the region Ω_{γ} is being varied, that is to find a function and a boundary segment that minimises

$$I_3[\varphi, \gamma] = \int_{\Omega_{\gamma}} |\nabla\varphi|^2 \, d\Omega \tag{7.11}$$

over φ belonging to some set D_3 and γ is continuous. For a variational method of this type, used to solve a free boundary problem, see Betts (1979).

A typical weighted-residual method is to find a function φ belonging to D_1 such that

$$\int_{\Omega} \{k\, \nabla\varphi\, \nabla\Psi - f\Psi\}\, d\Omega = 0 \tag{7.12}$$

for all Ψ belonging to D_1 (or some other set $\tilde{D}_1$). To find approximate solutions we substitute a trial function φ into (7.12) and determine the parameters of φ by taking test functions Ψ from D_1 . The solution of this weighted-residual method can be the same as that of the variational method (7.6). If the trial function Φ is given by

$$\Phi = \gamma_1\varphi_1 + \gamma_2\varphi_2 + \ldots + \gamma_N\varphi_N \tag{7.13}$$

and if the N generalised coordinates $\{\gamma_i\}_{i=1}^{N}$ of (7.13) are determined by using the N test functions $\{\varphi_i\}_{i=1}^{N}$, then the approximate solutions by the weighted-residual and variational methods are identical. Thus the weighted-residual method (7.12) can be considered as a generalisation of the variational method (7.6). The third variational method (7.11) has been given to refute the often quoted remark that every variational method can be written more generally as a weighted-residual method.

Finally, an example of a variational inequality is given, namely to find a function φ belonging to some set K such that

$$\int_{\Omega} \nabla\varphi.\nabla(\varphi - \Psi)\, d\Omega \leqslant 0 \tag{7.14}$$

for all Ψ belonging to K where K is, for example, the set of functions that have square-integrable first derivatives in Ω, are zero on $\partial\Omega$ and satisfy

$$\text{if} \quad \varphi \in K \quad \text{then} \quad \varphi \geqslant w \quad \text{in} \quad \Omega \tag{7.15}$$

for some function w.

4. Assembling the equations

For the sort of problems described above the test and trial functions appear in integrals over the whole domain of the problem. These integrals are represented as a sum of integrals over elements. In a finite-element package, the program will work through the elements setting up the non-zero contributions from a particular element to the overall integrals and then adding them in. In general the integrals over elements are performed numerically and suitable quadrature formulae will be supplied by the package. This process of going through the elements is a highly repetitive part of the algorithm which contributes to the computational efficiency of the method.

This part of the algorithm is often phased in with the linear solver in such a way as to try and minimise the overall storage requirement. Solving a differential equation accurately by the finite-element method inevitably involves the solution of a large system of equations and various solution schemes have been put forward, for example, skyline (Jennings 1966), frontal (Irons 1970), sparse, etc. . It depends on the size of grid and the architecture of the computer which is the best method of approach for a given problem. It is likely that the new generation of vector processors will provide more scope for initiative in this area.

One simple way to reduce the amount of work is to remove all nodal coordinates at nodes within an element at the element assembly stage; such a process is called 'static condensation'. This is possible because of the nature of the local support of the functions φ_i in the trial function (7.5).

For example, the nodal coordinate at the centroid of the element shown in Fig. 1 is determined by the other nodal coordinates in that element.

5. Simple element types

Probably the most widely used elements are called 'Lagrange elements' for which all the nodal coordinates are values of the local functions and there is only one nodal coordinate at each node. For example, four such elements are illustrated in Fig. 3.

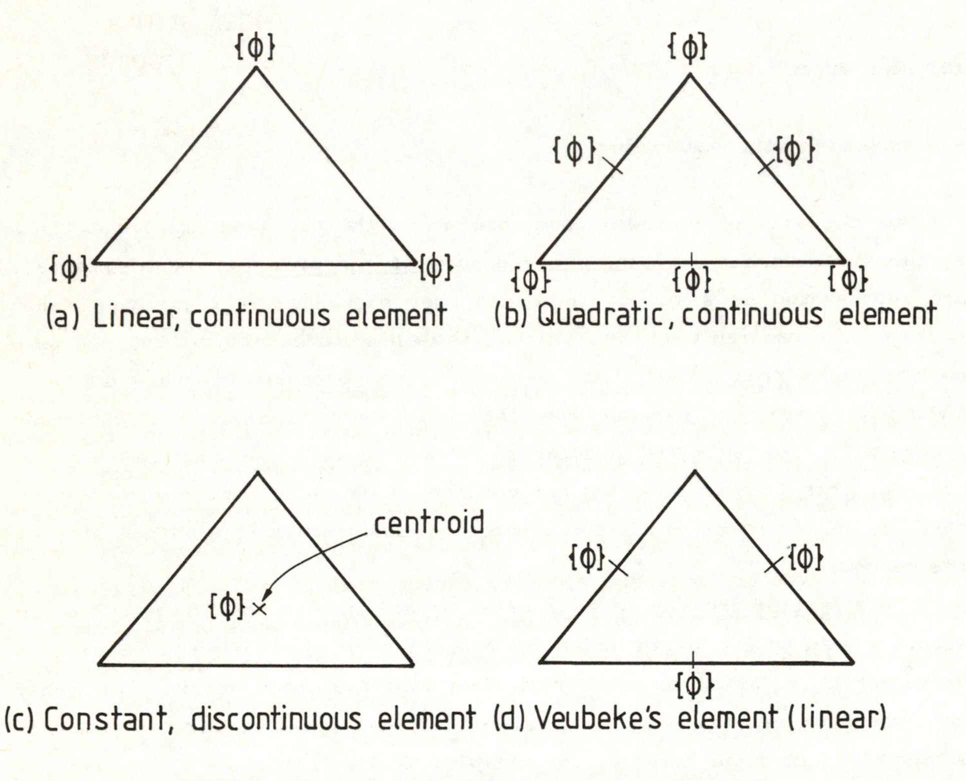

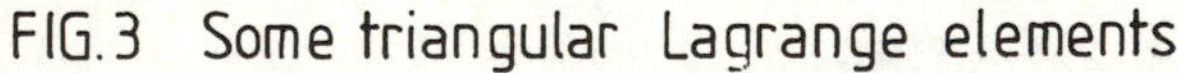
FIG. 3 Some triangular Lagrange elements

It is noted that elements (a) and (b) provide continuous functions but (c) and (d) provide discontinuous functions. It is also noted that although (d) gives a discontinuous function it is not fully discontinuous. A fully discontinuous, piecewise linear function is provided by the element illustrated in Fig. 4.

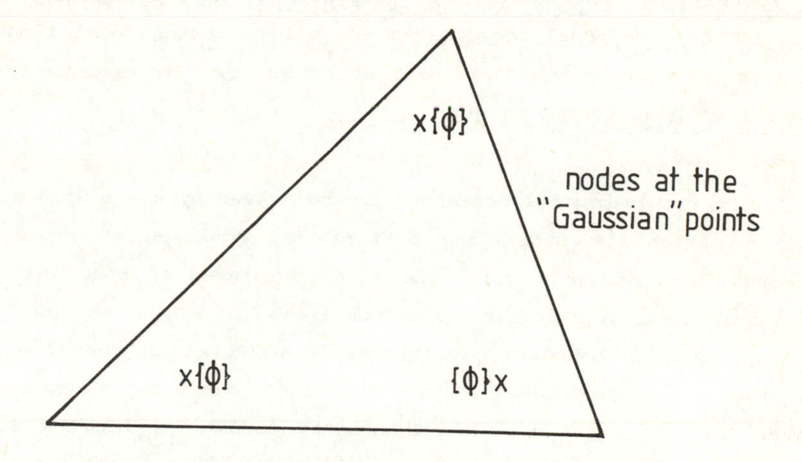

FIG. 4 Linear, fully discontinuous element

There are other triangular elements in common use, most of them based on Hermite interpolation for which some nodal coordinates represent derivatives and there may be more then one nodal coordinate per node.

Much work is now being done on problems in three-dimensional domains where the natural analogue of the triangle is the tetrahedron. This is in theory a perfectly suitable element, except that it is very difficult to visualise and hence difficult to use. A three-dimensional brick element, a distorted cube, is much easier to visualise and is almost exclusively used for three-dimensional problems. Such an element is the analogue of a two-dimensional quadrilateral element - a distorted square. Thus although it is arguable that triangles are the best two-dimensional elements, because of the need to generalise two-dimensional codes to three dimensions many finite-element users prefer quadrilateral elements for two-dimensional problems. This is not the only reason for preferring quadrilateral elements. When a region is triangulated it is often first split into quadrilaterals which are divided by constructing one of the two diagonals. There are then problems over which diagonal to construct, and over the effect of possibly giving some preferential direction to the approximate solution.

A simple bilinear transformation from (ξ, η) coordinates in the unit square to the (x, y) coordinates of the problem exists for each quadrilateral element, and provided the Jacobian is bounded away from zero within the element there are no major difficulties. The nodal interpolating functions are simple polynomials in the (ξ, η) coordinates and the bilinear transformation preserves the polynomial form along a boundary and thus makes the required inter-element continuity conditions easy to satisfy. This idea is simple to generalise for the three-dimensional brick.

6. Finite differences and finite elements

All the problems mentioned above can be solved by the finite-difference method in which the underlying differential equation is replaced by a difference equation on a grid. The main advantages of the finite-element method over the finite-difference method are:

(i) handling boundary conditions; we will not discuss how boundary conditions are imposed but a feature of the various methods mentioned above is that most of the more awkward boundary conditions for the underlying differential equation turn out to be or are made to be natural conditions and do not have to be satisfied by trial or test functions.

(ii) handling arbitrarily shaped domains; clearly if the domain is polygonal it can easily be covered with triangles. If there are curved boundaries we can either use elements with curved sides (this usually means approximating the boundary with polynomial segments of the same order as the local trial function) or solve an approximate problem; that is, replace the original problem with one in a polygonal domain.

(iii) local grid refinement; this is probably the most widely quoted advantage and is one of the main reasons for the popularity of the finite-element method.

7. Local grid refinement

If it is known that the solution is changing rapidly in one part of the domain and less rapidly in another then it is natural to put more effort into the former region to try to resolve the true behaviour of the solution. This can be done in the finite-difference method but does not fall neatly into the process. In the finite-element method however all that is necessary is to put more (and smaller) elements in the region of rapid change and larger elements in other regions. Such a modification requires no change in the program, only a change in the data file.

Another aspect of grid refinement is that, if after solving the problem on a given grid the quality of the solution is assessed to be deficient in some sub-region, a local refinement in the grid can be adopted in such a way that only minor changes in the data file are required. This provides a lot of flexibility to the finite-element user.

A note of warning on grid refinement: the error term in the finite-element method usually involves a term ($\frac{1}{\sin \alpha}$) where α is the largest angle in the triangulation; see for example Strang and Fix (1973). Thus, on refining the grid care must be taken not to produce angles approaching π. For this reason, a refinement such as that illustrated in Fig. 5a should be avoided in favour of the refinement illustrated in Fig. 5b, for the former refinement inevitably increases the value of α.

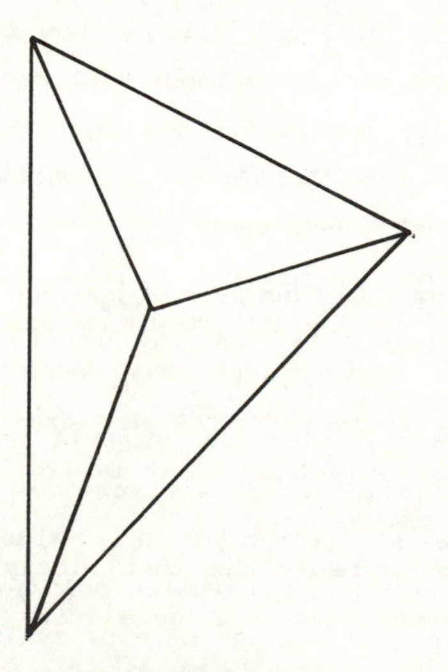

(a) A poor refinement

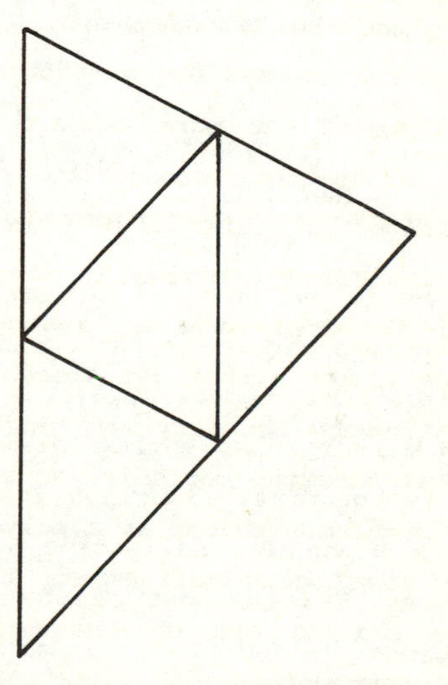

(b) A good refinement

FIG. 5 Refining in a triangular element.

8. Conforming and non-conforming elements

There is often a certain amount of confusion about conforming and non-conforming elements. If the variational form, e.g. (7.6), (7.8), (7.11), (7.12) or (7.14), involves (p+1)th order derivatives a conforming trial function must have continuous derivatives up to order p. It is tempting to define a non-conforming trial function by the converse of this rule but this can be misleading. It is true that a non-conforming trial function is one that does not have continuous derivatives up to order p but such trial functions are not necessarily non-conforming.

For example, the Veubeke element given in Fig. 3d provides a function that is continuous only at the mid-points of the element. Strictly speaking such an element cannot be used, for example, for minimising the functional (7.6) when Ω is a two-dimensional region. But the elemental contributions (taking the case $k = 1$) are:-

$$\int_{\text{element}} \nabla N_i \, \nabla N_j \, d\Omega \qquad \text{and} \qquad \int_{\text{element}} f \, N_i \, d\Omega \,. \tag{7.16}$$

These can be evaluated in the usual manner in each Veubeke element and a set of equations may be constructed, motivated by the algorithm to find an approximate minimum of (7.6).

Similarly the Morley element, illustrated in Fig. 6, provides a piecewise quadratic function that is continuous at the vertices and has a continuous normal derivative at the mid-points of the sides of the triangles. This element can be used to find approximate solutions of the variational method (7.8) for many but not all values of c by evaluating the integrals of W over the Morley elements.

In both these examples the approximate solutions have been constructed in the spirit of a non-conforming approach and illustrate how non-conforming elements are used. However, in both cases the process adopted can be understood as a straightforward complementary energy (or dual solution) approach; see Werner (1981). In this sense these two particular processes are not strictly non-conforming methods of solution and can be justified and analysed in the usual way without using any special techniques for non-conforming methods.

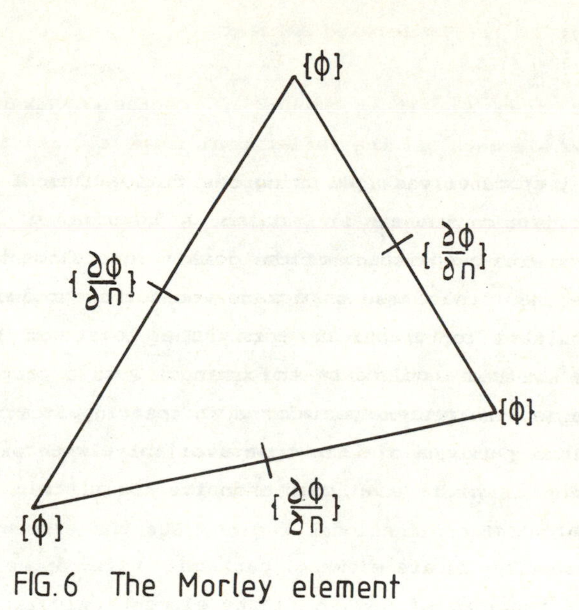

FIG.6 The Morley element

9. Using mixed interpolation

The solution of a set of simultaneous differential equations may involve the use of different elements for different unknowns. In practice, the same splitting of the domain would usually be made but different local functions for the different parameters may be used. For example, to solve the set of differential equations

$$\left.\begin{aligned} F(u, v) &= f , \\ G(u, v) &= g , \end{aligned}\right\} \qquad (7.17)$$

where F and G are differential operators and u and v the unknown functions, if it is known that u varies a lot more than v it would be advisable to use a higher order finite-element interpolation for u than for v on the same grid, for example a cubic element for u and a linear one for v.

This sort of choice may be at the discretion of the user or it may be forced by the equations themselves. Thus to solve the Stokes equations (or more generally the Navier-Stokes equation) for velocities and pressures in a fluid it is advisable and, arguably essential, to use a lower order polynomial for pressure than for the velocities; see for example Jackson and

Cliffe (1981).

10. Software

The early computer packages using the finite-element method were aimed at solving particular classes of problem. A description of the problem was given together with a splitting of the domain into elements and the type of element to be used; the computer package would then produce the answer in a suitably tabulated or graphical form. The level of knowledge of the finite-element method required by the user of such a package was not high, but the packages were rather specialised in their application. There are now major commercial packages of this type available, with extensive input and output sections, which give comprehensive facilities for a range of engineering problems.

More recently finite-element packages have appeared which are a collection of routines of use in finite-element calculations, for example the NAG package, based on work by Smith (1982), further developed by Greenough and Robinson (1982). The level of knowledge of the finite-element method required by a user of this type of package is quite high.

CHAPTER 8

OTHER METHODS FOR ELLIPTIC EQUATIONS

J. Walsh and D. Kershaw

1. Finite-difference methods

For many types of elliptic equations, finite-difference formulae give a simple and convenient way of constructing an approximating system of equations, which can then be solved numerically. Solution techniques for these systems have been improved in recent years by the development of multigrid and cyclic reduction algorithms, and the calculation can be done very efficiently in many cases.

To illustrate the form of the approximating system, we consider the linear second-order elliptic equation

$$a\frac{\partial^2 u}{\partial x^2} + b\frac{\partial^2 u}{\partial x \partial y} + c\frac{\partial^2 u}{\partial y^2} + p\frac{\partial u}{\partial x} + q\frac{\partial u}{\partial y} + ru = f, \tag{8.1}$$

given in a closed region R, with boundary conditions for u at all points of the boundary C. The basic formulae for approximating the partial derivatives on a rectangular mesh are given by Mitchell and Griffiths (1980, Ch. 3). The finite-difference approximation to (8.1) at an interior mesh-point gives a linear equation connecting the approximate values of u at neighbouring mesh-points. The complete system of equations is obtained by using this formula at all interior points of the region, with a suitable representation of the boundary conditions at mesh-points on the boundary. Suppose the resulting algebraic system is

$$A\,\underline{u} = \underline{b}, \tag{8.2}$$

where $\underline{u}$ is the vector of approximate solution values at mesh-points. The form of the matrix A is determined by the geometry of the region and also by the particular finite-difference formulae used.

Taking the simplest approximations to the derivatives in (8.1), we obtain a five-point formula at interior points if $b = 0$, and a nine-point formula if $b \neq 0$. The matrix A may be represented in partitioned form as

$$A = \begin{bmatrix} Q_1 & R_1 & & & & \\ P_2 & Q_2 & R_2 & & & \\ & P_3 & Q_3 & R_3 & & \\ & & \cdot & \cdot & \cdot & \\ & & & \cdot & \cdot & \cdot \\ & & & & \cdot & \cdot \; \cdot \end{bmatrix}, \tag{8.3}$$

where the submatrices Q_i are tridiagonal, and P_i, R_i are diagonal if $b = 0$, and tridiagonal otherwise. For a rectangular region R, the bandwidth of the matrix A is constant and the submatrices are of fixed order (equal to the number of mesh-points in one of the coordinate directions). For non-rectangular regions, the shape of the band is less regular and the submatrices are of variable order. Further irregularities may be produced in the matrix by mesh deletions or by special approximations to the boundary conditions on a curved boundary.

A comprehensive survey of methods for solving elliptic finite-difference equations is given by Fox (1977). The order of the algebraic system (8.2) is generally very large, so that it is essential to use efficient methods of solution which take account of the special structure of the matrix A. This matrix has very few non-zero elements in each row, and the system can be stored compactly; however, if we use direct methods of solution based on elimination, there is considerable 'fill-in' within the band, leading to much larger storage requirements. Iterative methods are generally far more economical in storage.

Algorithms for simple direct methods of solution are readily available in program libraries. We give below the main types of matrix for which special algorithms are appropriate. Suppose first that the region is rectangular, with $m \times n$ mesh-points where $m \leqslant n$.

(a) Symmetric positive definite matrix. This arises when the elliptic equation is self-adjoint, with $b = 0$, $r < 0$ in (8.1). The matrix can be factorised symmetrically using a band Cholesky algorithm. The storage needed is $O(m^2n)$, and the number of arithmetic operations is $O(m^3n)$.

(b) Unsymmetric matrix with diagonal dominance, given by $b = 0$, $r < 0$ in (8.1). The equations can be solved by direct elimination without interchanges. For a standard band solver, the storage requirement is $O(2m^2n)$, number of operations $O(2m^3n)$.

(c) Band matrix, no special properties. A band solver with interchanges should be used, requiring storage of $O(3m^2n)$, number of operations $O(3m^3n)$.

For more general regions with variable-length rows and possibly with mesh deletions, the main types of matrix which need to be considered are the following:

(d) Matrix with variable band. This can be factorised with or without interchanges. The storage and operation count depend on the particular case.

(e) Sparse matrix. This may be treated by a general sparse algorithm, or various special devices may be used to 'dissect' the problem; the latter approach is useful for a composite region such as a T or an H shape, which is composed of simple rectangular subregions. The storage and computation required are problem-dependent.

None of these direct algorithms makes full use of the structure of A. If we have a very simple problem such as Poisson's equation in a rectangle, it is reasonable to assume that a more efficient method can be found by using the special form of the submatrices in (8.3). Algorithms developed along these lines (reduction methods) will be discussed in section 2. Often the problem can be reformulated by the use of finite Fourier transforms; then the Fast Fourier transform (FFT) algorithm may be used to give a very efficient method (see section 5).

In the early days of computers, iterative methods were used in the solution of (8.2) in order to avoid large storage requirements. The classical point successive over-relaxation (SOR) method gives a very compact algorithm, but more complicated methods are generally preferred for efficiency, now that storage is less restricted. If the eigenvalues of the iteration matrix are real, the convergence can be accelerated by Chebyshev extrapolation or conjugate gradients. In the case of a symmetric positive definite matrix A, the eigenvalues are real when we use Jacobi iteration or symmetric SOR, and the convergence can then be accelerated; however, these techniques do not extend to unsymmetric matrices. A comparison of iterative methods is given by Kincaid and Grimes (1977).

To take account of the character of the solution of (8.1), which is generally a smooth function except on the boundary, we can use line iteration or alternating directions. These methods are often applied to the solution of large parabolic problems, with an implicit time integration. The most promising recent development in the area of iterative solution is the multigrid method, which uses a sequence of successively finer grids to represent the problem. This will be considered in section 3.

The solution of the system of linear equations (8.2) is only a part of the numerical treatment of the elliptic problem. The most difficult part is often the specification of the region and the mesh, and the construction of the approximating equations at interior and boundary points. An algorithm for setting up the equations in general regions involves a great deal of programming simply to handle the geometry, but this is quite unnecessary in the case of a simple region. It is desirable therefore to separate the construction of the algebraic equations from the solution of the system (8.2), and to have a standard interface between them. Not all algorithms can be separated in this way, however, and in some cases it is more efficient to handle the setting-up and solution phases together.

The choice of method is often dependent on the type of results required. Finite-difference methods produce a detailed solution at all points of the region, and there is no way of obtaining only part of it. The user may be interested only in the values on the boundary, say, and in such a case it may be more convenient to use integral equation methods, which give approximate solutions over boundary elements. This approach is also appropriate for exterior problems. It will be considered briefly in section 4.

2. Reduction methods

For the case of Poisson's equation over a rectangle, with Dirichlet boundary conditions, the matrix A in (8.2) takes a very simple form. Suppose there are $m \times n$ mesh-points as before; then the linear system may be written as

$$A\,\underline{u} = \begin{bmatrix} Q & I & & & \\ I & Q & I & & \\ & I & Q & I & \\ & & \cdot & \cdot & \cdot \\ & & & I & Q \end{bmatrix} \begin{bmatrix} \underline{u}_1 \\ \underline{u}_2 \\ \underline{u}_3 \\ \cdot \\ \cdot \\ \cdot \\ \underline{u}_n \end{bmatrix} = \begin{bmatrix} \underline{b}_1 \\ \underline{b}_2 \\ \underline{b}_3 \\ \cdot \\ \cdot \\ \cdot \\ \underline{b}_n \end{bmatrix}, \quad (8.4)$$

where the submatrices are of order m, and the given boundary values are incorporated in the vectors $\underline{b}_k$. Because the submatrices in A are constant, it is easy to eliminate alternate rows of the solution by a 'reduction' process. For the rows k-1, k, k+1 of (8.4), the first step of the reduction is to eliminate $\underline{u}_{k-1}$, $\underline{u}_{k+1}$, giving

$$\underline{u}_{k-2} + (2I - Q^2)\underline{u}_k + \underline{u}_{k+2} = \underline{b}_{k-1} - Q\underline{b}_k + \underline{b}_{k+1} . \quad (8.5)$$

This process may be applied for k = 2, 4, 6, ..., and we obtain another block tridiagonal system of the same form as (8.4), but only about half the size. Suppose $n = 2^p-1$ say; then we can repeat the reduction p-1 times to obtain finally a single block equation, which is equivalent to m scalar equations which can be solved directly. The complete solution is obtained by back-substitution.

The essential idea of the method is contained in (8.5), but it has to be implemented in a slightly different form because the calculation of the right-hand side in (8.5) is unstable. The details are given by Swartztrauber (1977). We note that the diagonal matrices in the reduced equations form the sequence

$$Q_1 = Q , \quad Q_r = 2I - Q_{r-1}^2 , \quad r = 2,3,\ldots . \quad (8.6)$$

The matrices Q_r are polynomials of degree 2^{r-1} in Q, which can be expressed in an explicit factorised form without storing the full matrices. The complete process is very efficient and economical in storage, with an operation count of $O(3mn \log_2 m)$.

The method described above can be extended to more general equations (including the Helmholtz equation), and to Neumann boundary conditions. However, it is essentially dependent on the very simple form of the matrix, which derives from the regularity of the finite-difference equations on a rectangular region.

An alternative way of solving Poisson's equation in a rectangle is to use a finite Fourier transform in one direction. This is essentially a linear transformation of the vectors $\underline{u}_k$; if we put $\underline{u}_k = V\underline{w}_k$, the kth equation of (8.4) becomes

$$\underline{w}_{k-1} + (V^{-1}QV)\underline{w}_k + \underline{w}_{k+1} = V^{-1}\underline{b}_k . \tag{8.7}$$

In the case of the Laplace operator the matrix Q is very simple, and we can find the explicit form of V such that

$$V^{-1} Q V = \mathrm{diag}\,[\lambda_1\ \lambda_2\ \ldots\ \lambda_m] . \tag{8.8}$$

Then the equations (8.7), for $k = 1,2,\ldots,n$, reduce to m independent tridiagonal systems of order n, which can be solved for the $\underline{w}_k$. The values of $\underline{u}_k$ are then easily obtained.

The detailed calculation is slightly different according to whether the boundary conditions are of Dirichlet or Neumann type. In all cases, however, the operations may be carried out efficiently by means of the FFT algorithm (see section 5). Swartztrauber (1977) shows that the operation count is $O(2mn\log_2 m)$, which makes it slightly faster than the reduction method. The most efficient algorithm of this type appears to be a combination of reduction and Fourier transformation (FACR).

It would be attractive to extend the fast reduction methods to regions of general shape. One approach is the capacitance matrix method, for which an algorithm is given by Proskurowski (1983). Suppose the region of interest R, with boundary C, is embedded in a rectangle R_1 . The elliptic problem is extended over R_1 with suitable boundary conditions, and approximated by finite-difference equations which can be solved by a reduction method. This solution does not satisfy the boundary conditions on C, however, and we have to calculate corrections to it. Suppose the boundary conditions are of Dirichlet type, and there are p boundary mesh-points on C. Then we have to solve in effect p additional systems over the rectangle in order to obtain the correction for each point; by superposition we then have the solution of the original problem. This method is not economical unless p is relatively small.

3. Multigrid methods

The basic idea of the multigrid method goes back to the development of relaxation methods for elliptic problems, when it was noted that the low-frequency error components on a fine mesh could be effectively reduced by moving temporarily to a coarser mesh. This behaviour can be expressed in terms of the eigenfunctions of the iteration matrix. Suppose the linear equations (8.2) are written in the iterative form

$$\underline{u}^{(k+1)} = B\underline{u}^{(k)} + \underline{c} \, . \tag{8.9}$$

The rate of convergence is governed by the eigenvalues of B, and the error modes are the corresponding eigenfunctions. Suppose B has been chosen so that its eigenvalues are all real. For simple point iterative methods, the error modes can be divided into two sets, the low-frequency modes which can be represented on the current mesh and also on coarser meshes, and the high-frequency modes which can be represented only on the current mesh and not on coarser ones. If the matrix B has good 'smoothing' properties, a few steps of the iteration (8.9) will effectively remove the higher modes from the calculation, and we can then restrict the data to a coarser mesh in order to annihilate the lower modes.

A detailed discussion of the basic algorithms and their application to model problems is given by Stüben and Trottenberg (1982). Briefly, we have a sequence of grids $G_0, G_1, \ldots, G_M$ covering the region, in which the points of G_q are a subset of the points of G_{q+1}. We start by constructing an approximate solution $\underline{u}_M^{(0)}$ on the finest grid G_M. Suppose the finite-difference equations on this grid are

$$A_M \underline{u}_M = \underline{b}_M \, . \tag{8.10}$$

We carry out a few steps of the iteration, say k, on G_M to obtain a reasonably smooth approximation $\underline{u}_M^{(k)}$. The residual is given by

$$\underline{r}_M = \underline{b}_M - A_M \underline{u}_M^{(k)} \, , \tag{8.11}$$

and $\underline{r}_M$ is also smooth. The calculation is now transferred to the grid G_{M-1} in order to accelerate the convergence. The number of points is reduced by applying a restriction operator R_{M-1} to $\underline{r}_M$, and the new equation to be solved on G_{M-1} is

$$A_{M-1}\,\underline{e}_{M-1} = \underline{r}_{M-1} = R_{M-1}\,\underline{r}_M . \quad (8.12)$$

Note that this is an equation for a correction to the previous solution, and not for the solution itself. We take $\underline{e}^{(0)}_{M-1} = \underline{0}$, carry out a few iterations as before, and transfer the results to the next grid G_{M-2} and so on.

Eventually the procedure arrives at the coarsest grid G_0 . This has only a small number of points and so the equations can be solved exactly, using a band matrix method or some other suitable technique. The result obtained is a correction to the solution on G_1 , but first it has to be extended to the finer grid using a prolongation operator P_1 say. This is an interpolation step which is likely to re-introduce the high-frequency errors, so we carry out a few smoothing steps on G_1 before moving to G_2. Proceeding in this way, the final stage gives a solution on G_M which is expected to provide an improved solution of the original problem.

This strategy can be varied in many ways, and one difficulty in writing algorithms is to find suitable criteria for moving from one grid to the next. Theoretical results about the speed of the calculation are of some interest but limited applicability. With simple model analysis, an operation count of O(mn), where mn is the number of points, is indicated; however, as with all iterative methods the actual time taken will depend on the accuracy required. Bank and Sherman (1981) give some practical comparisons (actually for finite-element equations); these showed that the multigrid method was very competitive for 2-3 figure accuracy, but for 6-7 figures a direct method using a sparse solver was better, if one could discount the time needed to work out the elimination strategy.

In the case of nonlinear problems it is often convenient to start the calculation on the coarsest grid, in order to obtain a reasonable first approximation for the finer grids. It is necessary to keep the solution values throughout, rather than just the corrections to the solution, in order to calculate the nonlinear terms. If the equations are linearised globally, (8.12) is replaced by

$$A_{m-1}\,\underline{u}_{M-1} = A_{M-1}\,(R_{M-1}\,\underline{u}^{(k)}_M) + R_{M-1}\,(\underline{b}_M - A_M\,\underline{u}^{(k)}_M) = \underline{b}_{M-1}, \quad (8.13)$$

and the transfer to coarser grids proceeds analogously. Brandt (1982, section 8.3) suggests that local (pointwise) linearisation is more effective, but it is difficult to analyse the nonlinear convergence, and the best method probably depends on the problem.

A number of algorithms have been developed for the multigrid method, of varying degrees of generality (see section 6). There will undoubtedly be further algorithms of this type for the solution of linear problems, but the design of a good general strategy for nonlinear problems is likely to prove more difficult.

4. Other special methods

When the elliptic equation has the simple Laplace form

$$\frac{\partial^2 u}{\partial x^2} + \frac{\partial^2 u}{\partial y^2} = 0 , \tag{8.14}$$

it can be reformulated as an integral equation, leading to some useful methods of solution. A summary of the relevant theory is given by Symm (1974). If the Laplace equation (8.14) is satisfied in a closed region R with smooth boundary C, Green's formula gives the value of u at a point P on the boundary in the following form

$$u(P) = \frac{1}{\pi}\int_C \ln |P - Q| \frac{\partial}{\partial n} u(Q)\, ds - \frac{1}{\pi}\int_C \frac{\partial}{\partial n} \ln |P - Q|\, u(Q)\, ds , \tag{8.15}$$

where $\frac{\partial}{\partial n}$ denotes differentiation along the inward normal to C at Q. For the Dirichlet problem the values of u are specified at all points of the boundary, and (8.15) is an integral equation for $\frac{\partial u}{\partial n}$ on C. Similarly when $\frac{\partial u}{\partial n}$ is specified on the boundary, we have an equation for the boundary values of u. In either case the solution of (8.15) may be obtained numerically by dividing the contour C into a number of boundary elements, taking u and $\frac{\partial u}{\partial n}$ to be constant over each element, and evaluating the resulting integrals by numerical quadrature.

This simple method has been extended in many directions, to treat regions with corners and other singularities, multiply-connected regions, and more general equations. When the boundary values have been found, the interior values of u may be calculated by a formula similar to (8.15), if necessary. But the method is often used for problems where the interior values are not required, and it is particularly economical for such cases.

The advantages of the boundary formulation are more evident for three-dimensional equations

$$\frac{\partial^2 u}{\partial x^2} + \frac{\partial^2 u}{\partial y^2} + \frac{\partial^2 u}{\partial z^2} = 0 \ . \tag{8.16}$$

Formulae analogous to (8.15) are given by Symm (1974), and the general method is the same as before. The boundary elements are now two-dimensional segments of the bounding surface, and convenient forms have been developed for the local approximation and numerical integration. Compared with finite-difference methods, the number of variables is considerably less, but the matrix of the system does not have the simple band or sparse form of section 2.

A survey by Machura and Sweet (1980) gives a brief description of some symbolic languages for specifying elliptic equations, to assist in setting up the approximating system. However, in practical computation the most difficult part of the program is usually the treatment of the geometry, the input of the region and mesh, and the output of the results in a form related to the region. The major engineering packages already provide extensive facilities for input and output, but it is likely that better ways of handling the geometry interactively will be produced as more powerful work-stations become available.

5. The Fast Fourier transform

Fourier transforms and series are widely used in the solution of boundary-value problems in ordinary differential equations, and also in data analysis and time-series. The Fast Fourier transform (FFT) is an important algorithm for calculating approximate Fourier transforms in all these contexts, and we present it here in general terms, and not simply for the case of elliptic equations.

Let f_0, f_1, ... , f_{n-1} be a sequence of complex numbers; then the finite Fourier transforms $\hat{f}_0, \hat{f}_1, \ldots, \hat{f}_{n-1}$ are given by

$$\hat{f}_r = \frac{1}{n} \sum_{s=0}^{n-1} f_s \exp(-2\pi i \frac{rs}{n}) , \quad r = 0,1,\ldots,n-1 . \tag{8.17}$$

(In some definitions the factor $1/n$ is replaced by unity or $1/\sqrt{n}$; clearly this is not of vital importance.)

The transform can be thought of as arising as follows. Suppose the function f has period 1, so that

$$f(x+1) = f(x) , \tag{8.18}$$

and that it has the complex Fourier series

$$f(x) = \sum_{r=-\infty}^{+\infty} c_r \exp(2\pi irx) . \tag{8.19}$$

Since the following orthogonality relations are satisfied

$$\int_0^1 \exp(2\pi irt) \exp(-2\pi ist)\, dt = \begin{cases} 0 , & r \neq s, \\ 1 , & r = s, \end{cases}$$

we have, formally at least,

$$c_r = \int_0^1 f(t) \exp(-2\pi irt)\, dt . \tag{8.20}$$

Consequently if the Fourier coefficients $\{c_r\}$ are known the function f can be recovered.

On the other hand, suppose that f is known (or has been sampled) at the equally-spaced points

$$0 , \frac{1}{n} , \frac{2}{n} ,\ldots, \frac{n-1}{n} . \tag{8.21}$$

Then we shall only be able to find $\{c_r\}$ approximately, and the integral in

(8.20) will be approximated by a quadrature formula. The points (8.21) may be regarded as being equally spaced around a circle, which suggests that the quadrature weights should be equal, implying the use of the trapezoidal rule. Since $f(0) = f(1)$, we take

$$\hat{f}_r = \frac{1}{n} \sum_{s=0}^{n-1} f\left(\frac{s}{n}\right) \exp\left(-2\pi i \frac{rs}{n}\right) \tag{8.22}$$

as the approximation to c_r. To simplify this expression we introduce the notation

$$f_s = f\left(\frac{s}{n}\right), \quad e(x) = \exp(-2\pi i x),$$

from which it follows that

$$\left.\begin{aligned} &e(x+y) = e(x).e(y), \\ &e(n) = 1, \; e(n + \tfrac{1}{2}) = -1, \text{ for integer } n. \end{aligned}\right\} \tag{8.23}$$

Lemma 1 If p is an integer, then

$$\sum_{s=0}^{n-1} e\left(\frac{ps}{n}\right) = \begin{cases} n & \text{if } p = 0, \pm n, \pm 2n, \ldots, \\ 0 & \text{otherwise}. \end{cases}$$

This may easily be verified.

We can now calculate the error in the approximation of c_r by $\hat{f}_r$. From (8.22) and (8.19), we have

$$\hat{f}_r = \sum_{t=-\infty}^{+\infty} c_t \left[\frac{1}{n} \sum_{s=0}^{n-1} e\left(\frac{rs}{n}\right) e\left(\frac{-st}{n}\right) \right].$$

Using (8.23) and Lemma 1, we see that the inner sum is zero except when $t-r$ is an integral multiple of n, so that

$$\hat{f}_r = \sum_{p=-\infty}^{+\infty} c_{r+np} .$$

Hence the error is

$$\hat{f}_r - c_r = \sum_{p\neq 0} c_{r+np} .$$

The inability of the finite Fourier transform to distinguish between frequencies which differ by a multiple of n is called aliasing.

We note a certain duality between the sequences f_0 , f_1 , ... , f_{n-1} and $\hat{f}_0$, $\hat{f}_1$, ... , $\hat{f}_{n-1}$, which may be stated as follows.

Lemma 2 If

$$\hat{f}_r = \frac{1}{n} \sum_{s=0}^{n-1} f_s \, e\left(\frac{rs}{n}\right) , \quad r = 0,1,\ldots,n-1 ,$$

then

$$f_s = \sum_{r=0}^{n-1} \hat{f}_r \, e\left(\frac{-rs}{n}\right) , \quad s = 0,1,\ldots,n-1 .$$

Thus f_0 , f_1 , ... , f_{n-1} can be regarded as transforms of $\hat{f}_0$, $\hat{f}_1$, ... , $\hat{f}_{n-1}$.

To calculate $\hat{f}_r$ for only a few values of r, we can use nested multiplication to evaluate (8.17), since it may be regarded as a complex polynomial in $z_r = e(r/n)$:

$$\hat{f}_r = \frac{1}{n} \sum_{s=0}^{n-1} f_s z_r^s .$$

Then for each value of r, the calculation of $\hat{f}_r$ will require a total of n-1 complex multiplications and n-1 complex additions, followed by one division. So if all the n transforms are required this will give a total of $O(n^2)$ complex multiplications and additions, which is very time-consuming for large n.

An efficient method for reducing the computation was given by Cooley and Tukey (1965), and another by Gentleman and Sande (1966). For ease of presentation we consider the case when n is even, n = 2N say. Suppose we are given $f_0, f_1, \ldots, f_{2N-1}$, and we wish to calculate

$$\hat{f}_r = \frac{1}{2N} \sum_{s=0}^{2N-1} f_s \, e\left(\frac{rs}{2N}\right), \quad r = 0,1,\ldots,2N-1 . \tag{8.24}$$

The method of Cooley and Tukey takes (8.24) as two sums, one over the even numbers and one over the odd numbers,

$$2N\,\hat{f}_r = \sum_{s=0}^{N-1} f_{2s}\, e\left(\frac{rs}{N}\right) + e\left(\frac{r}{2N}\right) \sum_{s=0}^{N-1} f_{2s+1}\, e\left(\frac{rs}{N}\right). \tag{8.25}$$

Now since e(s) = 1, we have

$$2N\,\hat{f}_{r+N} = \sum_{s=0}^{N-1} f_{2s}\, e\left(\frac{rs}{N}\right) + e\left(\frac{r+N}{2N}\right) \sum_{s=0}^{N-1} f_{2s+1}\, e\left(\frac{rs}{N}\right).$$

Since $e\left(\frac{1}{2}\right) = -1$ it follows that

$$2N\,\hat{f}_{r+N} = \sum_{s=0}^{N-1} f_{2s}\, e\left(\frac{rs}{N}\right) - e\left(\frac{r}{2N}\right) \sum_{s=0}^{N-1} f_{2s+1}\, e\left(\frac{rs}{N}\right). \tag{8.26}$$

Thus we need to calculate only the transforms

$$\sum_{s=0}^{N-1} f_{2s}\, e\left(\frac{rs}{N}\right), \quad \sum_{s=0}^{N-1} f_{2s+1}\, e\left(\frac{rs}{N}\right), \quad r = 0,1,\ldots,N-1, \qquad (8.27)$$

which can then be combined to give $\hat{f}_r$, $\hat{f}_{r+N}$ for $r = 0,1,\ldots,N-1$.

Since each sum in (8.27) requires $O(N)$ complex multiplications and additions, the calculation of all the values $\hat{f}_0$, $\hat{f}_1$, ..., $\hat{f}_{2N-1}$ requires $O(2N^2)$ complex multiplications and additions. This compares with $O(4N^2)$ for the simple nested multiplication method.

The method of Gentleman and Sande arranges the summation in (8.24) as

$$2N\,\hat{f}_r = \sum_{s=0}^{N-1} f_s\, e\left(\frac{rs}{2N}\right) + \sum_{s=0}^{N-1} f_{s+N}\, e\left(\frac{rs}{2N}\right) e\left(\frac{r}{2}\right).$$

We now evaluate $\hat{f}_r$ at even and odd values of r to give

$$2N\,\hat{f}_{2r} = \sum_{s=0}^{N-1} [\, f_s + f_{s+N}\,]\, e\left(\frac{rs}{N}\right), \quad r = 0,1,\ldots,N-1,$$

$$2N\,\hat{f}_{2r+1} = \sum_{s=0}^{N-1} [\, f_s - f_{s+N}\,]\, e\left(\frac{rs}{N}\right) e\left(\frac{s}{2N}\right), \quad r = 0,1,\ldots N-1.$$

Again the problem is that of finding two transforms of length N, and the operational count is the same as in the Cooley-Tukey method.

We now assume that $n = 2^q$ and state the complete Cooley-Tukey algorithm as given by Hockney and Jesshope (1983), with a slight change of notation. Given f_s for $s = 0,1,2,\ldots,2^q-1$, the problem is to calculate

$$\hat{f}_r = 2^{-q} \sum_{s=0}^{2^q-1} e\left(\frac{rs}{2^q}\right) f_s, \quad r = 0,1,\ldots,2^q-1. \qquad (8.28)$$

Define

$$\hat{f}_0(i,\ 0) = f_i \quad , \quad i = 0,1,2,\ldots,2^q-1 \ ,$$

and calculate for level $\ell = 0,1,\ldots,q-1$,

$$\left.\begin{aligned} \hat{f}_r(i,\ \ell+1) &= \hat{f}_r(i,\ \ell) + e\left(\frac{r}{2^{\ell+1}}\right) \hat{f}_r(i+2^{q-\ell-1},\ \ell) \ , \\ \hat{f}_{r+2^\ell}(i,\ \ell+1) &= \hat{f}_r(i,\ \ell) - e\left(\frac{r}{2^{\ell+1}}\right) \hat{f}_r(i+2^{q-\ell-1},\ \ell) \ , \end{aligned}\right\} \tag{8.29}$$

where $r = 0,1,2,\ldots,2^{\ell-1}$, $i = 0,1,2,\ldots,2^{q-\ell-1}-1$. Then when $\ell = q-1$, we have

$$\hat{f}_r\,(0,\ q) \quad , \quad \hat{f}_{r+2^{q-1}}\,(0,\ q) \quad , \ r = 0,1,2,\ldots,2^{q-1} \ ,$$

and we can evaluate

$$\hat{f}_r = 2^{-q}\,\hat{f}_r(0,\ q) \quad , \quad \hat{f}_{r+2^{q-1}} = 2^{-q}\,\hat{f}_{r+2^{q-1}}(0,\ q) \quad . \tag{8.30}$$

To prove the algorithm we shall show that if

$$\hat{f}_r(i,\ \ell) = \sum_{s=0}^{2^\ell-1} e\left(\frac{rs}{2^\ell}\right) f_{i+s.2^{q-\ell}} \ , \tag{8.31}$$

$$r = 0,1,\ldots,2^\ell-1 \ ; \ i = 0,1,\ldots,2^{q-\ell}-1 \ ,$$

then this satisfies the relations.

Clearly when $\ell = q$ we have

$$\hat{f}_r(0, q) = \sum_{s=0}^{2^q-1} e\left(\frac{rs}{2^q}\right) f_s ,$$

and so $\hat{f}_r = 2^{-q}\,\hat{f}_r(0, q)$, which is the required transform.

Now when $\ell = 0$ we see that (8.31) becomes

$$\hat{f}_0(i, 0) = f_i , \quad i = 0,1,\ldots,2^q-1 .$$

It remains to show that (8.31) satisfies (8.29). From (8.31) we have

$$\hat{f}_r(i, \ell+1) = \sum_{s=0}^{2^{\ell+1}-1} e\left(\frac{rs}{2^{\ell+1}}\right) f_{i+s.2^{q-\ell-1}} , \quad \begin{array}{l} r = 0,1,\ldots,2^{\ell+1}-1 , \\ i = 0,1,\ldots,2^{q-\ell-1}-1. \end{array}$$

Take the sum over even and odd values of s to give

$$\hat{f}_r(i, \ell+1) = \sum_{s=0}^{2^\ell-1} e\left(\frac{rs}{2^\ell}\right) f_{i+s.2^{q-\ell}}$$

$$+ \sum_{s=0}^{2^\ell-1} e\left(\frac{rs}{2^\ell} + \frac{r}{2^{\ell+1}}\right) f_{i+2^{q-\ell-1} + s.2^{q-\ell}}$$

$$= \hat{f}_r(i, \ell) + e\left(\frac{r}{2^{\ell+1}}\right).\hat{f}_r(i+2^{q-\ell-1}, \ell) ,$$

for $r = 0,1,\ldots,2^{\ell+1}-1$; $i = 0,1,\ldots,2^{q-\ell-1}-1$.

Finally, subdivide the range of r into $r = 0,1,\ldots,2^\ell-1$ and $r = 2^\ell, 2^\ell+1,\ldots,2^{\ell+1}-1$ to give

$$\hat{f}_r(i, \ell+1) = \hat{f}_r(i, \ell) + e\left(\frac{r}{2^{\ell+1}}\right) \hat{f}_r(i+2^{q-\ell-1}, \ell) ,$$

$$\hat{f}_{r+2^\ell}(i, \ell+1) = \hat{f}_r(i, \ell) - e\left(\frac{r}{2^{\ell+1}}\right) \hat{f}_r(i+2^{q-\ell-1}, \ell) ,$$

each for $r = 0,1\ldots,2^{\ell}-1$, since $\hat{f}_{r+2^{\ell}}(i, \ell) = \hat{f}_r(i, \ell)$.
Thus the algorithm is confirmed .

With regard to the number of arithmetic operations involved, it can be shown that this is $O(q.2^q)$, or equivalently $O(n \log_2 n)$, instead of $O(n^2)$ for the simple method based on nested multiplication.

The algorithm is highly efficient when n is a power of 2, and this should be arranged whenever possible. In the case where n is not a power of 2, suppose it is the product of the factors $n_1, n_2, \ldots, n_p$. Then a similiar algorithm can be devised by recursively finding n_1 transforms of period $(n_2 n_3 \ldots n_p)$, which are themselves n_2 transforms of period $(n_3 \ldots n_p)$, etc.. If n is prime, no saving of this type is possible, but a device which is sometimes used is to add to the data enough zeros to make the number of points a power of 2. However, this leads to the problem of 'leakage' (compare Chapter 15, section 5).

The general algorithm given above is based on the full complex Fourier transform (8.17), and it uses complex arithmetic throughout. It is not difficult to see that economies can be made if the data sequence $\{f_r\}$ is real, or if it is complex but we require only the sine or the cosine transform. Special forms of the algorithm should be used in such cases.

For example, suppose $g_0, g_1, \ldots, g_{2N-1}$ is a real sequence and we require the real Fourier coefficients

$$\hat{a}_r = \frac{1}{N} \sum_{s=0}^{2N-1} g_s \cos\left(\frac{\pi rs}{N}\right), \quad r = 0,1,\ldots,N-1 ,$$

$$\hat{b}_r = \frac{1}{N} \sum_{s=0}^{2N-1} g_s \sin\left(\frac{\pi rs}{N}\right), \quad r = 1,2,\ldots,N-1 ,$$

which are respectively approximations to

$$2\int_0^1 g(t)\cos(2\pi rt)\,dt \quad , \quad 2\int_0^1 g(t)\sin(2\pi rt)\,dt .$$

Define the complex sequence $f_0, f_1, \ldots, f_{N-1}$ by

$$f_s = g_{2s} + i\, g_{2s+1} \quad , \quad s = 0,1,\ldots,N-1 \, ,$$

and calculate

$$\hat{f}_r = \frac{1}{N} \sum_{s=0}^{N-1} f_s \, e\left(\frac{rs}{N} \right) \quad , \quad r = 0,1,\ldots,N-1 \, . \qquad (8.32)$$

It is not difficult to show from (8.32) that

$$\hat{f}_r + \hat{f}_{N-r} = \frac{2}{N} \sum_{s=0}^{N-1} (g_{2s} + i\, g_{2s+1}) \cos\left(\frac{2\pi rs}{N} \right) ,$$

$$\hat{f}_r - \hat{f}_{N-r} = \frac{2i}{N} \sum_{s=0}^{N-1} (g_{2s} + i\, g_{2s+1}) \sin\left(\frac{2\pi rs}{N} \right) .$$

It follows that

$$\hat{a}_r = \frac{1}{2} \mathrm{Re}(\hat{f}_r + \hat{f}_{N-r}) + \frac{1}{2} \cos\left(\frac{\pi r}{N} \right) \mathrm{Im}(\hat{f}_r + \hat{f}_{N-r}) + \frac{1}{2} \sin\left(\frac{\pi r}{N} \right) \mathrm{Re}(\hat{f}_r - \hat{f}_{N-r}),$$

$$\hat{b}_r = -\frac{1}{2} \mathrm{Im}(\hat{f}_r - \hat{f}_{N-r}) - \frac{1}{2} \cos\left(\frac{\pi r}{N} \right) \mathrm{Re}(\hat{f}_r - \hat{f}_{N-r}) - \frac{1}{2} \sin\left(\frac{\pi r}{N} \right) \mathrm{Im}(\hat{f}_r + \hat{f}_{N-r}) .$$

For applications of Fast Fourier methods in a wide variety of contexts, reference should be made to Henrici (1979) and Hockney and Jesshope (1983). An alternative approach to the Fast Fourier transform is given by de Boor (1980).

6. Software for elliptic problems

The direct solution of band and sparse matrix problems is covered in many program libraries, particularly LINPACK (Dongarra, Bunch, Moler and Stewart 1979), SPARSPAK (George, Liu and Ng 1980), and the NAG Library. A selection of iterative methods for elliptic difference equations is given in ITPACK (Young and Kincaid 1980).

The ELLPACK system (Rice 1978) is a comprehensive collection of software for elliptic problems, constructed on the modular principle. The user has flexibility in selecting different types of approximation and different solution methods; these are linked through a standard interface.

One implementation of the multigrid method is the routine MGD1 produced by Wesseling (1982). This is designed for seven-point finite-difference equations in a rectangle, with uniform mesh spacing. The equations are supplied by the user; the seven-point form allows for cross-derivative terms and upwind differencing. The smoothing method is incomplete LU factorisation. Another multigrid routine is MGO1 (Stüben 1982) which treats the Helmholtz equation in a rectangle. The finite-difference equations are set up internally, and the user has a choice of smoothing method.

An extensive set of routines based on cyclic reduction and Fast Fourier transform is provided in FISHPAK (Adams, Swartztrauber and Sweet 1978). This has facilities for solving the Helmholtz equation over a rectangle in Cartesian, polar, or cylindrical polar coordinates, with extensions to more general equations in three dimensions. These algorithms are very efficient for the problems covered.

The NAG Library also has Fourier transform routines, less comprehensive than those in FISHPAK. It also has an implementation of the integral equation method for Laplace's equation, outlined in section 4.

CHAPTER 9

METHODS FOR PARABOLIC EQUATIONS

J. Walsh

1. Types of approximation

Parabolic problems are usually associated with time-dependent physical systems, and they have the character of initial-value problems in the time direction, and of boundary-value problems in space. To illustrate some of the methods of numerical solution, let us consider a parabolic equation with only one space variable,

$$\frac{\partial u}{\partial t} = a(u, x, t)\frac{\partial^2 u}{\partial x^2} + f\left(u, \frac{\partial u}{\partial x}, x, t\right). \tag{9.1}$$

For $a > 0$, we would expect to integrate forwards in time, from initial conditions given on $t = 0$. (If $a < 0$, or if we integrate backwards, the problem is very ill-posed.) In the space direction we usually have a finite interval $0 \leqslant x \leqslant 1$ say, with either u or $\frac{\partial u}{\partial x}$ given at the end-points. With the additional condition $\frac{\partial f}{\partial u} \leqslant 0$, and certain restrictions on the boundary conditions involving $\frac{\partial u}{\partial x}$, the solution satisfies a maximum principle (Friedman 1964). But a more important consideration in practical problems is the presence or absence of significant convective terms involving $\frac{\partial u}{\partial x}$. For example, in Burgers' equation

$$\frac{\partial u}{\partial t} = \nu\frac{\partial^2 u}{\partial x^2} - u\frac{\partial u}{\partial x}, \tag{9.2}$$

if ν is large the first (diffusion) term on the right is dominant and the solution is fairly simple, whereas if ν is small the second (convective) term is dominant. In the latter case the solution represents a steep wave-front moving across the region, and a very fine mesh is needed to obtain it accurately.

The physical distinction between the time and the space directions makes it natural to split the problem into two parts, the solution of the boundary-value problem in space, and the forward integration in time. Suppose we take equally-spaced mesh-points in the x-direction, $x_i = ih$ say, where $h = 1/(N+1)$, and approximate the space-derivatives in (9.1) by finite differences on three neighbouring points. Let $U_i(t)$ be the approximation to $u(ih, t)$; then we get an initial-value system of the form

$$\frac{dU_i}{dt} = F_i(U_{i-1}, U_i, U_{i+1}, t), \quad i = 1,2,\ldots,N, \tag{9.3}$$

with additional equations at $i = 0$, $N+1$ if the boundary conditions involve $\frac{\partial u}{\partial x}$. Thus the problem is reduced to one of step-by-step integration. Alternatively we can base the computation on the space direction. Let $V_j(x)$ be an approximation to $u(x, jk)$, where k is the time-step; then we approximate the time-derivative in (9.1) to obtain

$$a_j V_j'' + f(V_j, V_j', x, jk) - \frac{1}{k} V_j = -\frac{1}{k} V_{j-1} . \tag{9.4}$$

This is a boundary-value problem with conditions specified at $x = 0, 1$, which has to be solved at each time-step.

Recent developments in general algorithms for parabolic equations have been based on the initial-value problem in time (the method of lines) rather than the boundary-value problem (9.4). However the boundary-value method may be preferable in special cases, particularly when a very fast solver such as the Fast Fourier transform (see Chapter 8, section 5) can be used at each time-step.

It may be noted that a general parabolic solver can also be used for elliptic equations, or for a mixture of elliptic and parabolic equations. (The extension to hyperbolic equations is less straightforward, because the boundary for a well-posed problem has to be related to the characteristics.) Time-dependent problems may also include algebraic equations in the system, and very versatile algorithms are needed to allow a range of problems to be handled easily. For very large problems, for example in three space dimensions, it is unlikely that general algorithms will be satisfactory, because the code needs to be specially designed for efficiency. But it may still be useful to carry out exploratory calculations with a standard routine, even if it is rather slow.

We now consider in more detail the problem of the time integration, and the spatial approximations in one or more dimensions.

2. Solution of initial-value problems

The construction of general algorithms for ordinary differential equations is discussed in Chapter 3. Here we consider the special features of the initial-value problem which are associated with parabolic systems. The solution of a parabolic equation typically includes rapidly decaying transient terms, so that the approximating system of ODEs in the time direction is stiff (see Chapter 3). Therefore we need to use implicit integrators if we are to avoid severe restrictions on the time-step.

Let U_{ij} be the approximate value of $U_i(jk)$ in (9.3); then we can represent the equation by

$$U_{i,j+1} - k\theta\, F_i(U_{i+1,j+1}\, ,\, U_{i,j+1}\, ,\, U_{i-1,j+1}\, ,\, (j+1)k)$$

$$= U_{ij} + k(1-\theta)\, F_i(U_{i+1,j}\, ,\, U_{ij}\, ,\, U_{i-1,j}\, ,\, jk)\, , \qquad (9.5)$$

where $0 \leqslant \theta \leqslant 1$. This formulation (the θ-method) enables the user to select either the explicit method by taking $\theta = 0$, or an implicit method with $\theta > 0$, which requires the solution of a system of nonlinear equations at each time-step. The accuracy of (9.5) is limited, but it has the advantage of not needing the storage of earlier solution values.

To solve (9.5) we use Newton's method, with the first approximation obtained by extrapolating with $\theta = 0$, and successive corrections calculated in the usual way from the linearised equations. We note that the Jacobian of the nonlinear system in (9.5) is tridiagonal; if we have a system of parabolic equations it becomes block tridiagonal. In general the time-step is not constant, but is controlled by a suitable local accuracy test; it can also be reduced if there is any difficulty over the convergence of the Newton iteration.

To obtain higher accuracy it is natural to use Gear's formula, which is suitable for stiff problems with non-oscillatory solutions. We consider a more general system than (9.3) in order to cover a wider range of problems. Consider the implicit equations

$$\varphi(\underline{U}, \frac{d\underline{U}}{dt}, t) = \underline{0} \tag{9.6}$$

(Dew and Walsh 1981). In Gear's method the time-derivative is approximated by a backward-difference formula, leading to the following equations for $\underline{U}_{j+1} \simeq \underline{U}\{(j+1)k\}$:

$$\varphi\{\underline{U}_{j+1}, \frac{1}{k\beta_p}(\underline{U}_{j+1} - \sum_{i=0}^{p} \alpha_i \underline{U}_{j-i}), (j+1)k\} = \underline{0} . \tag{9.7}$$

The coefficients $\{\beta_p, \alpha_i, i = 0,1,\ldots,p\}$ are given by Lambert (1973, Ch. 8) and the method is locally of order p+1. The Jacobian matrix for this system is

$$J_p = \left[\frac{\partial\varphi}{\partial\underline{U}} + \frac{1}{k\beta_p} \frac{\partial\varphi}{\partial\dot{\underline{U}}} \right] . \tag{9.8}$$

If (9.6) represents the space discretisation of a system of parabolic equations, the matrix J_p will be of band form, usually block tridiagonal. The nonlinear equations (9.6) are solved by Newton's method as before.

It is usual in ODE integrators to vary the order of the formula and also the steplength as the integration proceeds, in order to maintain efficiency and accuracy (see Chapter 3). Initially only one line of solution values is available at t = 0, so we have to start with a first-order formula and a very small step. The order and step are gradually increased on the basis of local error estimates, but if there is a sudden disturbance in the solution, the integrator may drop back to the low-order starting procedure. Ideally we should like to have a facility for controlling the global error, but most integration routines provide only local error estimates, and the overall error has to be checked by repeating the calculation with a different tolerance.

For some evolutionary problems it may be possible to use non-stiff integrators over parts of the time interval, and this has the great advantage of avoiding the solution of implicit equations at each time-step. A suitable predictor-corrector method, such as one based on the Adams-Bashforth formula, provides local error estimates from which the order and step can be varied as before, and we can monitor the steplength as the solution proceeds. If it becomes unreasonably small relative to the changes in the solution, we assume that stiff terms are entering into the problem, and we can then switch to a stiff integrator.

However, if the general implicit formulation (9.6) is used, there is a problem over the definition of the system. For Adams-Bashforth it is more natural to take the standard form as

$$\frac{d\underline{U}}{dt} = \underline{F}(\underline{U}, t) , \qquad (9.9)$$

which excludes certain types of problem. A compromise form between (9.6) and (9.9) has been used by Hindmarsh (1981) in algorithm development; this is

$$A \frac{d\underline{U}}{dt} = \underline{F}(\underline{U}, t) , \qquad (9.10)$$

where A is a matrix which is allowed to be singular. This enables us to include algebraic equations and elliptic equations in the parabolic system.

3. Treatment of the space dimensions

It is relatively easy to control the local error in the time direction, using the initial-value formulation discussed above. But the result may be misleading if there is no way of checking the accuracy of the space discretisation (other than by repeating the whole calculation with more points). For a simple boundary-value problem, we can check the local accuracy by calculating higher difference terms and reducing the mesh-length if necessary, but in the parabolic case we have the added complication that the character of the solution may change radically as the solution proceeds, so that the mesh needs continual adjustment.

For example, in the case of Burgers' equation (9.2), the steep wave-front which occurs when ν is small requires a very fine mesh, so that we either have to take very many points over the range, or arrange for the mesh-points to be re-distributed as the wave-front moves. It is possible to insert and delete mesh-points by simply terminating the integration and restarting on a new mesh, but this is inefficient, because the time-integrator has to restart with a very small time-step, and all previous information is lost.

The monitoring of the space error takes different forms according to the discretisation used. For finite differences, the first term of the difference correction can be calculated on the mesh, with adapted formulae near the boundaries. The aim is to position the mesh-points so that the error is distributed uniformly across the range, but this is not easy because the high-order correction terms are rather unstable when the points move. To illustrate this simply, consider the formula for the second derivative on a uniform mesh, with $U_i \simeq u(ih)$,

$$\left(\frac{\partial^2 u}{\partial x^2}\right)_i = \frac{1}{h^2}\,(U_{i+1} - 2U_i + U_{i-1}) - \frac{1}{12}\,h^2 \left(\frac{\partial^4 u}{\partial x^4}\right)_i + \dots \;. \qquad (9.11)$$

Now suppose the points are perturbed so that the spacing becomes unequal; then the second derivative is still approximated on three adjacent points, but the leading error term is of order h instead of h^2, and it involves the third derivative instead of the fourth. If new mesh-intervals are to be calculated from high-order terms, it is necessary to 'smooth' them in some way to avoid sudden changes (Raith, Schnepf and Schonauer 1982).

If we can devise a strategy for re-mapping the mesh-points, the order of the time integrator can usually be maintained by re-mapping the data retained from previous time-steps as well as that on the current line, using the same interpolation formula throughout. If there are near-discontinuities in the time direction, the order will be automatically reduced by the integration routine, but this is not due to the mesh alteration.

In the case of finite-element approximation in space, a strategy for moving the mesh-points has been proposed by Miller (1981). An alternative approach is to keep the elements fixed and to increase the degree of the approximation in regions where the error appears to be large. This requires a suitable monitoring function to give an estimate of the error (Babuska and Rheinboldt 1978). The matrix of the finite-element equations does not have to be re-computed completely; the additional parameters can be incorporated in an updating procedure.

A third method of spatial discretisation, which has some advantages for error estimation, is the representation of the solution in terms of Chebyshev polynomials (Berzins and Dew 1981). Suppose the approximate solution of (9.1) is taken to be

$$U(x, t) = \sum_{i=0}^{N} c_i(t)\, T_i(2x-1) \tag{9.12}$$

in the interval $0 \leq x \leq 1$. Then we can obtain a set of ordinary differential equations for the coefficients $c_i(t)$ by collocation of the parabolic equation at N+1 points

$$x_s = \frac{1}{2}\left(1 + \cos \frac{s\pi}{N}\right), \qquad s = 0,1,\ldots,N, \tag{9.13}$$

incorporating the boundary conditions at the ends of the range. The error estimate in the space direction is obtained by monitoring the convergence of the coefficients $c_i(t)$, $i = 0,1,\ldots,N$, and it is easy to change the number of terms as the integration proceeds. In practice it is not satisfactory to use a single series (9.12) to represent the solution; usually the range is split into a number of sub-intervals, with a different polynomial representation in each. Additional matching conditions are then required at the interval boundaries, but the error estimation is still straightforward.

For problems in two or three space dimensions, it is important to distribute the space points efficiently, because the amount of calculation required to solve the implicit equations at each time-step is very large. For general regions, solution methods may be based on sparse solvers or on alternating direction methods. But if the region is a rectangle, and if we have a simple spatial operator (for example the Laplace or Helmholtz operator), the best method is probably a fast cyclic reduction technique (FACR), which is described briefly in Chapter 8 on elliptic problems.

4. Construction of general algorithms

It is difficult to specify the exact class of problems which can be solved by a general integrator based on the method of lines. The first parabolic solvers in the NAG Library (D03PGF and related routines) attempted to restrict the equations to those where the space terms gave a recognisable boundary-value problem, and the ODE system was assumed to be stiff. However, such routines are inevitably used for more general problems, and they cannot be made as robust as linear algebra routines, for example. The best approach is probably to provide a number of different methods for the various parts of the calculation, and to give advice on possible pitfalls in the documentation.

A general program for evolutionary problems is the SPRINT package (Berzins, Dew and Furzeland 1984), which aims for flexibility in the treatment of parabolic and other time-dependent systems. The main sections are as follows:

Control program

Discretiser, e.g. Skeel's finite-difference formulae

Time-integrator, e.g. Gear's method, θ-method

Linear equation solver, band and sparse forms

Monitoring routine.

The interfaces are designed so that sections can be easily replaced if alternative methods are required.

The monitoring routine is often needed in practical problems, for example to implement physical changes of state such as an explosive reaction occurring at a certain temperature. In general it may be necessary to determine when some function of the current solution reaches a given value, in order to terminate the calculation or to operate some switch in the definition of the system. This is essentially a root-finding problem, which is complicated in some cases by the condition that the solution cannot be continued past the critical point. It is possible to allow for such a constraint, but the user has to understand the problem very clearly so that he can define the critical condition in a suitable way.

5. Software

An implementation of the method of lines in one space dimension is given by Sincovec and Madsen (1975) in the routine PDEONE. This is designed for a system of parabolic equations; the mesh-points need not be equally spaced, and there is provision for coefficient discontinuities (corresponding to material interfaces). The boundary conditions have the form

$$\alpha u + \beta \frac{\partial u}{\partial n} = \gamma \quad , \tag{9.14}$$

and the approximating system is integrated by using a standard ODE solver.

The NAG Library has the fairly general routine D03PGF, for parabolic systems in one space dimension. The method used is similar to that of PDEONE, and the time integration is carried out by a Gear routine. The user interface is rather complicated, and two simpler calling routines (D03PAF, D03PBF) are provided for restricted classes of problems. These routines use the same underlying software as D03PGF and so their speed is not optimal, but they are suitable for test calculations on simplified problems.

The SPRINT package (Berzins, Dew and Furzeland 1984) is a much more general parabolic solver, which allows for different forms of discretisation, including the Chebyshev method outlined in section 3, and for various types of control and termination tests. It is intended for more experienced users, who can adapt the elements of the package to their particular problems.

CHAPTER 10

ADVANCES IN ALGORITHMS FOR CURVE FITTING

J.G. Hayes

1. Introduction

I shall be aiming to discuss a variety of algorithms which have appeared in the past few years for fitting spline curves to data either by least squares (with one variant) or by interpolation. In particular I shall be describing the routines we have produced at NPL since (in NAG parlance) Mark 8 of the NAG Library. Those in the Library at Mark 8 are, I think, covered by Hayes (1978), so in a way I am continuing from that paper. There is some fuzziness in between, though, as part of that paper was projecting availability into the future, and things never quite turn out as planned.

These new routines are part of the total of almost 50 new routines which comprise DASL, our Data Approximation Subroutine Library (Anthony, Cox and Hayes 1982). DASL is available from NAG. So, after covering notation and definitions in the next section, I shall describe in section 3 various additional features and facilities that have been provided in routines for the basic spline curve-fitting problem and in new auxiliary routines. Then in section 4 I shall deal with the periodic and parametric spline routines, which in particular provide capabilities for multi-valued functions, including closed curves. It will not be possible to cover all the background material for these DASL routines and where a specific reference is not given, further information may be found in Cox (1982).

DASL includes a simple knot-placement algorithm, intended to give the new user, particularly, at least a good start at a satisfactory choice of knots. In section 5, however, I shall describe a routine by a Belgian researcher which makes a more substantial attack on the problem of automatic knot-placement. It should give good results in many cases when there is plenty of data, usually without using many more knots than necessary.

In the final section I present some work by researchers in the US and the UK on monotonic interpolation. A spline fitted in a normal way to monotonic data will usually be monotonic itself, simply through following the data. But that will not always be so, especially where the slope of the data is small or zero. Thus there is sometimes a need for routines which ensure monotonicity, a need more readily satisfied by interpolation methods. Such routines, having been designed successfully to meet the monotonicity requirement, tend to be assessed and compared by applying them to data inadequately representing curves with rapidly changing slope. They are then pronounced as 'visually pleasing' (or not), perhaps reflecting the fact that one main motivation of this work is for drawing curves through points by computer.

Finally, I should comment on the fact that I have omitted any mention of fitting with ordinary polynomials. This omission is not because polynomials are no longer important in data fitting (indeed they are still very useful in many problems) but simply because the numerical methods were essentially determined long ago and in software terms there has been no very significant change for some years. A survey of polynomial methods is included in Hayes (1974) and software in Hayes (1978), already mentioned.

2. Notation and definitions

Let us denote the given data points by (x_i, f_i), $i = 1,2,\dots,m$, and some finite interval containing the x_i by $[x_{min}, x_{max}]$. Commonly in practice x_{min} and x_{max} will be set equal respectively to the smallest and largest of the x_i. On this interval, a (polynomial) <u>spline</u> $s(x)$ of order n (degree $n-1$), with <u>knots</u> $\lambda_1, \lambda_2, \dots, \lambda_N$ (where $x_{min} < \lambda_1 < \lambda_2 < \dots < \lambda_N < x_{max}$) is a function with the following properties:

(a) In each of the intervals $\lambda_{j-1} \leqslant x \leqslant \lambda_j$, for $j = 1,2,\dots,N+1$, $s(x)$ is a polynomial of degree $n-1$ at most. (Here λ_0 and λ_{N+1} are to be interpreted as x_{min} and x_{max}.)

(b) $s(x)$ and its derivatives up to order $n-2$ are continuous.

The above is the case of a spline with <u>simple knots</u>, each knot being separate from the others. However, <u>coincident</u> or <u>multiple knots</u> can be allowed, reducing the continuity requirement. If r knots coincide at a particular value of x, the spline and its derivatives are taken to be continuous only up to order $n-r-1$ at that value of x.

The spline can be mathematically represented in many ways, for example by the N+1 polynomials comprising it. For our purposes, however, it is usually best represented in terms of the set of normalized B-splines $N_{n,j}(x)$, $j = 1,2,\ldots,q$, with $q = N+n$. Before this full set can be defined, 2n extra, artificial, knots must be added to the set of knots (the <u>interior knots</u>) already introduced. They are denoted as in the following inequalities, which they must satisfy:

$$\left.\begin{array}{l} \lambda_{1-n} \leqslant \lambda_{2-n} \leqslant \ldots \leqslant \lambda_0 \leqslant x_{min} , \\ x_{max} \leqslant \lambda_{N+1} \leqslant \lambda_{N+2} \leqslant \ldots \leqslant \lambda_q . \end{array}\right\} \qquad (10.1)$$

It is usual to take $\lambda_0 = x_{min}$ and $\lambda_q = x_{max}$. Indeed, for most purposes we take all equalities in (10.1), and can speak of <u>coincident end-knots</u>. Except for differences due to rounding error, the particular choice does not affect the value of a fitted spline in $[x_{min}, x_{max}]$.

The conventional definition of B-splines, in terms of divided differences, is not very illuminating and it is easier to visualise them from some of their properties. The <u>normalized B-spline</u> $N_{n,j}(x)$ defined on the knot set $\lambda_{1-n}, \lambda_{2-n}, \ldots, \lambda_q$ is a spline of order n which is zero outside the interval $[\lambda_{j-n}, \lambda_j]$, and satisfies

$$\sum_{j=1}^{q} N_{n,j}(x) \equiv 1 , \quad \text{for } x_{min} \leqslant x \leqslant x_{max} . \qquad (10.2)$$

In fact, that statement is a complete definition. $N_{n,j}(x)$ is strictly positive everywhere in the interval $(\lambda_{j-n}, \lambda_j)$ and has a single maximum. The continuity conditions for a spline imply that, if λ_{j-n} and λ_j are simple knots, $N_{n,j}(x)$ joins the x-axis with all derivatives, up to order n-2, having the value zero; if they are not simple knots, the continuity with the x-axis is correspondingly reduced.

The B-splines satisfy a recurrence relation (Cox 1972, de Boor 1972), which provides the best means of computing their values at particular x-values. It is

$$N_{n,j}(x) = \left\{ \frac{x - \lambda_{j-n}}{\lambda_{j-1} - \lambda_{j-n}} \right\} N_{n-1,j-1}(x)$$

$$+ \left\{ \frac{\lambda_j - x}{\lambda_j - \lambda_{j-n+1}} \right\} N_{n-1,j}(x) , \qquad (10.3)$$

starting with

$$N_{1,j}(x) = \begin{cases} 1, & \text{for } \lambda_{j-1} \leqslant x < \lambda_j , \\ 0, & \text{otherwise.} \end{cases}$$

(The choice of the left rather than the right inequality above to include equality is purely conventional. The right inequality must include equality when $j = N+1$.)

With the B-splines thus introduced, we have the result of Curry and Schoenberg (1966) that any spline $s(x)$ of order n with interior knots $\lambda_1 , \lambda_2 , \ldots, \lambda_N$ has a unique representation of the form

$$s(x) = \sum_{j=1}^{q} c_j N_{n,j}(x) , \text{ for } x_{min} \leqslant x \leqslant x_{max} . \qquad (10.4)$$

This is the <u>B-spline representation</u> and the c_j are <u>B-spline coefficients</u>. We note that q (= $N+n$) is the number of independent coefficients in a spline of order n with N interior knots.

In the standard spline-fitting problem with data points (x_i , f_i), we have the <u>observation equations</u>

$$s(x_i) = f_i , \quad i = 1,2,\ldots,m , \qquad (10.5)$$

or, when using the B-spline representation (10.4),

$$\sum_{j=1}^{q} c_j N_{n,j}(x_i) = f_i , \quad i = 1,2,\ldots,m . \qquad (10.6)$$

Then, if the problem is one of interpolation, we seek values for the coefficients c_j which satisfy (10.6) exactly. In this problem we have $m = q$, but even then there is in general no solution to (10.6). There is a unique solution, however, if the Schoenberg-Whitney conditions

$$\lambda_{j-n} < x_j < \lambda_j , \quad j = 1,2,\ldots,q , \tag{10.7}$$

are satisfied. (When coincident end-knots are used, the left-hand inequality may be relaxed to include equality when $j = 1$, and the right-hand one when $j = q$.)

In the least-squares problem, given weights w_i, we seek coefficients c_j such that the <u>weighted residual sum of squares</u>

$$\sum_{i=1}^{m} \{w_i[f_i - s(x_i)]\}^2 \tag{10.8}$$

is minimized. (Note that, as in the NAG routines, we define w_i to be inversely proportional to the standard deviation of f_i.) A solution always exists to this problem, but for a unique solution some subset of q of the x_j (all different) must satisfy the Schoenberg-Whitney conditions.

We shall also write the observation equations (10.6) in matrix form as

$$A\underline{c} = \underline{f} , \tag{10.9}$$

where $\underline{c}^T = [c_1 \ldots c_q]$, $\underline{f}^T = [f_1 \ldots f_m]$ and A is the $m \times q$ matrix with elements $a_{ij} = N_{n,j}(x_i)$.

3. Basic fitting problems

NAG routine E02BAF deals with the standard least-squares problem and E01BAF with the standard interpolation problem, as they are outlined towards the end of the previous section. Both routines use only cubic splines ($n = 4$), but those are very effective on most problems. The new routines in DASL do deal with general n, however. That generality is required in some applications, but its provision is also associated with the fact that we have written routines which give the first derivative or indefinite integral of a spline as a spline function of x. These splines are, of course, of different order from the parent spline. They can, however, be evaluated with the evaluation routine for general n. Moreover, by repeated application of the routines, higher derivatives or repeated integrals can be obtained.

An aspect of the new least-squares routine which users may find helpful is the fact that the data points can be accepted in any order. Previously they had to be in order of increasing x-value. That was to enable, first, the distinct x-values to be identified and then tested for satisfaction of the Schoenberg-Whitney conditions; second, to cause the matrix A of (10.9) to be of banded form, allowing a very efficient computational process. 'Banded form' here means (a) all but n adjacent elements in each row are zero, (b) column position of first non-zero element in each row is a monotonically increasing function of row number. The new routine carries out the same tests and gets the same benefit from the band form if it finds that the data points are ordered, but, with some loss in efficiency, will also deal with the disordered case. It does so by the method given by Cox (1981) specifically for a least-squares solution when the matrix A has the property (a) above but not property (b).

The design of the routine was also based on acceptance of the fact that a case in which the Schoenberg-Whitney conditions are satisfied, but only just (i.e. where slight changes in knots, moving them across data points, would result in the conditions being transgressed) can be numerically indistinguishable during the solution of the linear equations from cases which do not satisfy the conditions. The routine aims to identify such rank deficiency or numerical rank deficiency using some appropriate threshold, but still to determine one least-squares solution from the infinite number having the same or nearly the same residual sum of squares. We recall that the basic solution process we employ is to pre-multiply the weighted form of (10.9) by an orthogonal matrix Q which enables us to obtain the least-squares coefficient vector $\underline{c}$ from the equation

$$R\,\underline{c} = \underline{h}\ . \tag{10.10}$$

Here R is a q × q upper triangular matrix, indeed the first q rows of the matrix QWA, where W = diag[w_i], and $\underline{h}$ contains the first q elements of the vector $QW\underline{f}$. Numerical rank deficiency is taken as identified if there is a diagonal element of R less than the chosen threshold. Let us suppose, for simplicity, that there is only one such element and that it is in the jth position. This means that the coefficient c_j is (effectively) indeterminate, and to overcome this a resolving constraint having the form

$$\frac{c_{j+1} - c_j}{\lambda_j - \lambda_{j-n+1}} - \frac{c_j - c_{j-1}}{\lambda_{j-1} - \lambda_{j-n}} = 0 \tag{10.11}$$

is introduced as an extra observation equation. If the diagonal element had been an exact zero, (10.11) would be satisfied exactly and so c_j determined as a particular weighted mean of c_{j-1} and c_{j+1} (indeed the simple arithmetic mean if the knots are equi-spaced). The residual sum of squares would be unaffected by the extra equation. A small instead of zero diagonal element disturbs that situation very little. It has to be said, however, that when numerical rank deficiency is detected, a user would almost always be best advised to choose different, probably fewer, knots and start the computation again. This applies when curve fitting; surface fitting is a different matter.

As just implied, the new least-squares routine, like the old, requires the user to choose the knots. However, a simple routine which chooses the knot positions given the number of knots should normally give the user a good start. It arranges the knots so as to contain as near as possible the same number of data points in each knot interval. Thus it is particularly suited to those (commonly occurring) data sets which have closer x-values where the curve is changing shape most rapidly.

Finally, a most significant part of the basic spline-curve segment of DASL is the set of three routines which provide standard errors of values of the least-squares spline and its derivative and integral. Here, the computation follows from the fact that, given an arbitrary vector $\underline{b}$ with q elements, the standard error (SE) of the linear combination $\underline{b}^T\underline{c}$ of the computed coefficients c_j is given by (see for example Draper and Smith (1966, p. 121))

$$(SE)^2 = \underline{b}^T(A^T W^2 A)^{-1} \underline{b} \,(RMS)^2 , \tag{10.12}$$

where RMS is the root mean square of residuals of the fit. Because of the definitions following (10.10), this reduces to

$$(SE)^2 = \underline{v}^T\underline{v} \,(RMS)^2 , \tag{10.13}$$

where $\underline{v}$ can be obtained as a solution of the triangular system

$$R^T \underline{v} = \underline{b} \; . \tag{10.14}$$

Values of the fitted spline and its derivative and integral are all linear combinations of the computed c_j's in which the multipliers are the $N_{n,j}(x)$ or their derivatives or integrals evaluated at a particular x. Thus the computation given by (10.13) and (10.14) applies in all cases.

To summarize the new features of the basic spline curve segment of DASL, they are (a) arbitrary order spline, (b) arbitrary ordering of points in least-squares routine, (c) application of resolving constraints in rank-deficient cases in the same routine, (d) derivative and indefinite integrals as spline functions of x, (e) simple knot-placement routine, and (f) computation of standard errors.

4. Periodic and parametric curves

Sometimes it is appropriate to fit a periodic function to data. We may, for example, be given a value of some variable in each of a number of directions from a base point and wish to derive a fit approximating the variable round the whole 360°. This can be achieved, using splines, by first taking the set of knots interior to the interval (x_{min}, x_{max}), which would likely be $(0,360^o)$ in the example, and repeating them moved a distance $d = (x_{max} - x_{min})$ to the right and also to the left. Then these new knots are used as the extra knots needed for defining the B-splines. Thus, for $k < 1$, $\lambda_k = \lambda_{k+n} - d$ and, for $k > N$, $\lambda_k = \lambda_{k-N} + d$. This cyclic choice of knots ensures that the B-splines which are non-zero in the interval $[\lambda_N, x_{max}]$ are the same as those in the interval $[x_{min}, \lambda_1]$ shifted on a distance d. In other words, $N_{n,N+j}(x+d) = N_{n,j}(x)$, $j = 1,\ldots,n$. Thus, if also we insist that in the B-spline representation (10.4) we must have $c_{j+N} = c_j$, the polynomial in the interval $[\lambda_N, x_{max}]$ will be the same polynomial as in the interval $[x_{min}, \lambda_1]$ shifted a distance d. So the spline and all its derivatives will have the same values at each end of the range. We may note that setting $c_{j+N} = c_j$ disturbs the band form of the matrix A in the observation equations (10.9): in effect, the first and last n columns are laid one on top of the other, leaving only N columns. These 'non-band' columns can be placed at the end, another arrangement for which Cox (1981) provides an efficient algorithm. For the periodic spline, as for the basic spline, DASL provides a least-squares fitting and an interpolation routine, routines for computing values, derivatives and integrals, and standard errors of all three, together with a simple knot-placement routine.

So far we have discussed only single-valued functions but it is also possible to treat multi-valued functions if we use parametric splines. There is no longer a dependent/independent relationship between the two variables and I shall use x and y rather than x and f to indicate the essential symmetry in their roles. General curves in the (x, y) plane can then be represented in parametric form:

$$x = s_1(t) \, , \qquad y = s_2(t) \, , \tag{10.15}$$

where s_1 and s_2 are spline functions of some suitable parameter t. Arc-length, the distance measured along the curve from some base point, is usually satisfactory. Then, as the parameter value moves over its range, the point (x, y) given by (10.15) describes a curve in the plane. In the fitting problem, the main need is to give a parameter value t_i to each data point $(x_i\ , y_i)$. Arc-length is not available to us, but, at any rate in the interpolation problem, we can compute cumulative chord-length, that is to say, distance along the polygon formed by joining adjacent points by straight lines. For the approximation (smoothing) problem, Earnshaw and Yuille (1971) used this same parameterization, though it is no longer cumulative chord-length, of course. Provided the errors in the data are reasonably small compared with the distances between adjacent points, it works quite well in providing a smooth approximant. Once the parameter values have been established, $s_1(t)$ is determined by interpolation or by a least-squares fit, as appropriate, to the 'data points' $(t_i\ , x_i)$. Similarly $s_2(t)$ is derived from the points $(t_i\ , y_i)$. Note that a closed curve results if the splines used are periodic splines. DASL contains interpolation and least-squares fitting routines, together with an evaluation routine, for both open and closed curves. All the fitting routines select their own knots, using simple strategies.

5. An almost automatic algorithm

Dierckx (1981a) has given a spline-fitting algorithm which requires the user to choose only one parameter, to control the smoothness of the spline in a certain sense. It uses a fitting criterion akin to that for the well-known 'smoothing spline' (Reinsch 1967), in that the spline is determined so as to minimize an intuitively acceptable measure of smoothness subject to the residual sum of squares for the given data being less than a constant S. It is the value of S (the 'smoothing factor') which the user has to choose. Some guidance is available if the size of the statistical errors in the data can be estimated; otherwise it is a matter of experiment. Reinsch's smoothness measure involves the kth derivative of the spline, where k is half the (even) order of the spline. Dierckx uses the measure

$$D = \sum_{j=1}^{N} d_j^2 , \qquad (10.16)$$

where d_j is the discontinuity in the (n-1)th derivative of the spline at λ_j.

The smoothing spline has the disadvantage of having a knot at every data point, and so a very bulky representation. Dierckx aims to be reasonably frugal with the number of knots, choosing their positions in a simple ad-hoc fashion. He starts with no knots and adds a few at a time. At each stage, he computes the least-squares spline with the current knot set and computes the weighted sum of squares of the residuals in each interval $[\lambda_{j-1}, \lambda_j]$. The intervals with the largest sums of squares each receive a new knot. The formula for the total number of knots to be added, aimed at progressing as rapidly as possible while avoiding various pitfalls, is a little involved and I shall not describe it. The process continues until a knot set is found for which the total weighted residual sum of squares is less than S. Then the algorithm goes on to determine a spline (shown to exist and be unique) which has this knot set and which satisfies the full fitting criterion. We may write this criterion:

$$\text{minimize } D , \quad \text{subject to } \delta \leqslant S , \qquad (10.17)$$

where D is as given in (10.16) and δ is the weighted residual sum of squares defined in (10.8). The minimization is with respect to the coefficients in the B-spline representation, which the algorithm employs. We note that both D and δ are quadratic functions of these coefficients.

Solving (10.17), using the method of Lagrange, involves the repeated solution of the sub-problem:

$$\text{find the spline } s_p(x) \text{ to minimize } D + p\delta , \text{ for given p.} \qquad (10.18)$$

The solution to that problem is in fact the least-squares solution of a set of equations, linear in the B-spline coefficients c_j, which consists simply of the observation equations (10.9) together with N other equations relating to the derivative discontinuities at the knots. The observation equations have their usual weights and the other equations each have a weight $1/\sqrt{p}$ (which is the only dependence on p).

Having obtained $s_p(x)$ by this means, we may define

$$F(p) = \sum_{i=1}^{m} w_i^2 \, [f_i - s_p(x_i)]^2 \tag{10.19}$$

and then the final piece of the computation is to find the value of p such that

$$F(p) - S = 0 \, . \tag{10.20}$$

In other words, the computation is a root-finding process in which the function evaluation stage involves the least-squares solution just described. The function F(p) is monotonic and convex and so the root, within known bounds, can be found efficiently. The corresponding $s_p(x)$ is the required result. The least-squares solutions involved are derived by the use of Givens rotations without square roots, advantage being taken of the special structure of the equations.

6. Monotonic interpolation

In this problem, assuming $x_1 < x_2 < \ldots < x_m$, our data ordinates f_i are monotonic and we wish to find a function which passes through the points (x_i , f_i) and is monotonic. We may look in particular for a monotonic function, p(x) say, which is a cubic polynomial in each interval (x_i, x_{i+1}) and has just first-derivative continuity at the x_i (and so could be described as a cubic spline with two knots at each x_i). A cubic segment is completely specified by its value and derivative at the two ends of its range, and Fritsch and Carlson (1980) provide necessary and sufficient conditions on these quantities for p(x) to be monotonic. Thus our problem can be seen as one of providing a formula for the slopes $d_i = p'(x_i)$ which causes the conditions always to be satisfied. Brodlie (1980, p. 37) gives such a formula, which has the advantage that each d_i is derived very simply from information local to x_i . Denoting the slope of the line joining data points j and j+1 by m_j , so that $m_j = (f_{j+1} - f_j)/h_j$, with $h_j = x_{j+1} - x_j$, Brodlie's formula is

$$\frac{1}{d_i} = \frac{\alpha}{m_{i-1}} + \frac{1-\alpha}{m_i} , \quad \text{if} \quad m_{i-1}m_i > 0 ,$$

and

$$d_i = 0 , \quad \text{otherwise.}$$

Here

$$\alpha = \frac{h_{i-1} + 2h_i}{3(h_{i-1} + h_i)} .$$

Fritsch and Butland (1982) report 'visually pleasing' results with it in extensive tests and it has been incorporated in PCHIP (Fritsch 1982). This is a package of about a dozen routines to deal with monotonic interpolation and various associated tasks, including evaluation, differentiation and integration of the interpolant.

In this same area, Gregory and Delbourgo (1982) describe a method based on rational quadratics which gives a monotonic interpolant for any choice of derivative values which have the correct sign, and suggests two particular choices. The rational quadratic in the interval $[x_i, x_{i+1}]$ is

$$\frac{f_{i+1}\theta^2 + m_i^{-1}(f_{i+1}d_i + f_i d_{i+1})\theta(1-\theta) + f_i(1-\theta)^2}{\theta^2 + m_i^{-1}(d_i + d_{i+1})\theta(1-\theta) + (1-\theta)^2} ,$$

with $\theta = (x - x_i)/h_i$. They also have a development of this method (Delbourgo and Gregory 1983) which chooses the derivatives so as to make the interpolant have second derivative continuity. Of course it is not a local method, in fact involving the solution of m-2 nonlinear equations. Both algorithms apparently produce 'visually pleasing' results.

CHAPTER 11

L_1 AND L_∞ ALGORITHMS AND APPLICATIONS

I. Barrodale and C. Zala

1. Introduction

For the past several years reliable and efficient programs based on the simplex methods of linear programming have existed for curve fitting in the L_1 and L_∞ norms, thereby offering an alternative to traditional L_2 (least-squares) methods. Following a review of available algorithms we focus attention on several practical applications where L_1 and L_∞ techniques appear to have some advantages over least-squares methods. The main emphasis is on applications of the L_1 norm in geophysical processing, where examples involving overdetermined and underdetermined systems of equations occur. Brief remarks are also included on the efficacy of solving linear programming problems via L_1 algorithms, on dealing with erratic data and complex-valued data, and on some applications of the L_∞ norm in signal processing.

2. Linear programming

Linear programming (LP) is a subject in which the aim is to maximise or minimise a linear function of several variables, where these variables are subject to given constraints expressed as linear equations or inequalities. The <u>simplex method</u>, introduced in 1947 by G.B. Dantzig, is a direct method for solving any LP problem, although several variants of the basic method have since been developed for handling particular classes of problems more efficiently.

Actually, the simplex method deals only with LP problems involving linear equality constraints in nonnegative variables. Problems involving linear inequality constraints, or variables which are unrestricted in sign, are transformed as follows.

Any inequality constraint can be written as an equality constraint by simply introducing a new nonnegative slack variable; thus, inequalities such as

$$x_1 + 2x_2 \leqslant 7 \qquad \text{and} \qquad 3x_1 - 4x_2 \geqslant 1$$

can be rewritten as

$$x_1 + 2x_2 + x_3 = 7 \quad \text{and} \quad 3x_1 - 4x_2 - x_4 = 1 \; .$$

Secondly, a variable x_1 which may assume negative values can be replaced by setting $x_1 = x_1' - x_1''$, where x_1' and x_1'' are new nonnegative variables.

Also, it is convenient to assume that every LP problem is a minimisation problem, since the maximum of a function f can always be determined by minimising the function (-f).

Any LP problem can be expressed in the following form:

$$\text{minimise} \qquad z = c_1x_1 + c_2x_2 + \ldots + c_nx_n \tag{11.1}$$

$$\text{subject to} \quad \begin{cases} a_{11}x_1 + a_{12}x_2 + \cdot \;\; \cdot \;\; \cdot + a_{1n}x_n = b_1 \\ a_{21}x_1 + a_{22}x_2 + \cdot \;\; \cdot \;\; \cdot + a_{2n}x_n = b_2 \\ \quad \cdot \qquad\quad \cdot \qquad\qquad\quad \cdot \qquad \cdot \\ \quad \cdot \qquad\quad \cdot \qquad\qquad\quad \cdot \qquad \cdot \\ \quad \cdot \qquad\quad \cdot \qquad\qquad\quad \cdot \qquad \cdot \\ a_{m1}x_1 + a_{m2}x_2 + \cdot \;\; \cdot \;\; \cdot + a_{mn}x_n = b_m \end{cases} \tag{11.2}$$

$$\text{and} \qquad x_j \geqslant 0 \quad \text{for} \quad j = 1,2,\ldots,n \tag{11.3}$$

where $m < n$ and a_{ij} , b_i and c_j are given real numbers. The simplex method also requires that each b_i is nonnegative; this can always be arranged by multiplying equations in (11.2) by (-1) where necessary.

The constraints (11.2) and (11.3) form a system of m linear equations in n nonnegative variables x_j . Since $m < n$ this system usually has infinitely many solutions, and the problem is to select an optimal solution which minimises the objective function (11.1).

Using matrix notation, the general LP problem above can be stated as follows:

$$\text{minimise} \qquad z = \underline{c}\,\underline{x}$$

$$\text{subject to} \qquad A\,\underline{x} = \underline{b}$$

$$\text{and} \qquad \underline{x} \geqslant \underline{0}$$

where $\underline{c} = [c_1 \; c_2 \; \ldots \; c_n]$ is a row vector, $\underline{x} = [x_1 \; x_2 \; \ldots \; x_n]^T$ and $\underline{b} = [b_1 \; b_2 \; \ldots \; b_m]^T$ are column vectors, and $A = [a_{ij}]$ is an $m \times n$ matrix.

Geometrically, the domain defined by (11.2) and (11.3) is a convex hyperpolyhedron, which may be bounded or unbounded. In some cases the nature of the constraints may cause this domain to be empty; then the problem has no solution. For suitable values of z, the equation $\sum_{j=1}^{n} c_j x_j = z$ defines a family of hyperplanes which intersects the hyperpolyhedron. As z decreases, the intersection eventually contracts to a corner point of the domain, and this represents the optimal solution. (Strictly speaking, the optimal intersection may be a face or edge of the hyperpolyhedron containing more than one corner point; in this case the optimal solution is not unique.)

The simplex method first identifies a corner point and then proceeds successively to other corners, making sure that z decreases at each step. The initial identification of a corner point requires that an $m \times m$ identity matrix can be formed from the columns of A. This is usually arranged by introducing artificial variables into the constraints (11.2) which have a large positive coefficient M in the objective function (11.1). Corner points are characterised as nonnegative solutions to $A\underline{x} = \underline{b}$ which have at most m positive components; thus, at least n-m of their components are zero. A movement from a given corner to a neighbouring corner (i.e. a simplex iteration) is accomplished by a pivoting operation which exchanges a zero component with a non-zero component and vice versa, while generating a new solution to the constraints. Thus, each simplex iteration involves a pivot selection followed by one step of Gauss-Jordan or Gauss elimination (about mn multiplications and additions). As a rule, all the artificial variables are set to zero during the first few iterations, and the simplex method then proceeds from corner to corner of the original domain until an optimal solution is found.

The total number of simplex iterations required to solve an LP problem is approximately two or three times the number m of constraints involved (excluding the nonnegativity constraints). These computations can sometimes be reduced considerably by applying the simplex method to the dual of the 'primary' LP problem instead.

Given the primary problem $\{\min \underline{c}\ \underline{x} \mid A\underline{x} = \underline{b},\ \underline{x} \geqslant \underline{0}\}$, its dual unsymmetric problem is $\{\max \underline{b}^T\underline{u} \mid A^T\underline{u} \leqslant \underline{c}^T\}$; notice that the m-dimensional column vector $\underline{u}$ is not constrained to be nonnegative here. Similarly, for the primary problem $\{\min \underline{c}\ \underline{x} \mid A\underline{x} \geqslant \underline{b},\ \underline{x} \geqslant \underline{0}\}$ the dual symmetric problem is $\{\max \underline{b}^T\underline{u} \mid A^T\underline{u} \leqslant \underline{c}^T,\ \underline{u} \geqslant \underline{0}\}$. For both cases we have $\min \underline{c}\ \underline{x} = \max \underline{b}^T\underline{u}$, and the optimal solutions to the primal and dual problems can be obtained from the final stage of the simplex method applied to either form. LP problems which contain far more inequality constraints than variables should first be converted to their dual form before solution by the simplex method.

The standard form of the simplex method transforms the complete matrix A at each iteration, and so it is suitable for dense, medium-size LP problems. The revised form of the simplex method is most suitable for large sparse problems; in this form the matrix A remains unchanged, but each iteration generates and stores sufficient auxiliary information to proceed through exactly the same sequence of corner points as in the standard simplex method. The auxiliary information is retained at each step as the inverse of an $m \times m$ matrix; the inverse can be kept explicitly or in factored form.

During the past thirty years the simplex method has been used to solve hundreds of different types of optimisation problems; we concentrate here on applications of linear programming to curve fitting alone.

3. L_1 and L_∞ curve fitting

Consider the set of measurements y_i, corresponding to a variable t_i, which is recorded and graphed in Fig. 1.

Suppose that we wish to 'fit' these data by a straight line $Y(t) = x_1 + x_2 t$, where x_1 and x_2 are to be chosen so as to make the residuals $r_i = y_i - Y(t_i)$ as small as possible. The popular least-squares (L_2 norm) approach to this problem is to define the 'size' of the residuals to be $\sqrt{(\sum_i r_i^2)}$, and then use differential calculus to minimise this expression and so obtain x_1 and x_2. In this chapter we shall discuss curve-fitting problems in which the size of the residuals is defined as $\sum_i |r_i|$ or as $\max_i |r_i|$. These expressions are minimised by using the simplex method of linear programming, as explained later.

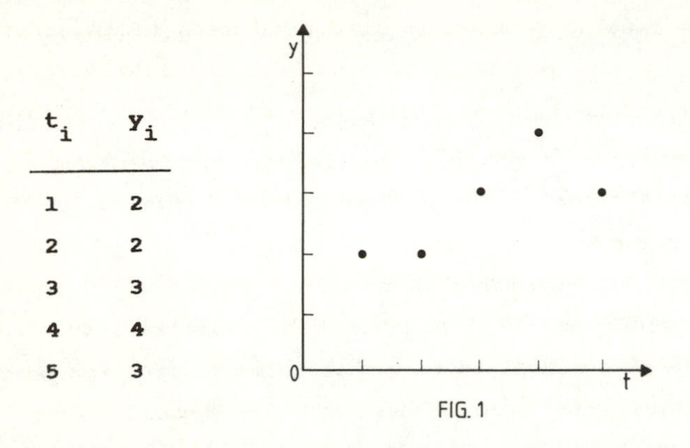

t_i	y_i
1	2
2	2
3	3
4	4
5	3

FIG. 1

It is convenient to adopt matrix notation when describing curve-fitting problems and algorithms. Thus, the above problem can be stated as an overdetermined system of linear equations $A\underline{x} = \underline{b}$, where

$$A = \begin{bmatrix} 1 & 1 \\ 1 & 2 \\ 1 & 3 \\ 1 & 4 \\ 1 & 5 \end{bmatrix}, \quad \underline{b} = \begin{bmatrix} 2 \\ 2 \\ 3 \\ 4 \\ 3 \end{bmatrix}, \quad \text{and} \quad \underline{x} = \begin{bmatrix} x_1 \\ x_2 \end{bmatrix}, \qquad (11.4)$$

for which the objective is to 'solve' the system by appropriate choice of $\underline{x}$. This is accomplished by defining the residual vector $\underline{r}$ as $\underline{r} = \underline{b} - A\underline{x}$, and then minimising some norm of $\underline{r}$. The norms considered here are:

$$L_1 \text{ norm:} \quad ||\underline{r}||_1 = \sum_i |r_i| = \sum_i |b_i - \underline{A}_i \, \underline{x}| \qquad (11.5)$$

$$L_\infty \text{ norm:} \quad ||\underline{r}||_\infty = \max_i |r_i| = \max_i |b_i - \underline{A}_i \, \underline{x}| \qquad (11.6)$$

where $\underline{A}_i$ denotes the ith row of A. In the simple problem above expression (11.6) has a minimum value of 2/3.

It is worth emphasising at this point that algorithms for solving an overdetermined system $A\underline{x} = \underline{b}$ can be applied to more general problems than might be assumed. For example, expressions (11.5) or (11.6) could represent a multidimensional curve-fitting problem involving hundreds of parameters x_j. In addition, each of the residuals r_i could be assigned different weights. Furthermore, the original curve-fitting problem could have involved complex-valued data. Of course, there are also many situations which call for the solution of systems of linear equations which do not involve curve-fitting problems at all.

We shall assume in most of what follows that the problem at hand is that of solving an overdetermined system of real linear equations $A\underline{x} = \underline{b}$. However, we also include remarks on complex-valued data fitting, underdetermined systems of real equations, and linearly constrained problems. We shall make no reference here to curve-fitting problems where the parameters x_j appear nonlinearly.

The basic problems with which we are concerned can be stated as follows. Given $\underline{b} = [b_1\ b_2\ \ldots\ b_m]^T$ and an $m \times n$ matrix $A = [a_{ij}]$, let $\underline{A}_i$ denote the ith row of A. The unconstrained L_1 linear approximation problem is to minimise

$$||\underline{b} - A\underline{x}||_1 = \sum_{i=1}^{m} |b_i - \underline{A}_i\,\underline{x}| \qquad (11.7)$$

where $\underline{x} = [x_1\ x_2\ \ldots\ x_n]^T$ is a vector of unknown parameters. The unconstrained L_∞ linear approximation problem is to minimise

$$||\underline{b} - A\underline{x}||_\infty = \max_{1 \leqslant i \leqslant m} |b_i - \underline{A}_i\,\underline{x}| \,. \qquad (11.8)$$

An obvious extension to these two problems involves the imposition of linear constraints on the choice of $\underline{x}$. Thus, the constrained L_1 or L_∞ linear approximation problem is to minimise expression (11.7) or (11.8), respectively, subject to given linear constraints

$$C\underline{x} = \underline{d} \qquad \text{and} \qquad E\underline{x} \leqslant \underline{f} \,. \qquad (11.9)$$

For example, the unconstrained L_1 solution to the 5×2 system (11.4) yields a minimum value for expression (11.7) of 2 when $x_1 = 1.5\lambda + 1.75(1-\lambda)$

and $x_2 = 0.5\lambda + 0.25(1-\lambda)$, for any λ satisfying $0 \leqslant \lambda \leqslant 1$ (i.e. the L_1 solution is not unique). However, suppose now that the straight line $Y(t)$ has to satisfy the constraints $Y(6) = 5$ and $Y(0) \leqslant 1.2$. The constrained L_1 problem is to minimise (11.7) subject to the constraints (11.9), where $C = [1 \;\; 6]$, $\underline{d} = [5]$, $E = [1 \;\; 0]$ and $\underline{f} = [1.2]$. The unique solution is given by $x_1 = 1$ and $x_2 = 2/3$, and the minimum value for expression (11.7) is increased to 7/3.

In practice, none of the matrices A, C or E should be assumed to be full rank nor should it be assumed beforehand that vectors $\underline{x}$ satisfying the constraints (11.9) actually exist.

It is usually the case in curve-fitting problems that rank(A) = n, but the linear programming algorithms referred to in the next section also handle the case rank(A) < n.

4. Solution by simplex method

The constrained L_1 linear approximation problem of minimising (11.7) subject to (11.9) can be restated in a form suitable for solution by linear programming methods as follows:

$$\text{minimise} \quad \underline{e}(\underline{u} + \underline{v}) + M\underline{e}'(\underline{u}' + \underline{v}') + M\underline{e}''\,\underline{v}'' \tag{11.10}$$

$$\text{subject to} \quad \begin{cases} A(\underline{x}' - \underline{x}'') + \underline{u} - \underline{v} = \underline{b} \\ C(\underline{x}' - \underline{x}'') + \underline{u}' - \underline{v}' = \underline{d} \\ E(\underline{x}' - \underline{x}'') + \underline{u}'' - \underline{v}'' = \underline{f} \end{cases}$$

$$\text{and} \quad \begin{cases} \underline{x}', \underline{x}'' \geqslant \underline{0}\,, \quad \underline{u}, \underline{v} \geqslant \underline{0}\,, \quad \underline{u}', \underline{v}' \geqslant \underline{0}\,, \\ \underline{u}'', \underline{v}'' \geqslant \underline{0}\,. \end{cases}$$

Here we have put $\underline{x} = \underline{x}' - \underline{x}''$, introduced column vectors $\underline{u}$ and $\underline{v}$, and $\underline{u}''$ is a column vector of slack variables. The column vectors $\underline{u}'$, $\underline{v}'$ and $\underline{v}''$, all have artificial variables as components. Finally, $\underline{e}$, $\underline{e}'$ and $\underline{e}''$ are row vectors of appropriate dimensions with each component unity, and M is a large positive number denoting the cost of each artificial variable.

A fast and compact algorithm for solving problem (11.10) by a suitably tailored version of the simplex method is described in Barrodale and Roberts (1978), and a Fortran implementation is supplied in Barrodale and Roberts (1980).

In the absence of the constraints (11.9) this algorithm and software produce identical results to the method given in Barrodale and Roberts (1973) and the program supplied in Barrodale and Roberts (1974).

The unconstrained L_∞ linear approximation problem of minimising (11.8) can also be restated as a linear programming problem by first setting

$$\max_{1 \leqslant i \leqslant m} |b_i - \underline{A}_i \underline{x}| = w .$$

Then the problem is to

$$\begin{array}{ll} \text{minimise} & w \\ \text{subject to} & -w \leqslant b_i - \underline{A}_i \underline{x} \leqslant w, \quad \text{for } i = 1,2,\ldots,m \end{array}$$

or

$$\begin{array}{ll} \text{minimise} & w \\ \text{subject to} & \begin{bmatrix} A & \underline{e} \\ -A & \underline{e} \end{bmatrix} \begin{bmatrix} \underline{x} \\ w \end{bmatrix} \geqslant \begin{bmatrix} \underline{b} \\ -\underline{b} \end{bmatrix} \end{array}$$

where $\underline{e}$ is an m-dimensional column vector of 1's. This is an LP problem with considerably more inequality constraints than its (n+1) variables $\underline{x}$ and w, and so the following dual problem is solved instead:

$$\begin{array}{ll} \text{maximise} & \underline{b}^T(\underline{s} - \underline{t}) \\ \text{subject to} & \begin{cases} A^T(\underline{s} - \underline{t}) = \underline{0} \\ \underline{e}^T(\underline{s} + \underline{t}) \leqslant 1 \end{cases} \\ \text{and} & \underline{s}, \underline{t} \geqslant \underline{0} . \end{array} \qquad (11.11)$$

A fast and compact algorithm for solving (11.11) by a specialised form of the simplex method is described in Barrodale and Phillips (1974), and a Fortran implementation is given in Barrodale and Phillips (1975).

Roberts and Barrodale (1980) generalised this algorithm to handle the constrained L_∞ problem of minimising (11.8) subject to (11.9), and a Fortran implementation is given in Roberts and Barrodale (1978). In addition, Madsen and Powell (1975) supplied a more efficient Fortran code for the special case of minimising (11.8) subject to upper and lower bounds on the variables x_j .

If the matrix and vector elements in the equation $A\underline{x} = \underline{b}$ are complex numbers, then in the norm definitions (11.5) and (11.6) the quantity $| \,.\, |$ should be interpreted as a modulus sign. Unfortunately, this causes the corresponding complex L_1 and L_∞ problems to become nonlinear.

However, on substituting the following norms instead of the complex analogues of (11.5) and (11.6), the algorithms for solving the unconstrained L_1 and L_∞ problems can be adapted to treat the complex case. Consider a vector $\underline{v}$ with complex components $v_t = x_t + iy_t$, and let $|| \,.\, ||_\dagger$ be defined as

$$||\underline{v}||_\dagger = \sum_t (|x_t| + |y_t|) .$$

Barrodale (1978) shows how to use the NAG routine E02GAF (see section 9) to minimise $||\underline{b} - A\underline{x}||_\dagger$ rather than $||\underline{b} - A\underline{x}||_1$, and provides bounds and a numerical example which demonstrate that the norm $|| \,.\, ||_\dagger$ is a reasonable estimate of $|| \,.\, ||_1$.

Similarly, in Barrodale, Delves and Mason (1978) the norm $|| \,.\, ||_*$ defined as

$$||\underline{v}||_* = \max_t \{ \max(|x_t| , |y_t|) \}$$

is shown to be a reasonable estimate of $|| \,.\, ||_\infty$, and details of how to use the NAG routine E02GCF (see section 9) to minimise $||\underline{b} - A\underline{x}||_*$ are provided. Barrodale et al. (1978) also describe an algorithm for minimising $||\underline{b} - A\underline{x}||_\infty$.

5. L_∞ applications in signal processing

The L_∞ norm is a natural criterion to use in several different applications where the objective is to design a model to within specified tolerance limits. A very familiar example to computer users is the determination to within machine accuracy of polynomial approximations to standard mathematical functions. In this section we present two applications of L_∞ techniques in signal processing: the design of digital band-pass filters and the suppression of sidelobes in beamforming techniques.

A common problem in digital filtering is the design of zero-phase band-pass filters. The desired frequency response is defined by specifying a pass-band and stop-band, and then a time domain filter is sought which, when transformed to the frequency domain, yields a frequency response within an acceptable tolerance of these desired bands. This can be posed as the unconstrained L_∞ problem of determining filter coefficients h_0, h_1, ..., h_{n-1} to minimise

$$\max_{1 \leq k \leq m} |D(f_k) - H(f_k)| \qquad (11.12)$$

where

$D(f_k)$ = desired frequency response at frequency f_k,

$H(f_k) = h_0 + 2\sum_{j=1}^{n-1} h_j \cos(2\pi j f_k)$, frequency response of the filter,

f_k = specified frequency for $k = 1,2,\ldots,m$.

Notice that in view of the zero-phase requirement, no complex numbers occur in this problem.

In practice, it is important to keep the order n of the filter as low as possible while achieving satisfactory behaviour in its frequency response. An efficient computational scheme which gradually increases n until expression (11.12) is less than a specified value is described by Blommers (1978); it makes use of an LP-based algorithm of Roberts (1976).

The familiar equi-ripple property of the L_∞ approximation can easily be modified by choice of appropriate weights to yield different size deviations in the pass-band and stop-band. For example, it is possible to design filters with strongly enforced stop-bands, provided that larger ripples can be tolerated in the pass-bands.

A frequent requirement in beamforming (or spectral estimation) applications is the detection of weak signals in the presence of much stronger signals. In order to achieve this aim it is common practice to preprocess the data by multiplication with a set of weights (a data window). In this way, contaminating sidelobes due to leakage of power from the strong signal into other spectral regions may be suppressed, at the expense of broadening the main lobe. If the detection of a weaker signal outside the main lobe of the strong signal is the central aim, this broadening is of little concern. Thus, the problem at hand is the determination of a set of weights which minimises the maximum modulus of the sidelobes. Only for very simple array geometries is it possible to arrive at an analytic solution to this problem. However, a constrained L_∞ complex minimisation procedure could be applied to calculate an appropriate set of weights for any array configuration and beamformer.

Here we illustrate the method using the simple case of an equispaced linear array with the distance between elements equal to half the wavelength of the signal. Under these conditions the problem is to determine a set of weights $w_1, w_2, \ldots, w_n$ to minimise

$$\max_{1 \leq k \leq m} \left| \sum_{j=1}^{n} w_j \exp(i\pi j u_k) \right| \tag{11.13}$$

$$\text{subject to} \quad \sum_{j=1}^{n} w_j = 1.0 ,$$

where n = number of elements in the array,

$u_k = \sin\theta_0 - \sin\theta_k$ for $k = 1,2,\ldots,m$,

θ_0 = direction of arrival of the signal,

θ_k = look direction of the array $(-\pi/2 \leq \theta_k \leq \pi/2)$,

and the range of u_k is restricted to that region where sidelobe suppression is desired.

Whether the weights w_j are real or complex will depend on the application and capability of the beamformer technique. Computationally, the minimisation problem (11.13) can be solved approximately (to within a factor of $\sqrt{2}$) by the use of the norm $\|\cdot\|_*$ defined in section 4.

6. L_1 applications, erratic data

The best L_1 approximation by a constant to a set of measurements $y_1, y_2, \ldots, y_m$ is the median of $\{y_i\}$, which is a robust estimator. In contrast, the corresponding L_2 approximation is the mean of $\{y_i\}$, which is more sensitive to erratic data values. Generalising from this simple case, in the presence of data containing 'outliers' or 'wild points' the use of the L_1 norm for curve fitting is usually recommended over the L_2 norm. We summarise here the results of an extensive set of computational experiments designed to compare the effectiveness of the L_1 and L_2 norms in solving overdetermined systems $A\underline{x} = \underline{b}$ which contain erratic data in both A and $\underline{b}$.

Barrodale and Chow (1982) compared the use of L_1 and L_2 methods in solving model curve-fitting problems of the form $A\underline{x} = \underline{b}$, in which first $\underline{b}$, and then A and $\underline{b}$ together, had been contaminated with noise. Small levels of noise were introduced into all the data, and larger noise was added to 5% - 30% of the data (thereby simulating the occurrence of outliers).

The outcome of these numerical experiments confirmed that when noise occurs in $\underline{b}$ only, the L_1 solutions were far more accurate than the L_2 solutions, until the ratio of mean outlier noise to mean background noise fell below 10:1 . For ratios less than 5:1 the L_1 and L_2 solutions were of comparable accuracy.

When noise occurred in both A and $\underline{b}$ it was found that L_1 solutions were still far more accurate under favourable conditions (noise ratios of 100:1, no more than 10% outliers in the data), although no advantage over L_2 solutions could be observed when the outliers were less prominent compared to the background noise.

In addition, these experiments showed that examination of L_1 residuals is an effective means of identifying rows of A and $\underline{b}$ that contain outliers, and that automatic rejection of rows with large L_1 residuals is an acceptable method for trimming erratic data.

7. L_1 applications, solving LP problems

As has been explained earlier, any L_1 problem is an LP problem which can be solved by the simplex method; or better still, by a special version of the simplex method adapted to the structure of the L_1 problem. Interestingly, Bartels, Conn and Sinclair (1978) pointed out that the converse result holds true: any LP problem can be re-posed as an L_1 problem.

An obvious question is whether or not an L_1 code might offer any computational advantage over the simplex method in solving general LP problems. Steiger (1980) claimed that this is indeed the case, and we have recently confirmed that the NAG routine E02GAF (see section 9) for solving the unconstrained L_1 problem described in section 4, consistently requires fewer iterations than the NAG LP routine H01ADF (see section 9) on a variety of medium-size LP problems ranging from dense to sparse. Indeed, in some of these examples E02GAF required only about half the number of iterations required by H01ADF; the savings in computation time were proportional too.

8. L_1 applications, deconvolution of seismic data

8.1 Scope of the problem

In seismic methods of geophysical prospecting, information about the layered structure of the earth below is obtained by measuring the reflections of seismic waves produced by an impulsive source near the surface. Data is recorded by an array of receivers on the surface and is digitised and processed to yield a series of seismic traces, which contain information about the reflections.

In the convolutional model for this process, a seismic trace t is assumed to result from the convolution of a sparse 'spike train' of reflections s of length n (also known as a reflectivity series or impulse response) with a wavelet w of length k, together with additive noise r:

$$\text{i.e.} \qquad t = w * s + r \tag{11.14}$$

where t and r are of length $m = n+k-1$ and '*' denotes convolution. In matrix notation (11.14) may be written as

$$\underline{t} = W\underline{s} + \underline{r}\ , \tag{11.15}$$

where W is an $m \times n$ band matrix formed from the elements of w as follows:

$$W_{ij} = \begin{cases} w_{i-j+1} & , \quad 1 \leqslant i-j+1 \leqslant k \\ 0 & , \quad \text{otherwise.} \end{cases}$$

The effect of this convolution is to distort and blur the underlying spike train, which contains the required information. The purpose of deconvolution is to recover the spike train and wavelet from the trace.

In the absence of constraints on the form of the spike train and wavelet, the problem (11.15) is ill-posed. However, the following constraints may be imposed, to yield a more tractable problem:

(a) The wavelet may be known to a high degree of accuracy, especially in marine applications. Even if the wavelet is not known a priori, a reasonable estimate may frequently be obtained, based on a knowledge of the source characteristics and the processing history of the data.

(b) In many seismic applications, an estimate of the spike train is available for one or more selected trace(s). This is obtained from acoustic measurements made from drilled wells (well logs) at selected sites.

(c) The wavelet and spike trains are of different statistical character, the wavelet being deterministic and of short duration, and the spike train being sparse, random and of length comparable to the trace.

Thus, in many cases, the problem reduces to one of obtaining sparse spike trains from a series of traces, where the wavelet is already available or may be reliably estimated. The unique properties of the L_1 norm have given rise to a number of methods for the effective solution of this problem, for which the corresponding L_2 methods are either inappropriate or less effective. These geophysical applications of the L_1 norm are presented below.

The impetus for these applications was in large part due to Claerbout and Muir (1973), who outlined a number of potential advantages of L_1 methods over L_2 procedures in several areas of geophysical analysis, including deconvolution, filter design and earthquake epicentre location. L_1 methods should have advantages in situations where the data are erratic (e.g. presence of outliers, asymmetric or long-tailed error distribution), or where the distribution of residuals is non-Gaussian or sparse (e.g. a spike train).

8.2 The method of Taylor et al. (time domain)

A deconvolution scheme using the L_1 criterion was proposed by Taylor, Banks and McCoy (1979). The formulation suggested by them was to minimise the expression

$$\sum_{i=1}^{m} |r_i| + \lambda \sum_{j=1}^{n} |s_j| \qquad (11.16)$$

where λ is an adjustable parameter. For large values of λ (~100) the output spike train is zero, and as λ is decreased the number of spikes obtained increases. This particular formulation was chosen in analogy with the familiar L_2 case where the expression to be minimised is

$$\sum_{i=1}^{m} |r_i|^2 + \lambda \sum_{j=1}^{n} |s_j|^2 . \qquad (11.17)$$

The second term in (11.17) is a stabilising or pre-whitening condition, and the parameter λ is interpreted as the level of pre-whitening additive noise. In the case of (11.16) Taylor et al. (1979) concluded from their simulation study that a suitable value usually exists for this parameter within the range $15 < \lambda < 40$; this was determined by varying λ and comparing the trace with the result of convolving w with the extracted spike train. There is no guarantee, however, that a value of λ chosen within these bounds will be suitable for other traces, and in practice the choice of an 'optimal' λ involves considerable computation and experimentation.

8.3 One-at-a-time L_1 deconvolution (time domain)

A common assumption in geophysical modelling is that the underlying spike train is sparse. In conjunction with co-workers, we have developed a fast and compact L_1 deconvolution algorithm which yields a sparse spike train by extracting spikes one-at-a-time. This algorithm proceeds as follows. Initially, we set $\underline{s} = \underline{s}^{(0)} = \underline{0}$ and $||\underline{r}^{(0)}|| = ||\underline{t} - W\underline{s}^{(0)}|| = ||\underline{t}||$. In the first iteration all n possible spike positions are considered in determining the spike train $\underline{s} = \underline{s}^{(1)}$, with one non-zero spike, which minimises $||\underline{t} - W\underline{s}||$. We then set $||\underline{r}^{(1)}|| = ||\underline{t} - W\underline{s}^{(1)}||$ and note that $||\underline{r}^{(1)}|| \leq ||\underline{r}^{(0)}||$. In the second iteration the remaining n-1 spike positions are considered in determining the spike train $\underline{s} = \underline{s}^{(2)}$, with two non-zero spikes, which minimises $||\underline{t} - W\underline{s}||$ subject to the constraint that the spike position chosen in $\underline{s}^{(1)}$ (but not necessarily its amplitude) be preserved in $\underline{s}^{(2)}$. Setting $||\underline{r}^{(2)}|| = ||\underline{t} - W\underline{s}^{(2)}||$, it follows that $||\underline{r}^{(2)}|| \leq ||\underline{r}^{(1)}||$. Continuing in this manner, $\underline{s}^{(j+1)}$ is determined during the (j+1)th iteration by minimising $||\underline{t} - W\underline{s}||$ where $\underline{s}$ is constrained to have j+1 non-zero spikes, of which j have the same position as in $\underline{s}^{(j)}$ and the extra spike is positioned optimally from the remaining n-j choices. Iterations continue until an upper limit for the number of non-zero spikes is reached, or until the quantity $||\underline{r}^{(j)}||$ becomes sufficiently small compared to $||\underline{r}^{(0)}||$ thereby yielding a spike train whose convolution with the wavelet closely approximates the trace. See Chapman and Barrodale

(1983).

This algorithm is locally optimal, in the sense that both the new spike position and all the amplitudes of $\underline{s}^{(j+1)}$ are chosen to yield the maximum possible decrease in moving from $||\underline{r}^{(j)}||$ to $||\underline{r}^{(j+1)}||$.

In a comprehensive series of experiments with numerous model wavelets and spike trains, we have found that the L_1 norm generally yields superior results to those obtained with the corresponding one-at-a-time algorithm based on the L_2 norm, even when up to 10% RMS Gaussian noise was added to the traces. These experiments also revealed that in general the L_1 procedure performs better with smooth, band-limited wavelets of the type common in seismic data, while the L_2 procedure yields better results only with spiky, broad-band wavelets. (A band-limited wavelet possesses frequencies in the Fourier transform for which the amplitudes are zero or near-zero.)

In attempting to understand these results we applied singular value decomposition (SVD) analysis to wavelet matrices W formed from a number of model wavelets, in order to measure their numerical condition. A high condition number for W implies that substantially different spike trains may yield, when convolved with the wavelet, a close fit to the trace. We observed that the L_1 procedure gave superior results to the L_2 algorithm for all but exceptionally well-conditioned wavelets (i.e. condition number less than 50). See Barrodale, Zala and Chapman (1984).

In further work (Milinazzo, Barrodale and Zala 1984) an upper bound on the condition number of W was identified as the ratio of the maximum to the minimum value of the Fourier amplitude spectrum of the wavelet. We note that a wavelet for which W is ill-conditioned will be band-limited.

In summary, the one-at-a-time L_1 spike extraction algorithm has proven to be an effective means of deconvolving seismic data, even when the wavelet is poorly conditioned or band-limited. Further details on applications of this algorithm are given in Chapman, Barrodale and Zala (1984).

8.4 Underdetermined L_1 deconvolution (frequency domain)

Another application of L_1 methods to deconvolution of band-limited seismic data has been described by Levy and Fullagar (1981). In this method a sparse spike train is obtained by solving an underdetermined system of equations which arises when the problem is formulated in the frequency domain. An L_1 solution to an underdetermined m × n system of equations has the property that at most m of the values of s (the spike train) will be non-zero. (In contrast, every element will in general be non-zero in the corresponding L_2 solution.)

The system of equations may be expressed as follows:

$$\left.\begin{aligned} \text{Real } \{S_j\} &= 1/N \sum_{k=0}^{N-1} s_k \cos(k\omega_j) \\ \text{Imag } \{S_j\} &= -1/N \sum_{k=1}^{N-1} s_k \sin(k\omega_j) \end{aligned}\right\}, \qquad (11.18)$$

where S_j = the Fourier transform of the spike train (obtained by frequency domain division of the trace by the wavelet) at frequency ω_j ;

ω_j = $2\pi j/N$ for $j = 0,1,\ldots,N/2$;

N = the number of points in the Fourier transform ($N \geqslant m$);

s_k = the time domain spike series for which a solution is sought.

For band-limited data the system (11.18) is effectively underdetermined, as those rows corresponding to frequencies with near-zero amplitude for the Fourier transform of the trace may be disregarded and only those rows corresponding to reliable spectral data be included. In the underdetermined L_1 solution to (11.18), a sparse spike train with the required characteristics is obtained directly.

8.5 L_1 wavelet estimation

In the above section the problem of deconvolution of a trace using a known wavelet was considered. In the case of data for which a wavelet is not known, even approximately, it must be estimated. The use of L_2 methods in wavelet estimation, though capable of yielding good estimates of the amplitude characteristics of the wavelet, cannot provide information about its phase properties. In applying these L_2 methods it has been conventional practice to make restrictive assumptions (such as minimum phase) about the phase of the wavelet to be estimated. Alternative methods of wavelet estimation which do not require such assumptions are clearly desirable.

We have recently confirmed and extended an idea due to Scargle (1977), who suggested that using L_1 methods instead of L_2 methods may allow recovery of both the amplitude and the phase of the wavelet, provided the underlying spike series is sparse and the wavelet is not band-limited. The method is based on linear inverse filter theory and may be outlined as follows.

Consider a trace as a noise-free convolution of a wavelet and a spike train:

$$t = w * s. \tag{11.19}$$

For a broad-band (not band-limited) w there exists an inverse filter of w, denoted as w^{-1}, such that $w^{-1} * w = [0 \ldots 0\ 1\ 0 \ldots 0]$, i.e. a spike at some delay. Convolving both sides of (11.19) with w^{-1}, we obtain:

$$w^{-1} * t = w^{-1} * w * s$$

$$w^{-1} * t = s.$$

Thus, if we have an inverse filter w^{-1}, we may obtain the wavelet by inverting w^{-1}, and the spike train by convolving w^{-1} with the trace. This inverse filter w^{-1} is known to be equivalent (to within a scaling factor) to a prediction error filter, where a number of independent forwards and/or backwards coefficients are used to predict the trace values based on previous and future trace values. This overdetermined system of equations has the following form:

$$t_i = \sum_{j=1}^{NF} a_j t_{i-j} + \sum_{j=1}^{NB} a_{-j} t_{i+j} \tag{11.20}$$

where t_i are the trace elements for $NF < i \leqslant m-NB$; a_j are the forwards coefficients for $j = 1,2,\ldots,NF$; a_{-j} are the backwards coefficients for $j = 1,2,\ldots,NB$.

As the residuals of (11.20) correspond to the sparse spike train, it is natural to apply the L_1 norm. In a large number of numerical experiments we have found that if appropriate numbers of forwards and backwards coefficients are chosen, and the assumptions of a broad-band wavelet and a sparse spike train are valid, a good estimate of the wavelet, complete with phase information, may be obtained. This information cannot be extracted from these equations using L_2 methods.

9. Software

The NAG Library currently provides, as routine H01ABF, a version of the standard simplex method which handles '$\leqslant$' constraints. It also provides routine H01ADF; this is a revised simplex code which does not however permit the user on entry to compress the (sparse) matrix A to its non-zero elements only. Both of these routines are scheduled for replacement in the near future. The NAG Library also includes a version of the simplex method in which row interchanges are performed during the pivoting operation so as to help preserve numerical accuracy. This is routine H01BAF; however this code does not provide special treatment for LP problems where A is a sparse matrix.

In addition the NAG Library includes routines E02GAF (Barrodale and Roberts 1974) and E02GCF (Barrodale and Phillips 1975) for solving respectively the unconstrained L_1 and L_∞ problems of section 4. The constrained L_1 problem (subject to inequality constraints, see section 4) is implemented in NAG routine E02GBF (Bartels, Conn and Sinclair 1978).

CHAPTER 12

Nonlinear Optimisation : Unconstrained

P.J. Harley

1. Introduction

The development of efficient algorithms for solving unconstrained minimisation problems is an important area of research. This importance is derived not only from the desire to solve unconstrained problems but also from the use made of these algorithms in constrained optimisation; many constrained optimisation problems are solved as a sequence of unconstrained ones. Thus unconstrained minimisation lies at the heart of the whole of nonlinear optimisation.

The material in this chapter has been collected together from two main sources. The first is the book by Gill, Murray and Wright (1981) which is an excellent and detailed text covering the whole area of nonlinear optimisation. The second consists of the books by Fletcher (1980, 1981) of which the first covers unconstrained optimisation. Together with the paper by Brodlie (1977), these constitute good background reading for anyone new to this area.

2. Notation

We will consider the problem of minimising the nonlinear function $f(\underline{x})$ where $\underline{x}$ is a vector in R^n, the n-dimensional space of real numbers. We will assume that the function is sufficiently smooth for our purposes noting those conditions where problems can arise due to lack of smoothness. We define the gradient $\underline{g}(\underline{x})$ of $f(\underline{x})$ to be

$$\underline{g}(\underline{x}) = \underline{\nabla} f(\underline{x}) = \left[\frac{\partial f}{\partial x_1} \; \frac{\partial f}{\partial x_2} \; \ldots \; \frac{\partial f}{\partial x_n}\right]^T$$

and the symmetric matrix of second derivatives, the Hessian $G(\underline{x})$, to be

$$G(\underline{x}) = \left[\frac{\partial^2 f}{\partial x_i \partial x_j}\right], \quad i = 1,\ldots,n, \quad j = 1,\ldots,n .$$

To simplify notation we will often drop reference to the vector $\underline{x}$, writing instead $f = f(\underline{x})$, $f^{(k)} = f(\underline{x}^{(k)})$, $G^{(k)} = G(\underline{x}^{(k)})$, etc. .

Before we consider the conditions for a minimum to exist we must consider the scaling of the problem. Most optimisation algorithms, in order to apply convergence tests, assume that the variables are of similar magnitude and usually of order unity in the region of interest. This makes it simple to make decisions about what is 'large' or 'small'. Consider an unscaled problem where one of the physical variables takes values of order 10^4 and another takes values of order 10^{-5}; this may result in the algorithm concentrating on the variable of higher order to the exclusion of the other.

Normally only linear transformations are considered in the re-scaling of problems and that which is most commonly used is of the form

$$\underline{x} = D\underline{y}$$

where D is a constant diagonal matrix. For the problem discussed above d_{11} could be 10^{-4} and d_{22} could be 10^5. This type of transformation has the disadvantage that accuracy may be lost; also gradients and Hessians are modified, possibly altering the convergence rate of an algorithm. A detailed discussion of scaling is given in Gill et al. (1981).

We now state conditions for $f(\underline{x})$ to have a unique local minimum.

A sufficient condition for $\underline{x}^*$ to be a local minimiser of $f(\underline{x})$ is that

$$\underline{g}^* = \underline{g}(\underline{x}^*) = \underline{0}$$

and

$$\underline{s}^T G^* \underline{s} = \underline{s}^T G(\underline{x}^*) \underline{s} > 0 \qquad (12.1)$$

for all $\underline{s}$.

To make any statement about <u>global</u> minima we need to specify further conditions on $f(\underline{x})$, strict convexity for example.

Matrices G^* satisfying (12.1) are said to be <u>positive definite</u> and satisfy the following, more easily verified, conditions

(a) all the eigenvalues of G^* are strictly positive.

(b) G^* has a unique Cholesky factorisation LL^T with all $\ell_{ii} > 0$.

(c) G^* has a unique factorisation LDL^T with all $\ell_{ii} = 1$, $d_{ii} > 0$.

If G^* is indefinite (eigenvalues of both signs) then the condition

$$\underline{g}^* = \underline{0}$$

implies only that we have reached a stationary point which is a saddle point as shown in Fig. 1.

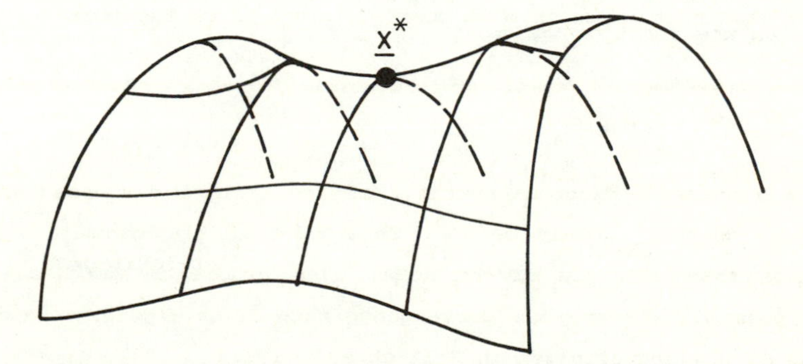

FIG.1 Saddle point

For the moment we will assume that $f(\underline{x})$ has a positive definite Hessian and outline the bones of a general method of determining its minimum.

3. The general algorithm

We assume that we have reached a point $\underline{x}^{(k)}$ at which the gradient is $\underline{g}^{(k)} = \underline{g}(\underline{x}^{(k)}) \neq \underline{0}$. The algorithm takes the form

(i) determine a <u>descent direction</u> $\underline{s}$.

A descent direction is one which satisfies the inequality

$$\underline{g}^{(k)T}\underline{s}^{(k)} < 0 ,$$

because $\underline{g}^{(k)}$ is a direction of increasing f.

(ii) determine a <u>steplength</u> α_k for which

$$f(\underline{x}^{(k)} + \alpha_k \underline{s}^{(k)}) < f(\underline{x}^{(k)})$$

(iii) set $\underline{x}^{(k+1)} = \underline{x}^{(k)} + \alpha_k \underline{s}^{(k)}$ and check for convergence.

We consider the three parts of this algorithm in reverse order; first the convergence test.

We have several choices of convergence test:

(a) on the function values

$$|f^{(k)} - f^{(k+1)}| \leqslant \epsilon_1 \qquad \text{(absolute)}$$

or

$$|f^{(k)} - f^{(k+1)}| \leqslant \epsilon_2 |f^{(k)}| , \qquad \text{(relative)}$$

(b) on the position

$$||\underline{x}^{(k)} - \underline{x}^{(k+1)}|| \leqslant \delta_1 \qquad \text{(absolute)}$$

or

$$||\underline{x}^{(k)} - \underline{x}^{(k+1)}|| \leqslant \delta_2 \, || \underline{x}^{(k)}|| , \qquad \text{(relative)}$$

(c) on the gradient

$$||\underline{g}^{(k+1)}|| \leqslant \gamma ,$$

where ϵ_i , δ_i and γ are suitably chosen small parameters. Practical convergence tests will usually contain elements of all three tests.

Conditions (a) and (b) are simple to apply although (b) does require the use of vector norms (the L_2 norm if n is small, the L_∞ norm if n is large) and it is not invariant to a re-scaling of the problem. These tests will usually work well for smooth functions where convergence is reasonably rapid.

4. The steplength α_k

The steplength α_k is chosen to ensure a 'sufficient' decrease

$$f^{(k)} - f^{(k+1)}$$

in the function value. This means that $\underline{x}^{(k+1)}$ should not be too close to the end-points of the set of $\underline{x}$ (the interval $[\underline{\beta}, \underline{x}^{(k)}]$ in Fig. 2) for which the function values are less than $f(\underline{x}^{(k)})$.

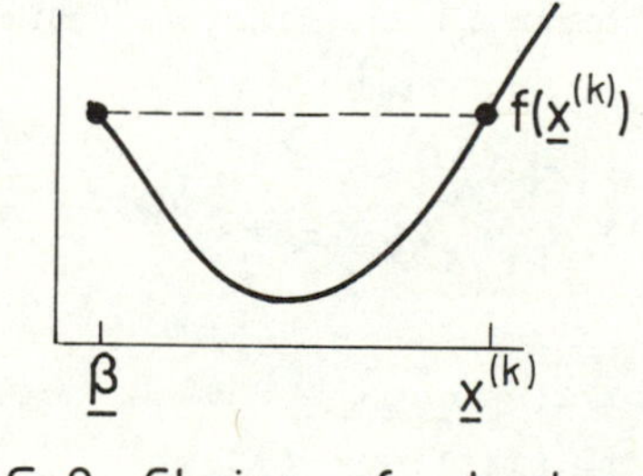

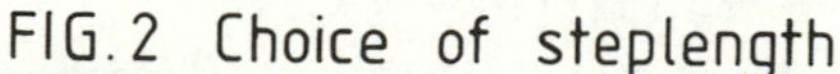
FIG. 2 Choice of steplength

The following restriction on α_k

$$0 < -\mu_1 \alpha_k \, \underline{g}^{(k)T} \underline{s}^{(k)} \leqslant f^{(k)} - f^{(k+1)} \leqslant -\mu_2 \alpha_k \, \underline{g}^{(k)T} \underline{s}^{(k)}, \quad (12.2)$$

$$0 < \mu_1 \leqslant \mu_2 < 1,$$

known as the Goldstein or Goldstein-Armijo condition, ensures that this does not happen. The upper and lower bounds in (12.2) prevent α_k from becoming 'too small' or 'too large'. Often μ_2 is chosen as $(1 - \mu_1)$, where $0 < \mu_1 < 1/2$; $\mu_1 = 0.1$ is a common choice.

This test is commonly allied to a scheme which provides trial values of α_k of the form $\epsilon^j\delta$, $j = 0,1,\ldots$, where δ is usually chosen to be unity. In general this way of determining α_k works well for methods where the initial choice δ tends to give a 'good' step so that only the first member of the set $\{\epsilon^j\delta\}$ needs to be evaluated.

An alternative choice of restriction on α_k is that it be chosen so that

$$\left|\underline{g}(\underline{x}^{(k)} + \alpha_k \underline{s}^{(k)})^T\underline{s}^{(k)}\right| \leqslant -\sigma\, \underline{g}^{(k)T}\underline{s}^{(k)}, \qquad 0 < \sigma < 1\ , \quad (12.3)$$

i.e. the gradient at $\underline{x}^{(k+1)}$ is more nearly orthogonal to $\underline{s}^{(k)}$ than is the gradient at $\underline{x}^{(k)}$. The choice of σ is important. If $\sigma = 0$ then (12.3) forces an exact line search; the minimum of f along $\underline{s}^{(k)}$ must be found. If σ is small (0.1 say) the line search is said to be <u>accurate</u>, and if σ is large (0.9 say) then a weak line search is being used. The smaller the value of σ the more work has to be performed to find α_k .

Condition (12.3) does not involve values of $f(\underline{x})$ and so cannot itself guarantee a sufficient reduction in the function value. It is usually used in conjunction with a test of the form

$$f(\underline{x}^{(k)}) - f(\underline{x}^{(k)} + \alpha_k \underline{s}^{(k)}) \geqslant -\mu\alpha_k\, \underline{g}^{(k)T}\underline{s}^{(k)} \qquad (12.4)$$

(cf. (12.2)) where $0 < \mu \leqslant 1/2$. If $\mu < \sigma$ then it can be shown that there is at least one point satisfying (12.3) and (12.4).

Current recommended practice in determining a suitable value for α_k is to use 'safeguard interpolation': a combination of a bracketing method (e.g. bisection) with a polynomial interpolation technique (usually quadratic). The bracket ensures that a suitable range of α-values is available and the interpolation gives an estimate of the 'optimal' α, that value which gives the greatest reduction in the function value. Details of these techniques can be found in Gill et al. (1981) and in Fletcher (1980).

5. The search direction

One obvious criterion which a descent direction must satisfy is that the function decreases significantly along it. This will not happen if the gradient and search directions are nearly orthogonal, i.e. if

$$\cos \theta^{(k)} = - \frac{\underline{g}^{(k)T}\underline{s}^{(k)}}{||\underline{g}^{(k)}||_2 \; ||\underline{s}^{(k)}||_2}$$

is close to zero. We can avoid this possibility by ensuring that

$$0 \leqslant \theta^{(k)} \leqslant \frac{1}{2}\pi - \rho \qquad (12.5)$$

for all k, for some $\rho > 0$.

With these restrictions on α_k and $\underline{s}^{(k)}$ we can prove that our model algorithm will give convergence to the true solution for reasonably well-behaved functions. The remainder of this chapter deals with the different ways currently used for choosing the search direction $\underline{s}^{(k)}$.

One fairly obvious choice of descent direction is

$$\underline{s}^{(k)} = - \underline{g}^{(k)},$$

the direction in which, locally, the function value decreases most rapidly; this yields the steepest descent method. This direction satisfies the descent property if $\underline{g}^{(k)} \neq \underline{0}$; we have

$$\underline{g}^{(k)T}\underline{s}^{(k)} = \underline{g}^{(k)T}(-\underline{g}^{(k)}) < 0 .$$

We can prove the theoretical convergence of the method but unfortunately it is possible for the method to generate arbitrarily small steps. Also, in practice, the method often fails due to the accumulation of round-off errors and so it is considered inefficient and unreliable.

The most successful methods base their choice of the search direction on a quadratic model of the function.

6. The quadratic model

Using n-dimensional Taylor series a smooth function can be represented by the series

$$f(\underline{x} + \underline{s}) = f(\underline{x}) + \underline{s}^T \underline{g}(\underline{x}) + \frac{1}{2}\, \underline{s}^T G(\underline{x}) \underline{s} + \ldots \qquad (12.6)$$

and its gradient by

$$\underline{\nabla} f(\underline{x} + \underline{s}) = \underline{g}(\underline{x}) + G(\underline{x}) \underline{s} + \ldots \quad . \qquad (12.7)$$

If we restrict ourselves to a sufficiently small neighbourhood of $\underline{x}$ we can neglect higher order terms and we see that a general function behaves like a quadratic. In fact we now make the approximation

$$f(\underline{x}^{(k)} + \underline{s}) \simeq q^{(k)}(\underline{s}) = f^{(k)} + \underline{s}^T \underline{g}^{(k)} + \frac{1}{2}\, \underline{s}^T G^{(k)} \underline{s} \qquad (12.8)$$

which is an approximation local to a neighbourhood of $\underline{x}^{(k)}$. It is a trivial exercise to show, provided $G^{(k)}$ is positive definite, that $q^{(k)}(\underline{s})$ has a minimum at the point $\underline{s}^{(k)}$ which is the solution of the equation

$$\underline{\nabla} q^{(k)}(\underline{s}) = \underline{g}^{(k)} + G^{(k)} \underline{s} = \underline{0} \ ,$$

that is

$$G^{(k)} \underline{s}^{(k)} = -\, \underline{g}^{(k)}. \qquad (12.9)$$

The method which chooses $\underline{s}^{(k)}$ to satisfy (12.9) is known as <u>Newton's method</u> and $\underline{s}^{(k)}$ is known as the <u>Newton direction</u>. The kth step of Newton's method can be written as

$$\left.\begin{array}{ll} \text{(i)} & \text{solve } G^{(k)} \underline{s} = -\, \underline{g}^{(k)} \quad \text{for } \underline{s}^{(k)} \ ; \\ \text{(ii)} & \text{set } \ \underline{x}^{(k+1)} = \underline{x}^{(k)} + \underline{s}^{(k)} \ . \end{array}\right\} \qquad (12.10)$$

It is immediate that Newton's method finds the minimum of a quadratic function in one step. In fact we can prove more:

If $\underline{x}^*$ is the minimiser of $f(\underline{x})$ and G^* is positive definite then, for

$\underline{x}^{(k)}$ sufficiently close to $\underline{x}^*$, Newton's method is well defined and converges with second order, i.e.

$$||\underline{s}^{(k+1)}|| \leq c\ ||\underline{s}^{(k)}||^2 \quad \text{for some constant } c.$$

Step (i) in (12.10) involves the solution of an n X n system of linear equations. We can use the LDL^T factorisation of $G^{(k)}$ to achieve this and also to check the positive definiteness condition. The basic Newton method described by (12.10) may not be useful far from the solution because it may, for example, take too large a step; as a result, the function value may even increase. This can be avoided by the use of a line search along the Newton direction

$$\underline{s}^{(k)} = -\ G^{(k)^{-1}}\ \underline{g}^{(k)},$$

which is a descent direction because

$$\underline{g}^{(k)T}\underline{s}^{(k)} = -\ \underline{g}^{(k)T}G^{(k)^{-1}}\ \underline{g}^{(k)} < 0$$

when $G^{(k)}$ is positive definite. We note in passing that Newton's method with a line search will converge with second order only if the steplengths α_k converge to the 'natural' value of unity, i.e. the method enters a region in which the quadratic model adequately fits the function.

This second-order convergence makes Newton's method extremely attractive whenever the derivative information ($\underline{g}$ and G) is available. Unfortunately problems do arise if $G^{(k)}$ is <u>not</u> positive definite; the function may not have a finite minimum or even a stationary point. It is also possible that any stationary point may be a saddle point.

There has been considerable research into techniques for resolving this problem. The resulting modifications fall into three general classes:

(a) techniques which operate directly on the eigensystem of the Hessian, $G^{(k)}$;

(b) techniques which modify the Hessian, $G^{(k)}$, by using a diagonal matrix, $D^{(k)}$;

(c) techniques which attempt to identify descent directions of negative curvature, i.e. where

$$\underline{s}^{(k)T} G^{(k)} \underline{s}^{(k)} < 0 \quad \text{and} \quad \underline{s}^{(k)T} \underline{g}^{(k)} < 0 .$$

We consider methods of type (a) first. They depend on the decomposition

$$G^{(k)} = P^T \Lambda P$$

where P is orthogonal and Λ is the diagonal matrix of the eigenvalues of $G^{(k)}$. The basic modification is to produce a new matrix $\bar{G}^{(k)}$ given by

$$\bar{G}^{(k)} = P^T \bar{\Lambda} P$$

where $\bar{\Lambda}$ is a diagonal matrix with $\bar{\lambda}_i$ equal to λ_i if λ_i is sufficiently positive; the other elements can be chosen in various ways outlined in Gill et al. (1981). However, as the calculation of the eigensystem of $G^{(k)}$ involves $O(n^3)$ arithmetic operations, these techniques have been superseded by more efficient methods.

Included in these are the methods of type (b) where a diagonal matrix $D^{(k)}$ is added to $G^{(k)}$ in order to make their sum positive definite. As a result the direction

$$\underline{s}^{(k)} = -(G^{(k)} + D^{(k)})^{-1} \underline{g}^{(k)}$$

is a descent direction. The difference between the various methods depends on the differing choices of $D^{(k)}$.

Perhaps the simplest, due to Levenberg (1944) and Marquardt (1963), is to choose $D^{(k)}$ as a multiple of the identity matrix so that

$$\underline{s}^{(k)} = -(G^{(k)} + \nu^{(k)} I)^{-1} \underline{g}^{(k)} , \qquad \nu^{(k)} > 0 .$$

We will not discuss this modification here; it plays an important part in least-squares techniques for the solution of nonlinear equations and is described in Chapter 14.

More generally the modification may be chosen whilst the matrix $G^{(k)}$ is being factorised into the product LDL^T. It might seem reasonable to proceed as follows: form the factorisation LDL^T of $G^{(k)}$ and then form the modified matrix $\bar{G}^{(k)}$ as $L\bar{D}L^T$ where $\bar{d}_{ii} = \max(d_{ii} , \delta)$ for some $\delta > 0$. However this approach has two major defects: the factorisation may not exist, and if it does exist its computation is usually a numerically unstable process. It is also possible that $\bar{G}^{(k)}$ can differ from $G^{(k)}$ by an arbitrarily large amount when $G^{(k)}$ is 'slightly' indefinite.

Gill and Murray (1974) developed a modified form of the Cholesky factorisation LDL^T. In this case the factors are modified in a numerically stable way to increase the diagonal elements, if necessary, as the factorisation proceeds. The end result is a matrix

$$\bar{G}^{(k)} = G^{(k)} + D^{(k)}$$

where $D^{(k)}$ is a nonnegative diagonal matrix; thus $G^{(k)}$ is modified only in its diagonal elements. This factorisation is discussed in some detail in Gill et al. (1981).

The final class of methods, (c), computes descent directions with negative curvature whenever $\underline{x}^{(k)}$ is close to a saddle point. These methods also depend on the LDL^T factorisation of $G^{(k)}$ and Fletcher and Freeman (1977), for example, have described a technique for computing such directions stably. The implementation of these methods requires some care, however, and further research is needed in this area.

7. Quasi-Newton methods

The major disadvantage of Newton's method, even when modified to ensure convergence, is that the user must supply second-derivative information to form the Hessian $G^{(k)}$. This is often an extremely 'expensive' process and so it is a disincentive to the method's use. However, methods with many of the features of Newton's method can be derived when only first-derivative information is available.

The most obvious example involves the use of finite differences to estimate the elements of $G^{(k)}$. For example, if h_i is an increment in the direction of the unit vector $\underline{e}_i$ then the matrix $\bar{G}$ with ith column given by the forward difference

$$\frac{\underline{g}(\underline{x}^{(k)} + h_i\underline{e}_i) - \underline{g}(\underline{x}^{(k)})}{h_i}$$

is evaluated and, to preserve symmetry, $G^{(k)}$ is approximated by

$$G^{(k)} \simeq \frac{1}{2}(\bar{G} + \bar{G}^T) .$$

There are disadvantages associated with this discrete Newton method; the matrix approximation to $G^{(k)}$ may not be positive definite (necessitating the use of modification techniques), n gradient evaluations are required in each iteration to estimate $\bar{G}$ and an $n \times n$ system of linear equations has to be solved at each iteration. There is one further disadvantage associated with the use of finite differences and that is the problem of choosing the increment h_i. To obtain convergence we want h_i to tend to zero but this introduces problems of numerical instability which are outlined in Gill et al. (1981).

Despite these difficulties the discrete Newton method can be useful especially for large sparse problems where advantage can be taken of the structure to reduce the amount of differencing necessary. Of more general applicability, however, is the class of quasi-Newton methods.

These methods are very similar to Newton's method with a line search except that the inverse Hessian ${G^{(k)}}^{-1}$ is approximated by a symmetric positive definite matrix $H^{(k)}$ which is corrected or updated after each iteration. Quasi-Newton methods have the general structure

(i) define the search direction as

$$\underline{s}^{(k)} = -H^{(k)}\underline{g}^{(k)}$$

rather than

$$\underline{s}^{(k)} = -{G^{(k)}}^{-1}\underline{g}^{(k)} ;$$

(ii) perform a line search giving

$$\underline{x}^{(k+1)} = \underline{x}^{(k)} + \alpha_k \underline{s}^{(k)} ;$$

(iii) update $H^{(k)}$ to give $H^{(k+1)}$.

The starting matrix $H^{(1)}$ is usually taken to be the identity matrix, I, unless better information is available. This means, of course, that

$$\underline{s}^{(1)} = -\underline{g}^{(1)}$$

and so the method's initial direction is that of steepest descent.

Possible advantages of the method (against Newton's method) are the following

(a) only first derivatives are required (instead of second);

(b) $H^{(k)}$ positive definite implies search directions $\underline{s}^{(k)}$ are descent directions ($G^{(k)}$ may be indefinite);

(c) $O(n^2)$ multiplications per step ($O(n^3)$ for Newton's method).

(Not all quasi-Newton methods have property (b); those that do are often called <u>variable metric</u> methods.)

Externally, the main difference from Newton's method is in step (iii) and we concentrate initially on this step. The updating is an attempt to add to $H^{(k)}$ second-derivative information obtained at iteration k and, clearly, it is desirable if the process updates $H^{(k)}$ to yield $H^{(k+1)}$ as a better approximation to ${G^{(k)}}^{-1}$. To ensure this we impose a condition on successive estimates $H^{(k+1)}$. If we define

$$\underline{\delta}^{(k)} = \alpha_k \underline{s}^{(k)} = \underline{x}^{(k+1)} - \underline{x}^{(k)} \qquad (12.11)$$

and

$$\underline{\gamma}^{(k)} = \underline{g}^{(k+1)} - \underline{g}^{(k)}, \qquad (12.12)$$

then from (12.7)

$$\underline{g}^{(k+1)} = \underline{g}^{(k)} + G^{(k)}\alpha_k \underline{s}^{(k)} + \ldots$$

and we have

$$\underline{\gamma}^{(k)} = G^{(k)}\underline{\delta}^{(k)} + \ldots \quad .$$

The higher order terms are zero for a quadratic function and, given that our methods are based on a quadratic model, we neglect them, giving

$$\underline{\gamma}^{(k)} = G^{(k)}\underline{\delta}^{(k)} \quad . \tag{12.13}$$

Now, from (12.11) and (12.12), we can see that $\underline{\gamma}^{(k)}$ and $\underline{\delta}^{(k)}$ can only be calculated after we know $\underline{x}^{(k+1)}$ and so the existing estimate $H^{(k)}$ cannot, a priori, relate directly to them and satisfy (12.13), i.e.

$$H^{(k)}\underline{\gamma}^{(k)} \neq \underline{\delta}^{(k)} \quad .$$

However, we can force this behaviour on the <u>new</u> estimate of the Hessian so that

$$H^{(k+1)}\underline{\gamma}^{(k)} = \underline{\delta}^{(k)} \quad , \tag{12.14}$$

which is sometimes called the <u>quasi-Newton condition</u>.

Now we consider the method whereby we update the approximation $H^{(k)}$. The simplest method is additive :

$$H^{(k+1)} = H^{(k)} + E^{(k)}$$

and possibly the simplest correction is of the form

$$E^{(k)} = \beta\, \underline{u}\underline{u}^T$$

where $E^{(k)}$ is a <u>symmetric rank-one matrix</u>. Thus (12.14) takes the form

$$H^{(k+1)}\underline{\gamma}^{(k)} = H^{(k)}\underline{\gamma}^{(k)} + \beta\, \underline{u}\underline{u}^T\underline{\gamma}^{(k)} = \underline{\delta}^{(k)} \tag{12.15}$$

and so, because $\underline{u}^T\underline{\gamma}^{(k)}$ is a scalar, $\underline{u}$ is proportional to

$$\underline{\delta}^{(k)} - H^{(k)}\underline{\gamma}^{(k)} \quad .$$

Thus we set

$$\underline{u} = \underline{\delta}^{(k)} - H^{(k)}\underline{\gamma}^{(k)}$$

and impose the condition $\beta\underline{u}^T\underline{\gamma}^{(k)} = 1$ in order to satisfy (12.15). This gives the <u>symmetric rank-one formula</u>

$$H^{(k+1)} = H + \frac{(\underline{\delta} - H\underline{\gamma})(\underline{\delta} - H\underline{\gamma})^T}{(\underline{\delta} - H\underline{\gamma})^T\underline{\gamma}} \tag{12.16}$$

where the superscript (k) has been omitted from the right-hand side. This method has an important property known as <u>quadratic termination</u> (see Fletcher (1980)):

> If the rank-one method is well defined and if $\underline{\delta}^{(1)}, \ldots, \underline{\delta}^{(n)}$ are independent then the method terminates for a quadratic function in at most (n+1) line searches with $H^{(n+1)} \equiv G^{-1}$.

The proof of this result does not require the use of <u>exact</u> line searches, a very important feature.

Unfortunately, this method suffers from two major disadvantages: even when applied to a quadratic function it does not preserve the positive definiteness of $H^{(k)}$ and, in some cases, the denominator of (12.16) may become zero and so $H^{(k+1)}$ is not defined. Numerical difficulties also arise if the denominator becomes small.

More flexibility is introduced by allowing $E^{(k)}$ to become a matrix of rank two, giving the formula

$$H^{(k+1)} = H^{(k)} + a\,\underline{u}\underline{u}^T + b\,\underline{v}\underline{v}^T .$$

Again the quasi-Newton condition (12.14) must be satisfied giving

$$\underline{\delta}^{(k)} = H^{(k)}\underline{\gamma}^{(k)} + a\,\underline{u}\underline{u}^T\underline{\gamma}^{(k)} + b\,\underline{v}\underline{v}^T\underline{\gamma}^{(k)} . \tag{12.17}$$

$\underline{u}$ and $\underline{v}$ are no longer uniquely determined but an obvious choice, from the form of (12.17), is

$$\underline{u} = \underline{\delta}^{(k)} \quad \text{and} \quad \underline{v} = H^{(k)}\underline{\gamma}^{(k)} .$$

Then, finally, we require $a\,\underline{u}^T\underline{\gamma}^{(k)} = 1$ and $b\,\underline{v}^T\underline{\gamma}^{(k)} = -1$ to satisfy (12.17), thus determining a and b and resulting in the update formula

$$H_{DFP}^{(k+1)} = H + \frac{\underline{\delta}\underline{\delta}^T}{\underline{\delta}^T\underline{\gamma}} - \frac{H\underline{\gamma}\underline{\gamma}^T H}{\underline{\gamma}^T H\underline{\gamma}} \quad . \tag{12.18}$$

This is the well-known DFP method, based on work by Davidon (1959) and Fletcher and Powell (1963), which has been found to work well in practice. Initially the method was used with accurate line searches and performed well. However the current practice of using weak line searches exposes certain deficiencies in the method. The DFP method does have a number of important theoretical properties:

(a) for quadratic functions using exact line searches:

(i) it terminates in at most n iterations with $H^{(n+1)} = G^{-1}$ (quadratic termination) ;

(ii) previous quasi-Newton conditions are preserved in the sense that $H^{(k)}\underline{\gamma}^{(j)} = \underline{\delta}^{(j)}$ for $j = 1,2,\ldots,k-1$;

(iii) it generates so-called conjugate directions when $H^{(1)} = I$.

(b) for general functions

(i) it preserves positive definite $H^{(k)}$;

(ii) it is economical, requiring approximately $3n^2 + O(n)$ multiplications per iteration;

(iii) it yields a superlinear rate of convergence

$$||\underline{x}^{(k+1)} - \underline{x}^*|| \;/\; ||\underline{x}^{(k)} - \underline{x}^*|| \to 0 \; .$$

(c) for strictly convex functions, using exact line searches, it yields global convergence.

The positive definiteness condition is obviously important as it guarantees that the search directions are descent directions. This result depends on the condition

$$\underline{\delta}^{(k)T}\underline{\gamma}^{(k)} > 0$$

holding for all k and this is guaranteed provided the restriction of inequality (12.3),

$$\left|\underline{g}^{(k+1)T}\underline{s}^{(k)}\right| \leqslant -\sigma\, \underline{g}^{(k)T}\underline{s}^{(k)} ,$$

holds on the line search.

Another important formula derived from a rank-two update is the so-called BFGS method, derived independently in 1970 by Broyden, Fletcher, Goldfarb and Shanno (Fletcher 1970) in which

$$H_{BFGS}^{(k+1)} = H + \left[1 + \frac{\underline{\gamma}^T H\underline{\gamma}}{\underline{\delta}^T\underline{\gamma}}\right]\frac{\underline{\delta}\underline{\delta}^T}{\underline{\delta}^T\underline{\gamma}} - \frac{\underline{\delta}\underline{\gamma}^T H + H\underline{\gamma}\underline{\delta}^T}{\underline{\delta}^T\underline{\gamma}} . \qquad (12.19)$$

The BFGS formula has been found to work very well in practice, and in many instances better then the DFP formula. All the properties (a) - (c) of the DFP formula have been proved to hold for the BFGS formula with the addition that inexact line searches, satisfying conditions similar to (12.2) and (12.3), can be used and global convergence is still obtained.

There are many more such rank-two formulae; for example, there is a one-parameter family of methods, known as the Broyden family, which is given by the formula

$$H_{\varphi}^{(k+1)} = (1 - \varphi)\, H_{DFP}^{(k+1)} + \varphi\, H_{BFGS}^{(k+1)} . \qquad (12.20)$$

This important family shares many of the properties of the DFP and BFGS methods. There is a remarkable theoretical result due to Dixon (1972a, b) which specifies that, for a general function, if exact line searches are used then all the methods of the Broyden family generate identical iterates $\underline{x}^{(k)}$. This means that with exact line searches each method should take the same number of iterations to obtain a solution, although the number of function and derivative values used in the linear searches may vary from method to method. This is because each method generates identical directions, $\underline{s}_{\varphi}^{(k)}$, but the lengths of the vectors are different for varying values of φ.

It is important to note that there exist values of φ for which difficulties such as singular $H_{\varphi}^{(k+1)}$ can arise. These difficulties are avoided if, as Fletcher (1970) suggests, φ is restricted to lie in the interval [0,1] providing the formulae with some desirable stability properties.

In a practical situation it is virtually impossible to use an exact line search and so behaviour for inexact line searches is important. In this case the BFGS method ($\varphi = 1$) seems to be preferred.

We mention two further points concerning quasi-Newton methods. The first concerns the situation in which analytic gradients $\underline{g}^{(k)}$ are not available. Much work has been done in this area to develop stable algorithms; our comments earlier on finite differences are relevant here and there is a discussion of this in Gill et al. (1981).

The second point concerns the implementation and numerical stability of the methods. We have presented quasi-Newton methods from the point of view of determining the search direction using the formula

$$\underline{s}^{(k)} = - H^{(k)}\underline{g}^{(k)} \tag{12.21}$$

where $H^{(k)}$ is an approximation to the inverse Hessian $G^{(k)^{-1}}$. We did this because only matrix-vector multiplication ($O(n^2)$ multiplications) is involved in determining $\underline{s}^{(k)}$ rather than the $O(n^3)$ multiplications in solving a system of linear equations. However, Gill and Murray (1972) have pointed out that if $B^{(k)}$ is an approximation to $G^{(k)}$ and if the factors $L_kD_kL_k^T$ of $B^{(k)}$ are available then the system of equations

$$B^{(k)}\underline{s}^{(k)} = L_kD_kL_k^T\,\underline{s}^{(k)} = - \underline{g}^{(k)}$$

can be solved in $O(n^2)$ multiplications. It is also possible to update L_k, D_k to produce new factors L_{k+1}, D_{k+1} in $O(n^2)$ multiplications. There are two advantages claimed for this approach: it is easier to identify and correct an indefinite matrix by altering D_k, and also it is said that round-off errors are controlled better than if the approximation to the inverse Hessian is used. Discussion of these points can be found in Gill et al. (1981).

8. Conjugate direction methods

Given a positive definite matrix G and a set of vectors $\{\underline{s}^{(1)}, \underline{s}^{(2)}, \ldots, \underline{s}^{(n)}\}$ then the vectors are said to be <u>conjugate</u> if

$$\underline{s}^{(i)T} G \, \underline{s}^{(j)} = 0, \quad \text{for all } i \neq j \,. \tag{12.22}$$

A <u>conjugate direction method</u> is one which generates conjugate directions when applied to a quadratic function with Hessian G. Members of the Broyden family have this property provided exact line searches are used. Conjugate directions are particularly important because of the following result:

A conjugate direction method terminates for a quadratic function in at most n exact line searches, and each $\underline{x}^{(i+1)}$ is the minimiser in the subspace generated by $\underline{x}^{(1)}$ and the directions $\underline{s}^{(1)}, \underline{s}^{(2)}, \ldots, \underline{s}^{(i)}$, i.e. in the set of points

$$\{\underline{x} \mid \underline{x} = \underline{x}^{(1)} + \sum_{j=1}^{i} \alpha_j \, \underline{s}^{(j)}, \text{ for all } \alpha_j\} \,.$$

We can think of a conjugate direction method as a transformation which changes the problem into one in which a single search along each of the coordinate directions is sufficient to obtain the minimum.

The <u>conjugate gradient method</u> attempts to ally conjugacy properties with the method of steepest descent in order to produce efficient and reliable methods. The method is described as it could be applied to a quadratic function; it is assumed that exact line searches are used throughout.

We start with the steepest descent direction

$$\underline{s}^{(1)} = -\,\underline{g}^{(1)}$$

and for $k \geq 1$ we choose

$$\underline{s}^{(k+1)} = \text{component of } -\underline{g}^{(k+1)} \text{ conjugate to } \underline{s}^{(1)}, \underline{s}^{(2)}, \ldots, \underline{s}^{(k)}.$$

Because the function is quadratic we can rewrite the conjugacy condition (12.22) as

$$\underline{s}^{(i)T} G(\underline{x}^{(j+1)} - \underline{x}^{(j)}) = \underline{s}^{(i)T}(\underline{g}^{(j+1)} - \underline{g}^{(j)}) = \underline{s}^{(i)T} \underline{\gamma}^{(j)} = 0 \,, \quad \text{for all } i \neq j \,,$$

and so $\underline{s}^{(i)}$ and $\underline{\gamma}^{(j)}$ are orthogonal for all $i \neq j$. Thus the Gram–Schmidt method can be used to determine $\underline{s}^{(k+1)}$, giving

$$\underline{s}^{(k+1)} = -\underline{g}^{(k+1)} + \sum_j \beta_j \underline{s}^{(j)} .$$

In fact, for quadratic functions, $\beta_j = 0$ for all $j < k$ and so

$$\underline{s}^{(k+1)} = -\underline{g}^{(k+1)} + \beta_k \underline{s}^{(k)} \tag{12.23}$$

for $k \geqslant 1$, where

$$\beta_k = \frac{\underline{g}^{(k+1)T}\underline{g}^{(k+1)}}{\underline{g}^{(k)T}\underline{g}^{(k)}} . \tag{12.24}$$

This is the basis of the well-known conjugate gradient method of Fletcher and Reeves (1964) and has one obvious advantage over quasi-Newton methods; no matrix operations are required.

This method does have a disadvantage; for general functions the direction $\underline{s}^{(k+1)}$ derived from (12.23) and (12.24) may not be a descent direction. This can be overcome by choosing

$$\beta_k = -\frac{\underline{g}^{(k+1)T}\underline{g}^{(k+1)}}{\underline{g}^{(k)T}\underline{s}^{(k)}} .$$

An alternative choice is

$$\beta_k = \frac{(\underline{g}^{(k+1)} - \underline{g}^{(k)})^T\underline{g}^{(k+1)}}{\underline{g}^{(k)T}\underline{g}^{(k)}} \tag{12.25}$$

which is due to Polak and Ribiere (in Polak (1971)). Practical experience suggests that (12.25) may perform better than the Fletcher–Reeves choice (12.24); however, it requires four n-vectors of storage compared with three n-vectors for the Fletcher–Reeves method. Neither method is as efficient or robust as the quasi-Newton methods and so they should not be used in normal circumstances. However their simplicity, in that no matrix operations are required, does give them an advantage when very large problems, those with hundreds or thousands of variables, have to be solved.

9. Software

Programs implementing many of the methods that we have discussed are available in the following libraries : HARWELL, MINPACK, NAG (see Appendix). The NAG routines are in general based on those written for the National Physical Laboratory by P.E. Gill and W. Murray. In particular, the NAG Library includes routines EO4DFF and EO4EBF, which are based on modified-Newton methods, EO4CGF and EO4DEF based on quasi-Newton methods, and EO4DBF which uses the conjugate gradient method.

CHAPTER 13

NONLINEAR OPTIMISATION: CONSTRAINED

T.L. Freeman

1. Introduction

The past decade has seen real advances in software for constrained optimisation so that one can now attempt to solve nonlinearly constrained problems numerically, with confidence of obtaining an optimum. The decade has also seen the publication of very good textbooks which deal with constrained optimisation; the two fundamental references for this chapter are Fletcher (1981) and Gill, Murray and Wright (1981).

We consider the problem:

$$\min_{\underline{x} \in R^n} \quad f(\underline{x}) \tag{13.1}$$

$$\text{subject to} \quad c_i(\underline{x}) = 0, \quad i = 1,2,\ldots,t ,$$

$$\text{and} \quad c_i(\underline{x}) \geqslant 0, \quad i = t+1,\ldots,m .$$

Throughout we address the problem of finding a local, rather than a global, solution to (13.1). The characteristics of the problem and the most suitable method for its solution depend on the form of the objective function $f(\underline{x})$ and the constraint functions $c_i(\underline{x})$, $i = 1,2,\ldots,m$. For example when $f(\underline{x})$ is quadratic and $c_i(\underline{x})$, $i = 1,2,\ldots,m$, are linear then (13.1) is a <u>quadratic programming problem</u>. Special techniques should be used for the solution of this problem; see for example, Fletcher (1981, Ch. 10), and Gill, Murray and Wright (1981, section 5.3.2).

When the constraint functions $c_i(\underline{x})$ are linear, of the form

$$c_i(\underline{x}) = \underline{a}_i^T \underline{x} - b_i , \quad \underline{a}_i \in R^n, \quad b_i \in R , \tag{13.2}$$

and $f(\underline{x})$ is a general nonlinear function, then (13.1) is a linearly constrained problem. If each vector $\underline{a}_i$ is a unit vector $\underline{e}_j$, and all the constraints are inequalities, then the ith constraint is simply a bound on the jth variable and (13.1) is a bounds constrained problem. Techniques for the solution of these problems are described in section 3. These techniques are essentially techniques for unconstrained problems restricted to satisfy the linear constraints.

When the functions $f(\underline{x})$, $c_i(\underline{x})$, $i = 1,2,\ldots,m$, are both general nonlinear functions then (13.1) is the much more difficult nonlinearly constrained problem. In section 4 we describe two techniques for solving this problem. The first, the augmented Lagrangian function method, is a penalty function technique. The second involves the solution of a sequence of quadratic programming problems.

An understanding of these methods is not possible without considering first the conditions which define a solution to (13.1). These so-called optimality conditions are the subject of the next section.

Before proceeding we introduce the standard notation:-

$\underline{g}(\underline{x}) = \underline{\nabla}f(\underline{x})$, gradient vector of $f(\underline{x})$,

$G(\underline{x}) = \left[\dfrac{\partial^2 f}{\partial x_i \partial x_j} \right]$, Hessian matrix of $f(\underline{x})$,

$\underline{a}_i(\underline{x}) = \underline{\nabla}c_i(\underline{x})$, ith constraint normal.

2. Optimality conditions

We first consider the conditions which characterise a solution to the simple linear equality constrained problem:

$$\min_{\underline{x} \in R^n} f(\underline{x}) \tag{13.3}$$

$$\text{subject to} \quad A^T\underline{x} - \underline{b} = \underline{0} .$$

We assume that the n × m matrix $A = [\underline{a}_1\ \underline{a}_2\ \ldots\ \underline{a}_m]$, $m < n$, has full rank m. Note that $c_i(\underline{x}) = \underline{a}_i^T\underline{x} - b_i$. Necessary conditions for $\underline{x}^*$ to be a local minimiser of (13.3) are:

$$\text{(i)} \quad A^T\underline{x}^* = \underline{b} \,,$$

$$\text{(ii)} \quad \underline{g}(\underline{x}^*) = A\underline{\lambda}^* = \sum_{i=1}^{m} \lambda_i^* \underline{a}_i \,, \qquad (13.4)$$

$$\text{(iii)} \quad \underline{s}^T G(\underline{x}^*)\underline{s} \geqslant 0, \quad \text{for all } \underline{s} : A^T\underline{s} = \underline{0} \,.$$

The parameters λ_i^*, $i = 1,2,\ldots,m$, are the Lagrange multipliers. Note that $\underline{a}_i$ is a vector orthogonal to the hyperplane defined by the linear constraint $\underline{a}_i^T \underline{x} - b_i = 0$ and it is therefore referred to as a constraint normal. Hence (ii) shows that $\underline{x}^*$ is a local minimiser of (13.3) only if the gradient vector $\underline{g}(\underline{x}^*)$ of the objective function can be written as a linear combination of the constraint normals (see Fig. 1).

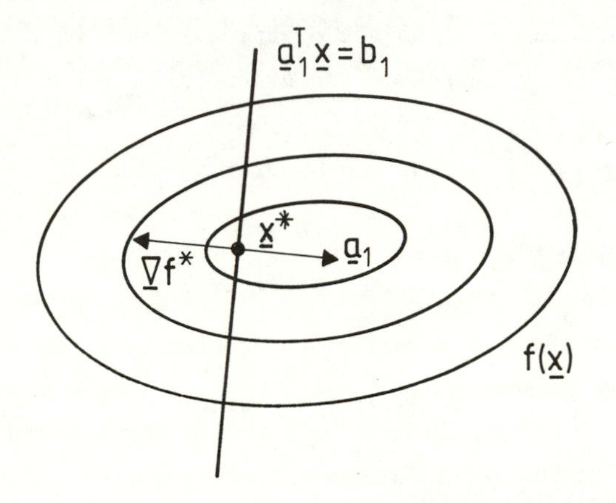

FIG.1 Equality constraints

Note also that the vector $\underline{s}$ in (iii) lies in the constraint space and hence that $\underline{x}^*$ is a local minimiser of (13.3) only if the projection of the Hessian matrix $G(\underline{x}^*)$ in any direction in the constraint space is nonnegative.

If we define any matrix Z whose columns span the null space of A,

$$A^T Z = 0 \,,$$

then conditions (ii) and (iii) of (13.4) can be rewritten as (ii)' $Z^T \underline{g}(\underline{x}^*) = \underline{0}$,

(iii)' $Z^T G(\underline{x}^*) Z$ is positive semi-definite.

Thus another view of these two conditions is that the projected gradient (projected onto the constraint space) is zero and the projected Hessian matrix is positive semi-definite.

Sufficient conditions for $\underline{x}^*$ to be a local minimiser of (13.3) are:

$$\begin{aligned} &(i) \quad A^T \underline{x}^* = \underline{b} , \\ &(ii) \quad \underline{g}(\underline{x}^*) = A\underline{\lambda}^* , \\ &(iii) \quad \underline{s}^T G(\underline{x}^*) \underline{s} > 0 , \text{ for all } \underline{s} : A^T \underline{s} = \underline{0} . \end{aligned} \qquad (13.5)$$

The optimality conditions are somewhat more complicated if the problem under consideration is the linear inequality constrained problem:

$$\min_{\underline{x} \in R^n} \quad f(\underline{x}) \qquad (13.6)$$

$$\text{subject to} \quad A^T \underline{x} - \underline{b} \geq \underline{0} .$$

We again assume that the n × m matrix A has full rank. We say that the ith constraint is <u>active</u> at some point $\hat{\underline{x}} \in R^n$ if $\underline{a}_i^T \hat{\underline{x}} - b_i = 0$. $\hat{M} \subset \{1,2,\ldots,m\}$ is used to denote the set of (indices of) active constraints at $\hat{\underline{x}}$.

Necessary conditions for $\underline{x}^*$ to be a local minimiser of (13.6) are:

$$\begin{aligned} &(i) \quad A^T \underline{x}^* \geq \underline{b} , \\ &(ii) \quad \underline{g}(\underline{x}^*) = \sum_{i \in M^*} \lambda_i^* \underline{a}_i , \text{ where } \lambda_i^* \geq 0 , \\ &(iii) \quad \underline{s}^T G(\underline{x}^*) \underline{s} \geq 0, \text{ for all } \underline{s} : \underline{a}_i^T \underline{s} = 0, \ i \in M^* . \end{aligned} \qquad (13.7)$$

Condition (ii) shows that $\underline{x}^*$ is a local minimiser of (13.6) only if the gradient vector $\underline{g}(\underline{x}^*)$ is a positive linear combination of the active constraint normals (see Fig. 2).

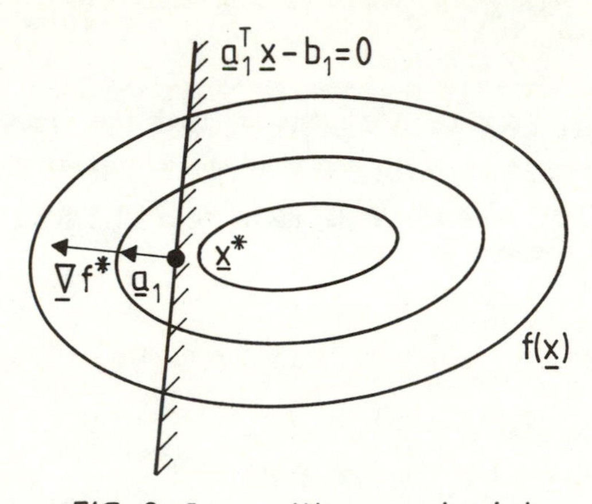

FIG. 2 Inequality constraints

Condition (iii) requires that the component of the Hessian matrix $G(\underline{x}^*)$ along any direction in the space of active constraints is nonnegative.

Sufficient conditions for $\underline{x}^*$ to be a local minimiser of (13.6) are:

$$\text{(i)} \quad A^T\underline{x}^* \geq \underline{b} \, ,$$

$$\text{(ii)} \quad \underline{g}(\underline{x}^*) = \sum_{i \in M^*} \lambda_i^* \underline{a}_i \, , \text{ where } \lambda_i^* > 0 \, , \tag{13.8}$$

$$\text{(iii)} \quad \underline{s}^T G(\underline{x}^*)\underline{s} > 0 \, , \text{ for all } \underline{s} \, : \, \underline{a}_i^T \underline{s} = 0 \, , \, i \in M^* .$$

Notice that we have now restricted the Lagrange multipliers to be positive. Zero multipliers add some complications to the sufficient conditions; see Fletcher (1981, p. 60).

Note that for the inequality constrained problem (13.6), it is the objective function and the active constraints which together define a local minimum.

The Lagrange multipliers, λ_i^* , provide useful information about the sensitivity of the solution $\underline{x}^*$ with respect to perturbations in the constraints. Consider for example a problem with a single linear equality constraint:

$$\min_{\underline{x} \in R^n} \quad f(\underline{x})$$

$$\text{subject to} \quad \underline{a}^T\underline{x} = b \ .$$

At the minimiser $\underline{x}^*$, $\underline{\nabla}f(\underline{x}^*) = \lambda^*\underline{a}$. Suppose that the right-hand side of the constraint is changed to $(b + \delta b)$ and that the minimiser changes accordingly to $(\underline{x}^* + \delta\underline{x}^*)$, so that $\underline{a}^T(\underline{x}^* + \delta\underline{x}^*) = b + \delta b$. Then the minimum value of the objective function becomes

$$\begin{aligned} f(\underline{x}^* + \delta\underline{x}^*) &= f(\underline{x}^*) + \delta\underline{x}^{*T}\underline{\nabla}f(\underline{x}^*) + O(\|\delta\underline{x}^*\|^2) \ , \\ &= f(\underline{x}^*) + \lambda^*\delta\underline{x}^{*T}\underline{a} + O(\|\delta\underline{x}^*\|^2) \ , \\ &= f(\underline{x}^*) + \lambda^*\delta b + O(\|\delta\underline{x}^*\|^2) \ . \end{aligned}$$

Hence, to first order, a change δb in the constraint induces a change $\lambda^*\delta b$ in the objective function. Hence if $\lambda_1 >> \lambda_2$ we could conclude that changes in the constraint $\underline{a}_1^T \underline{x} = b_1$ would have a much larger effect on $f(\underline{x}^*)$ than changes in the constraint $\underline{a}_2^T \underline{x} = b_2$. Clearly this conclusion is only valid if the constraints are suitably scaled relative to each other, since the Lagrange multipliers are scale-dependent.

We now consider the optimality conditions for the nonlinear equality constrained optimisation problem:

$$\min_{\underline{x} \in R^n} \quad f(\underline{x}) \tag{13.9}$$

$$\text{subject to} \quad c_i(\underline{x}) = 0, \quad i = 1,2,\ldots,t \ .$$

We define the Lagrangian function

$$L(\underline{x}, \underline{\lambda}) = f(\underline{x}) - \underline{\lambda}^T\underline{c}(\underline{x}) \ , \tag{13.10}$$

where $\underline{c}(\underline{x})\colon R^n \to R^t$, and $\underline{\lambda} \in R^t$. Necessary conditions for $\underline{x}^*$ to be a local minimiser of (13.9) are:

(i) $\underline{c}(\underline{x}^*) = \underline{0}$, or equivalently $\underline{\nabla}_{\underline{\lambda}}L(\underline{x}^*, \underline{\lambda}^*) = \underline{0}$,

$$\text{(ii)} \quad \underline{g}(\underline{x}^*) = \sum_{i=1}^{t} \lambda_i^* \underline{a}_i(\underline{x}^*) , \tag{13.11}$$

or equivalently $\underline{\nabla}_{\underline{x}} L(\underline{x}^*, \underline{\lambda}^*) = \underline{0}$,

$$\text{(iii)} \quad \underline{s}^T W(\underline{x}^*, \underline{\lambda}^*)\underline{s} \geqslant 0 \text{ , for all } \underline{s} : \underline{a}_i(\underline{x}^*)^T \underline{s} = 0 \text{ ,}$$

$$i = 1,2,\ldots,t \text{ ,}$$

where $\underline{a}_i(\underline{x}) = \underline{\nabla} c_i(\underline{x})$, $i = 1,2,\ldots,t$, are the constraint normals, assumed to be linearly independent, and $W(\underline{x}, \underline{\lambda}) = \left[\dfrac{\partial^2}{\partial x_i \partial x_j} L(\underline{x}, \underline{\lambda})\right]$ is the Hessian matrix of the Lagrangian function. Notice that these conditions reduce to (13.4) when $c_i(\underline{x}) = \underline{a}_i^T \underline{x} - b_i$ and $t = m$. Also note the fundamental role played by the Lagrangian function (13.10).

Sufficient conditions for $\underline{x}^*$ to be a local minimiser of (13.9) are (i) and (ii) of (13.11) together with

$$\text{(iii)} \quad \underline{s}^T W(\underline{x}^*, \underline{\lambda}^*)\underline{s} > 0, \text{ for all } \underline{s} : \underline{a}_i(\underline{x}^*)^T \underline{s} = 0 \text{ ,} \tag{13.12}$$

$$i = 1,2,\ldots,t \text{ .}$$

The extension of the optimality conditions (13.7) and (13.8) to the nonlinear inequality constrained problem is very similar to the extension of the optimality conditions for the equality constrained problem. We therefore refer to Gill, Murray and Wright (1981, p. 81), or Fletcher (1981, Ch. 9).

3. Linearly constrained problems

The inclusion of linear constraints in a minimisation problem simply has the effect of restricting the minimisation of $f(\underline{x})$ to a convex subset of R^n. Hence methods for dealing with linear constraints tend to be generalisations of unconstrained optimisation methods. We review the underlying ideas and refer to Gill, Murray and Wright (1981, Ch. 5), and Fletcher (1981, Ch. 11), for fuller details of the methods.

We first consider the equality constrained problem:

$$\min_{\underline{x} \in R^n} \quad f(\underline{x}) \tag{13.13}$$

$$\text{subject to} \quad A^T \underline{x} = \underline{b} \text{ ,}$$

where the $n \times m$ matrix A has full rank.

Essentially the equality constraints $A^T\underline{x} = \underline{b}$ have the effect of restricting the minimisation of $f(\underline{x})$ to an $n-m$ dimensional subset (the feasible region) of R^n. Provided that we can suitably parameterise this feasible region then we can solve (13.13) using unconstrained optimisation techniques in terms of the parameters of the feasible region.

Let Y be any $n \times m$ matrix whose columns span the range space of A and let Z be any $n \times (n-m)$ matrix whose columns span the null space of A, so that $A^T Z = 0$. Notice that there is some freedom in the choice of Y and Z. Any vector $\underline{x} \in R^n$ can be written as

$$\underline{x} = Y\underline{x}_Y + Z\underline{x}_Z \, . \qquad (13.14)$$

In order that $\underline{x}$ is feasible,

$$\begin{aligned} A^T\underline{x} &= A^T Y\underline{x}_Y + A^T Z\underline{x}_Z \\ &= A^T Y\underline{x}_Y = \underline{b} \, . \end{aligned} \qquad (13.15)$$

(13.15) uniquely determines $\underline{x}_Y$ and hence only the null-space part of $\underline{x}$, $\underline{x}_Z$, remains free to minimise $f(\underline{x})$. Hence the solution of (13.13) is essentially an unconstrained optimisation problem in the $(n-m)$ variables $\underline{x}_Z$. Details of these so-called null-space methods for the solution of (13.13) are given in section 5.1 of Gill, Murray and Wright (1981).

At the solution $\underline{x}^*$ to (13.13),

$$\underline{g}(\underline{x}^*) = A\underline{\lambda}^* ,$$

for some vector $\underline{\lambda}^* \in R^m$ of Lagrange multipliers. An estimate of this vector of multipliers is useful to assess the sensitivity of the equality constraints. At the approximate solution $\hat{\underline{x}}$ of (13.13), an estimate of the vector of Lagrange multipliers is given by the vector $\hat{\underline{\lambda}}$ which solves the least-squares problem

$$\min_{\underline{\lambda} \in R^m} \; \| A\underline{\lambda} - \hat{\underline{g}} \|_2 \, . \qquad (13.16)$$

Clearly the computation of this Lagrange multiplier estimate should take advantage of any matrix factorisation used earlier in the null-space method.

However the more interesting problem is the linear inequality constrained problem:

$$\min_{\underline{x} \in R^n} \quad f(\underline{x}) \tag{13.17}$$
$$\text{subject to} \quad A^T\underline{x} \geqslant \underline{b} ,$$

where, for simplicity, the $n \times m$ matrix $A = [\underline{a}_1\ \underline{a}_2\ \ldots\ \underline{a}_m]$ is assumed to have full rank. The most popular means of solving (13.17) is to use an 'active set' method. The aim is to predict the set of constraints $M \subset \{1,2,\ldots,m\}$ which are active at the solution of (13.17). If the true active set of (13.17) is M^*, then the solution of (13.17) is also a solution of the equality constrained problem:

$$\min_{\underline{x} \in R^n} \quad f(\underline{x})$$
$$\text{subject to} \quad \underline{a}_i^T\ \underline{x} = b_i\ , \quad i \in M^* .$$

The difficulty is to predict correctly the set of active constraints and any method must have strategies for both adding and deleting constraints from its current predicted set of active constraints.

If we let M_k denote the current set of active constraints and $\underline{x}^{(k)}$ denote the current approximation to $\underline{x}^*$, then the kth iteration of an active set method can be described as follows.

The first step of the iteration is to decide whether or not to delete a constraint from the active set M_k. We note that $\underline{x}^{(k)}$ is the solution to the currently active equality constrained problem

$$\min_{\underline{x} \in R^n} \quad f(\underline{x}) \tag{13.18}$$
$$\text{subject to} \quad \underline{a}_i^T\ \underline{x} = b_i\ , \ i \in M_k .$$

Then $\underline{g}(\underline{x}^{(k)}) = \sum_{i \in M_k} \lambda_i \underline{a}_i$. Now if $\underline{x}^{(k)}$ is also a solution to the inequality constrained problem (13.17), then $\lambda_i \geqslant 0$, $i \in M_k$. If any $\lambda_i < 0$, then $f(\underline{x})$

can be reduced by moving away from the ith constraint into the feasible region of (13.17). Hence the Lagrange multipliers can be used to decide whether or not to delete a constraint from M_k.

Having decided that the active set M_k is appropriate, a search direction $\underline{s}^{(k)}$ in the currently active constraint space is calculated. A new approximation $\underline{x}^{(k+1)}$ is then defined as

$$\underline{x}^{(k+1)} = \underline{x}^{(k)} + \alpha_k \underline{s}^{(k)} ,$$

where the steplength α_k is such that $f(\underline{x}^{(k+1)})$ is suitably less than $f(\underline{x}^{(k)})$ and also such that $\underline{x}^{(k+1)}$ does not violate the constraints. If $\underline{x}^{(k+1)}$ is such that the jth constraint becomes active then $M_{k+1} = M_k + \{j\}$, otherwise $M_{k+1} = M_k$ and the kth iteration of the method is completed. Clearly a convergence test must also be incorporated in an active set method.

The design of an efficient and robust active set method is a considerable task. We refer to Gill, Murray and Wright (1981) for details of active set methods.

A special linear inequality constrained problem is the bounds constrained problem:

$$\min_{\underline{x} \in R^n} f(\underline{x}) \tag{13.19}$$

$$\text{subject to} \quad \ell_j \leqslant x_j \leqslant u_j , \quad j = 1,2,\ldots,n .$$

Because each variable is either fixed at one of its bounds, or free, the active set method is particularly simple for (13.19). In fact (13.19) can be solved essentially just as efficiently as the corresponding unconstrained optimisation problem. This is reassuring since one would expect most practical optimisation problems to be in terms of variables which have some restrictions on size.

4. Nonlinearly constrained problems

We now address a more difficult problem in optimisation, to minimise a nonlinear objective function, where the variables are restricted by a set of nonlinear constraints:

$$\min_{\underline{x} \in R^n} \quad f(\underline{x}) \tag{13.20}$$

$$\text{subject to} \quad c_i(\underline{x}) = 0, \quad i = 1,2,\ldots,t ,$$

$$\text{and} \quad c_i(\underline{x}) \geq 0, \quad i = t+1,t+2,\ldots,m .$$

Gill, Murray, Saunders and Wright (1984) have recently produced an excellent survey of methods for the solution of (13.20), with special reference to the available software.

The difficulty of solving (13.20) stems from the conflicting aims of minimising the objective function and satisfying the constraints. With linear constraints it is relatively straightforward to satisfy the constraints and hence the methods described in section 3 can essentially concentrate on minimising the objective function. With nonlinear constraints it can be very difficult, if not impossible, to satisfy the constraints and hence there is a conflict between the difficult tasks of minimising $f(\underline{x})$ and satisfying the constraints. Different methods for the solution of (13.20) resolve this conflict in different ways. A common feature of the methods for the solution of this problem, which we will describe, is the reduction of the problem to a sequence of simpler sub-problems.

4.1 Augumented Lagrangian methods

For simplicity we consider the equality constrained problem:

$$\min_{\underline{x} \in R^n} \quad f(\underline{x}) \tag{13.21}$$

$$\text{subject to} \quad c_i(\underline{x}) = 0, \qquad i = 1,2,\ldots,t .$$

The earliest methods for the solution of (13.21) were the so-called penalty function methods. For example, Courant (1943) suggested the quadratic penalty function,

$$\varphi(\underline{x}, \sigma) = f(\underline{x}) + \frac{1}{2}\sigma \sum_{i=1}^{t} (c_i(\underline{x}))^2 , \tag{13.22}$$

where $\sigma > 0$. In (13.22) the first term is the objective function and the second term penalises constraint violations. The method consists of minimising $\varphi(\underline{x}, \sigma)$ for a sequence $\{\sigma_k\}$ of increasing values of σ. If we denote the minimiser of $\varphi(\underline{x}, \sigma_k)$ by $\underline{x}^*(\sigma_k)$ then it can be shown, under fairly mild conditions, that $\{\underline{x}^*(\sigma_k)\} \to \underline{x}^*$, a solution to (13.21), as $\{\sigma_k\} \to \infty$. Notice that the solution of (13.21) has been reduced to the solution of a sequence of unconstrained minimisation problems. The fundamental difficulty with the method is the inevitable ill-conditioning of the unconstrained problems as the sequence $\{\sigma_k\}$ tends to infinity.

The major step in overcoming this difficulty was taken independently by Powell (1969) and by Hestenes (1969). As originally proposed the two methods appeared to be quite different, but they are in fact very closely related. It would clearly be useful to find a penalty function $\varphi(\underline{x}, \sigma)$ which did not require the penalty parameter σ to tend to infinity. The Lagrangian function, defined in (13.10), is almost suitable in that $L(\underline{x}, \underline{\lambda}^*)$ has a stationary point at $\underline{x}^*$ and further has a local minimum at $\underline{x}^*$ within the subspace orthogonal to the constraint normals. Indeed if condition (iii), (13.12), were valid for all $\underline{s} \in R^n$ then $L(\underline{x}, \underline{\lambda}^*)$ would have a local minimum at $\underline{x}^*$. Powell and Hestenes suggested the use of an 'augmented Lagrangian' penalty function,

$$\varphi(\underline{x}, \underline{\lambda}, \sigma) = f(\underline{x}) - \underline{\lambda}^T\underline{c}(\underline{x}) + \frac{\sigma}{2}\,\underline{c}(\underline{x})^T\underline{c}(\underline{x}) , \qquad (13.23)$$

where for $\sigma > 0$ the term $(\frac{\sigma}{2}\,\underline{c}(\underline{x})^T\underline{c}(\underline{x}))$ introduces positive curvature in the subspace of constraint normals. It can be shown that there is a finite value $\hat{\sigma}$ such that, for any $\sigma > \hat{\sigma}$, an unconstrained local minimum of $\varphi(\underline{x}, \underline{\lambda}^*, \sigma)$ is also a local minimum of (13.21). Notice that $\underline{x}^*$ is not necessarily a local minimum of (13.23) except when $\underline{\lambda} = \underline{\lambda}^*$. In other words, to use the penalty function (13.23), we need to have estimates of both the solution $\underline{x}^*$ and the Lagrange multipliers $\underline{\lambda}^*$. Thus an augmented Lagrangian method has two sequences of approximations, $\{\underline{x}^{(k)}\} \to \underline{x}^*$ and $\{\underline{\lambda}^{(k)}\} \to \underline{\lambda}^*$. Typically the steps of the method are:

(i) Solve the unconstrained minimisation problem

$$\min_{\underline{x} \in R^n} \varphi(\underline{x}, \underline{\lambda}^{(k)}, \sigma_k) \text{ for fixed } \underline{\lambda}^{(k)}, \sigma_k \text{ to give}$$

a new estimate of the solution $\underline{x}^{(k+1)}$.

(ii) Update the estimates of the Lagrange multipliers

to give $\underline{\lambda}^{(k+1)}$.

(iii) If the constraint violations at $\underline{x}^{(k+1)}$ are not sufficiently less than those at $\underline{x}^{(k)}$ increase the penalty parameter to give σ_{k+1}.

(iv) Either terminate the iteration or return to (i).

Note that this method requires an unconstrained minimisation calculation in step (i).

The estimates $\underline{\lambda}^{(k)}$ of the Lagrange multipliers play a crucial role in this method. Indeed the method will converge to $\underline{x}^*$ only if $\{\underline{\lambda}^{(k)}\} \to \underline{\lambda}^*$. Further the rate of convergence of $\{\underline{x}^{(k)}\}$ to $\underline{x}^*$ cannot exceed the rate of convergence of $\{\underline{\lambda}^{(k)}\}$ to $\underline{\lambda}^*$ and hence high-order Lagrange multiplier estimates should be used (see Gill, Murray and Wright (1981), section 6.6).

The extension of the augmented Lagrangian penalty function to deal with inequality constraints was suggested by Rockafellar (1974) (see also Fletcher (1975)).

4.2 Sequential quadratic programming methods

Again we concentrate on the equality constrained problem (13.21), but now consider a method which replaces the nonlinearly constrained problem by a sequence of quadratic programming problems.

A Taylor expansion of the constraints about the current approximation $\underline{x}^{(k)}$ is

$$\underline{c}(\underline{x}) = \underline{c}(\underline{x}^{(k)}) + A(\underline{x}^{(k)})^T(\underline{x} - \underline{x}^{(k)}) + \ldots , \qquad (13.24)$$

where

$$A(\underline{x}^{(k)}) = [\underline{\nabla} c_1(\underline{x}^{(k)}) \;\; \underline{\nabla} c_2(\underline{x}^{(k)}) \;\; \ldots \;\; \underline{\nabla} c_t(\underline{x}^{(k)})] .$$

Thus a linear approximation to the constraints is given by

$$A(\underline{x}^{(k)})^T(\underline{x} - \underline{x}^{(k)}) = - \underline{c}(\underline{x}^{(k)}) ,$$

or by

$$A(\underline{x}^{(k)})^T \underline{s} = -\underline{c}(\underline{x}^{(k)}), \tag{13.25}$$

where $\underline{s}$ represents any step from $\underline{x}^{(k)}$.

Similarly a Taylor expansion of the objective function about $\underline{x}^{(k)}$ is

$$f(\underline{x}) = f(\underline{x}^{(k)}) + (\underline{x} - \underline{x}^{(k)})^T \underline{g}(\underline{x}^{(k)}) + \frac{1}{2}(\underline{x} - \underline{x}^{(k)})^T G(\underline{x}^{(k)})(\underline{x} - \underline{x}^{(k)}) + \ldots .$$

Hence a quadratic approximation to the objective function is given by

$$q^{(k)}(\underline{s}) = f(\underline{x}^{(k)}) + \underline{s}^T \underline{g}(\underline{x}^{(k)}) + \frac{1}{2}\underline{s}^T G(\underline{x}^{(k)})\underline{s}, \tag{13.26}$$

where $\underline{s}$ again represents some step from $\underline{x}^{(k)}$. This suggests that a suitable step $\underline{s}^{(k)}$ to progress from $\underline{x}^{(k)}$ to $\underline{x}^{(k+1)}$ would result from solving the quadratic programming problem

$$\min_{\underline{s} \in R^n} \quad q^{(k)}(\underline{s}) \tag{13.27}$$

$$\text{subject to} \quad A(\underline{x}^{(k)})^T \underline{s} = -\underline{c}(\underline{x}^{(k)}).$$

The major deficiency with this sub-problem is that it does not include information about the second derivatives of the constraints. A more suitable quadratic programming sub-problem is

$$\min_{\underline{s} \in R^n} \quad \underline{g}(\underline{x}^{(k)})^T \underline{s} + \frac{1}{2}\underline{s}^T W(\underline{x}^{(k)}, \underline{\lambda}^{(k)})\underline{s} \tag{13.28}$$

$$\text{subject to} \quad A(\underline{x}^{(k)})^T \underline{s} = -\underline{c}(\underline{x}^{(k)}).$$

The matrix $W(\underline{x}^{(k)}, \underline{\lambda}^{(k)})$, the Hessian matrix of the Lagrangian function, includes second-derivative information about both the objective function and the constraints. Note that the constant term of the objective function is omitted in (13.28) since it does not affect the resulting solution vector $\underline{s}^{(k)}$.

Thus a very simple quadratic programming method would solve (13.28) to find $\underline{s}^{(k)}$ and set

$$\underline{x}^{(k+1)} = \underline{x}^{(k)} + \underline{s}^{(k)} .$$

If the parameters $\underline{\lambda}^{(k)}$ in $W(\underline{x}^{(k)}, \underline{\lambda}^{(k)})$ of (13.28) are taken as the Lagrange multipliers of the preceding quadratic programming problem, then this simple method can be shown to have satisfactory local convergence properties.

The global behaviour of the method can be improved by including a line search, so that

$$\underline{x}^{(k+1)} = \underline{x}^{(k)} + \alpha_k \underline{s}^{(k)} ,$$

where α_k is chosen so that $\underline{x}^{(k+1)}$ is a 'better' approximation than $\underline{x}^{(k)}$. Clearly this requires a <u>merit function</u> to measure the quality of approximations to $\underline{x}^*$. One suggestion is to use the augmented Lagrangian function (13.23) as the merit function.

A sequential quadratic programming method based on (13.28) requires second derivatives of $f(\underline{x})$ and $c_i(\underline{x})$, $i = 1,2,...,t$. These derivatives are not usually assumed to be available and therefore much recent research has centred on using a quasi-Newton approximation to replace $W(\underline{x}, \underline{\lambda})$ in (13.28).

More details about sequential quadratic programming methods and the extension of the method to deal with inequality constraints can be found in section 6.5 of Gill, Murray and Wright (1981).

5. Software

The NAG Library contains a comprehensive set of bounds constrained minimisation routines. Also available, within the Library, are subroutines for dealing with nonlinear constraints using augmented Lagrangian function techniques, and a (new) subroutine, E04VCF, which uses a sequential quadratic programming technique, and which is based on the code NPSOL, written by Gill, Murray, Saunders and Wright (1983). Another code similar to NPSOL is VMCWD, which has been written by Powell (1982). This code is available in the Harwell Library and will soon be available in a modular version in IMSL.

CHAPTER 14

Nonlinear Least-Squares Problems

J.A. Belward

1. Introduction

Although the minimisation of a nonlinear function which is a sum of squares is one of the simpler problems in Mathematical Programming, techniques for its solution continue to develop steadily. Since the conference held at York in 1976 on the State of the Art in Numerical Analysis (Jacobs 1977), new methods have been proposed and several are now available in the form of coded algorithms. This review presents a survey of development over that period and summarises some recent tests of readily available algorithms. Particular techniques which take into account sparsity, separability, non-differentiability and linear constraints will also be mentioned briefly.

The nonlinear least-squares problem is

$$\underset{\underline{x}}{\text{minimise}}\ F(\underline{x}) = \frac{1}{2} \sum_{i=1}^{m} [f_i(\underline{x})]^2 \qquad (14.1)$$

where $\underline{x}$ belongs to R^n (usually $m > n$). It arises naturally in the context of finding a 'best fit' solution to the overdetermined system

$$\underline{f}(\underline{x}) = \underline{0} \qquad (14.2)$$

where

$$\underline{f} = [f_1\ f_2\ \dots\ f_m]^T ,$$

and the best fit solution is chosen to be that which minimises the Euclidean or L_2 norm of $\underline{f}$. (Unless stated to the contrary, $||\underline{a}||$ stands for the L_2 norm, $||\underline{a}||_2^2 = a_1^2 + a_2^2 + \dots + a_n^2$.) $F(\underline{x})$ is assumed to possess smoothness appropriate to the context in which it appears. A local minimiser of F will be denoted by $\underline{x}^*$ and assumed isolated. Corresponding quantities associated with the minimum are denoted by starred variables.

The quality of a best fit solution can be measured in many other norms; articles by Dennis (1977, 1982), Byrd (1982), Lemaréchal (1982) and Overton (1982), review the general problem of nonlinear fitting.

In addition to data-fitting in Statistics, nonlinear least-squares problems arise in Applied Mathematics, Engineering, Physics, etc., where numerical solutions of differential or integral equations are sought. Problems from these areas tend to fall into large-residual or small-residual problems, respectively (though the distinction is not at all clear cut). These terms describe the cases where the minimum value of $F(\underline{x})$ is large or small, in some sense.

The earliest methods, Gauss-Newton methods, were based on the assumption that the objective function $F(\underline{x})$ is small at optimality. These methods avoided the need to calculate second derivatives yet provided superlinear convergence in many cases. The modification of Levenberg (1944) and Marquardt (1963) was developed to overcome the inadequacy of this approximation at points remote from the optimum. It led to more robust algorithms, but in the large-residual problems deficiencies remained. In more recent times, work has focussed on retaining the basic philosophies of the Gauss-Newton (GN) and Levenberg-Marquardt (LM) methods, while attempting to make a good approximation to the second derivative terms without incurring the penalty of calculating them exactly.

Judgments of the relative efficiencies of several different codes are not easily reached and indeed objective testing in many areas of Computational Mathematics is very difficult. Nevertheless, numerical experiments are reported in the literature and some attempts have been made to evaluate the performance of the various codes available (Nazareth 1980, Hiebert 1981). In this review a summary of the theory appropriate to currently available software is given; we also mention some algorithms which are already listed or quite easily coded where the use of a major package is inconvenient or even impossible.

Space precludes a detailed presentation of the work summarised in this review; the interested reader is therefore urged to consult the source references given. Many good texts on Mathematical Programming are available, in particular those by Fletcher (1980) and Gill, Murray and Wright (1981), which present both the basic material and the latest developments in the area.

2. Gauss-Newton methods

From (14.1) the gradient of F is given by

$$\underline{\nabla}F(\underline{x}) = J^T\underline{f}(\underline{x}) \tag{14.3}$$

where J is the m × n Jacobian matrix of $\underline{f}$ with respect to $\underline{x}$, and the Hessian matrix H of F is given by

$$H(\underline{x}) = J^TJ + \sum_{i=1}^{m} f_i(\underline{x})\, G_i(\underline{x}) \tag{14.4}$$

where $G_i(\underline{x})$ is the Hessian matrix of $f_i(\underline{x})$.

From a point $\underline{x}^{(k)}$ we can calculate a Newton step $\underline{\delta x}^{(k)}$ from the equations

$$H(\underline{x}^{(k)})\underline{\delta x}^{(k)} = -\underline{\nabla}F(\underline{x}^{(k)})\ . \tag{14.5}$$

For small-residual problems (defined later) near an optimum point $\underline{x}^*$ the second term of the Hessian matrix H is small; thus we may approximate the second derivative by

$$H(\underline{x}) \simeq J^TJ\ . \tag{14.6}$$

On using (14.3) and (14.6) in (14.5), we obtain the equations for the Gauss-Newton step:

$$(J^{(k)})^T J^{(k)}\underline{\delta x}^{(k)} = -(J^{(k)})^T \underline{f}(\underline{x}^{(k)})\ . \tag{14.7}$$

This system may also be obtained by linearising each $f_i(\underline{x})$ about $\underline{x}^{(k)}$ with $\underline{x}^* = \underline{x}^{(k)} + \underline{\delta x}^{(k)}$; then

$$F(\underline{x}^{(k)} + \underline{\delta x}^{(k)}) \simeq \frac{1}{2}\,||\underline{f}(\underline{x}^{(k)}) + J^{(k)}\underline{\delta x}^{(k)}||^2\ , \tag{14.8}$$

and this quadratic in $\underline{\delta x}^{(k)}$ is minimised when equations (14.7) are satisfied. These are the normal equations for the overdetermined linear system

$$J^{(k)}\underline{\delta x}^{(k)} = -\underline{f}(\underline{x}^{(k)}) \; . \tag{14.9}$$

It can be shown (Gill, Murray and Wright 1981, hereafter referred to as GMW) that if

$$\left\| \sum_{i=1}^{m} f_i^* G_i^* \right\| < \epsilon \tag{14.10}$$

then the Gauss-Newton step $\underline{\delta x}_{GN}$ and Newton step $\underline{\delta x}_N$ satisfy

$$\frac{\|\underline{\delta x}_N - \underline{\delta x}_{GN}\|}{\|\underline{\delta x}_N\|} = O(\epsilon) \; . \tag{14.11}$$

Thus we see that the GN method will possess the same order of convergence (normally quadratic) as the unmodified Newton method.

An algorithm based on solving (14.7) directly would not be completely satisfactory, for it is not necessary to solve (14.7). Instead either the singular value decomposition or QR factorisation of $J^{(k)}$ can be used. A minimum norm vector $\underline{\delta x}^{(k)}$ can always be found which minimises

$$\underline{r}^{(k)} = J^{(k)}\underline{\delta x}^{(k)} + \underline{f}(\underline{x}^{(k)}) \tag{14.12}$$

and when $J^{(k)}$ has full rank the value of $\underline{\delta x}^{(k)}$ so found satisfies equation (14.7). Further, the condition number of the coefficient matrix in (14.7) is the square of that in (14.9); thus some loss in accuracy will often result.

Fletcher (1980) (hereafter referred to as F1) proves, for the full rank case, that the errors $\underline{e}^{(k)} = \underline{x}^* - \underline{x}^{(k)}$ satisfy

$$\|\underline{e}^{(k+1)}\| \leqslant \|B^*\| \, \|\underline{e}^{(k)}\| + o\left(\|\underline{e}^{(k)}\|^2\right) \tag{14.13}$$

where $B^* = (J^{*T}J^*)^{-1} \sum_{i=1}^{m} f_i^* G_i^*$. In this case the spectral radius ρ^* of B^*

distinguishes the zero residual case $\rho^* = 0$, from the non-zero residual case $\rho^* > 0$. The result (14.13) also suggests divergence for $\rho^* > 1$, and this is confirmed by a simple example; see F1, p. 94.

The GN method is intended to behave in a similar way to Newton's method, for which it is known that a Newton step does not in itself guarantee a function reduction. The approximate Hessian J^TJ is positive semi-definite and thus, as in Newton's method, when J has full rank the solution of (14.7) will provide a descent direction for F.

These considerations, together with Fletcher's result above, indicate the advisability of using a line search procedure along the direction of the solution of equation (14.7) to ensure a sufficient function reduction. The basic GN algorithm then becomes:

(i) Select $\underline{x}^{(0)}$.

(ii) For $k = 0,1,2,\ldots,k_{MAX}$ calculate $\underline{f}(\underline{x}^{(k)})$, $F(\underline{x}^{(k)})$ and $J^{(k)}$ and determine $\underline{s}^{(k)}$ (replacing $\underline{\delta x}^{(k)}$) which satisfies (14.7).

(iii) Minimise $F(\underline{x}^{(k)} + \alpha \underline{s}^{(k)})$ with respect to α.

(iv) Set $\underline{x}^{(k+1)} = \underline{x}^{(k)} + \alpha_k \underline{s}^{(k)}$.

(v) Test for convergence and maximum iteration count.

(vi) According to (v), either stop or return to (ii).

Suitable exit criteria are discussed later; see GMW or F1 for a good discussion of this point.

The above algorithm is based on the hope that J remains of full rank throughout the calculation. This will not always be the case and if the columns of J become linearly dependent, this may cause divergence of an otherwise convergent algorithm. Powell (1970) gives an example where the Jacobian loses rank, and convergence to a non-stationary point occurs. Fletcher (in F1) points out a test example (p. 95) in which J^* has rank less than n and a poor estimate of $\underline{x}^*$ is obtained.

In practice, a theoretical loss of rank will be characterised by severe instability; computationally a matrix will rarely be rank deficient, and the problem arises of making a decision on the rank. The most stable measure of the numerical rank of J is determined by its singular values (see Chapter 1, section 2.5; Chapter 2, section 4), and it is shown in the next section, that if the columns of J become linearly dependent, the search directions will be confined to a subspace of R^n - an effect which would seriously inhibit convergence.

3. The Levenberg-Marquardt method

The Levenberg-Marquardt method is one in which the step $\underline{\delta x}^{(k)}$ is determined as the solution to a system derived on the basis of a restriction on the magnitude of $\underline{\delta x}^{(k)}$. Specifically, $\underline{\delta x}^{(k)}$ is given by

$$((J^{(k)})^T J^{(k)} + \lambda_k I)\underline{\delta x}^{(k)} = -(J^{(k)})^T \underline{f}^{(k)} \tag{14.14}$$

where λ_k is a constant whose value implicitly controls the magnitude of $\underline{\delta x}^{(k)}$. From this point of view the method, proposed by Levenberg (1944) and Marquardt (1963), is called a trust region method; it is based on the observation that a certain approximation to the objective function is only accurate on a limited domain. The approximation to the function is then minimised on the restricted set; thus in (14.14) λ_k is the Lagrange multiplier for the problem

$$\underset{\underline{\delta x}^{(k)}}{\text{minimise}} \quad L(\underline{\delta x}^{(k)}) \equiv ||\underline{f}^{(k)} + J^{(k)}\underline{\delta x}^{(k)}|| \tag{14.15}$$

subject to

$$||\underline{\delta x}^{(k)}|| \leqslant \Delta_k . \tag{14.16}$$

$L(\underline{\delta x}^{(k)})$ is the linear approximation to $F(\underline{x}^{(k)} + \underline{\delta x}^{(k)})$ about $\underline{x}^{(k)}$. Now either

$$|| \underline{\delta x}^{(k)}|| < \Delta_k \quad \text{and} \quad \lambda_k = 0 ,$$

or

$$|| \underline{\delta x}^{(k)}|| = \Delta_k \quad \text{and} \quad \lambda_k \geqslant 0 .$$

Hence Δ_k determines λ_k .

A natural generalisation of (14.15) and (14.16) is to replace (14.16) by

$$||D\,\underline{\delta x}^{(k)}|| \leq \Delta_k \tag{14.17}$$

where $D = \text{diag}\,[\nu_1\ \nu_2\ \ldots\ \nu_n]$; then in (14.14), $\lambda_k I$ is replaced by $\lambda_k D^T D$. Notice that Δ_k and D act to restrict $\underline{\delta x}^{(k)}$, whereas in the GN algorithm the restriction is made more directly through the line search algorithm in which safeguarding procedures will not allow excessive values of $||\underline{\delta x}^{(k)}||$.

It can be observed that as $\lambda_k \to \infty$ in (14.14), $\underline{\delta x}^{(k)}$ tends to the steepest descent direction; thus it appears that convergence can be guaranteed by choosing a sequence of sufficiently large values of λ_k. In practice, this would be better avoided, since the choice of steepest descent directions often leads to slow convergence. Note that if λ_k is sufficiently large, this will ensure that $((J^{(k)})^T J^{(k)} + \lambda_k I)$ is always positive definite and problems of rank deficiency can thereby be avoided. On the other hand, it is desirable to have $\lambda_k \to 0$ ultimately in the small-residual case, so that the steps become GN steps near $\underline{x}^*$ with the consequent quadratic convergence. More precisely, it can be shown (Dennis 1977) that the LM step is related to the corresponding GN step by

$$||\underline{\delta x}^{(k)}_{LM} - \underline{\delta x}^{(k)}_{GN}|| \leq \frac{\lambda_k}{\sigma^2_{min}}\, ||\underline{\delta x}^{(k)}_{GN}|| \tag{14.18}$$

where σ_{min} is the smallest non-zero singular value of $J^{(k)}$.

As we show at the end of this section, the difficulty with restricted subspace steps in the GN method in the presence of loss of rank of the Jacobian is shared by the LM method. For large-residual problems the convergence rate can be linear at the best. Nevertheless experience has shown the LM method to be reasonably robust. Osborne (1976) and Moré (1978) have given algorithms which control the iteration through values of λ (Osborne) and Δ (Moré). Each gives a convergence theorem for the algorithm in question and Osborne's article includes a Fortran listing.

Moré's algorithm has several features which have been used in later development. It is the algorithm used in the MINPACK (for further details see section 8) routines LMDER and LMDIF, and appears to contribute substantially to the algorithm NL2SOL of Dennis, Gay and Welsch (1981). It proceeds as follows.

(i) Initialise with $\underline{x}^{(0)}$ and Δ_0. Calculate $\underline{f}(\underline{x}^{(0)})$, $J(\underline{x}^{(0)})$,

$D^{(0)} = \text{diag}\,[\nu_1\ \nu_2\ \dots\ \nu_n]$, $\nu_i = \frac{\partial F}{\partial x_i}(\underline{x}^{(0)})$ and set $\epsilon = 0.1$.

(ii) For $k = 0,1,\dots,k_{MAX}$, if

$$||D^{(k)}(J^{(k)})^{+}\ \underline{f}(\underline{x}^{(k)})|| \leqslant (1+\epsilon)\Delta_k\ , \tag{14.19}$$

set $\lambda_k = 0$ and $\underline{\delta x}^{(k)} = -(J^{(k)})^{+}\ \underline{f}(\underline{x}^{(k)})$,
otherwise find the minimum-norm solution of

$$\begin{bmatrix} J^{(k)} \\ \sqrt{(\lambda_k)}\ D^{(k)} \end{bmatrix} \underline{\delta x}^{(k)} = \begin{bmatrix} -\underline{f}(\underline{x}^{(k)}) \\ \underline{0} \end{bmatrix} \tag{14.20}$$

for which

$$\big|\ ||D^{(k)}\underline{\delta x}^{(k)}|| - \Delta_k\big| < \epsilon\Delta_k\ . \tag{14.21}$$

(iii) Compute

$$\rho_k = \frac{||\underline{f}^{(k)}||^2 - ||\underline{f}(\underline{x}^{(k)} + \underline{\delta x}^{(k)})||^2}{||\underline{f}^{(k)}||^2 - ||\underline{f}^{(k)} + J^{(k)}\underline{\delta x}^{(k)}||^2}\ . \tag{14.22}$$

(iv) Test for convergence.

(v) Update Δ_k, $\underline{x}^{(k)}$ and $D^{(k)}$ according to the value of ρ_k and return to (ii).

In (14.19) $(J^{(k)})^{+}$ is the generalised inverse of $J^{(k)}$. Step (ii) essentially involves solving an implicit functional relationship for λ_k. Equations (14.20) are the equations for which (14.14), with I replaced by $D^{(k)T}D^{(k)}$, are the normal equations. Conditions (14.21) and (14.19) are approximations to the precise requirement

$$||D^{(k)}\underline{\delta x}^{(k)}|| = \Delta_k\ .$$

Moré recommends the use of a QR decomposition in the solution of (14.20). Some economy can be made if several different values of λ have to be tried in (14.20) for the same J, D and $\underline{x}$. ρ_k evaluates the accuracy of the linear approximation to F near $\underline{x}^{(k)}$. The updating scheme sets $\underline{x}^{(k+1)} = \underline{x}^{(k)}$ if $\rho_k < 10^{-4}$, otherwise $\underline{x}^{(k+1)} = \underline{x}^{(k)} + \underline{\delta x}^{(k)}$. If $\rho_k < 0.25$, Δ_k is reduced by a factor of between 0.1 and 0.5; if $\rho_k > 0.75$ or $\lambda_k = 0$ and $\rho_k > 0.25$, $\Delta_{k+1} = 2\Delta_k$.

Finally in this section it is shown how the loss of rank of the Jacobian confines the steps in the Levenberg-Marquardt method to a subspace of R^n. Following the argument given by Nazareth (1980), let the columns of the matrices $V_1^{(k)}$ and $V_2^{(k)}$ span the range and null space of $J^{(k)T}J^{(k)}$. If the singular value decomposition of $J^{(k)}$ is

$$J^{(k)} = U^{(k)}\Sigma^{(k)}V^{(k)T} \tag{14.23}$$

where $U^{(k)}$ and $V^{(k)}$ are m × m and n × n orthogonal matrices and

$$\Sigma^{(k)} = \begin{bmatrix} S^{(k)} & 0 \\ 0 & 0 \end{bmatrix}$$

is an m × n matrix, with $S^{(k)} = \text{diag}\,[\sigma_1\ \sigma_2\ \dots\ \sigma_p]$, then p is the rank of $J^{(k)}$. The matrices $V_1^{(k)}$ and $V_2^{(k)}$ mentioned earlier can be taken to be the first p and remaining n-p columns of $V^{(k)}$:

$$V^{(k)} = [V_1^{(k)}\quad V_2^{(k)}] \; . \tag{14.24}$$

Since $S^{(k)}$ has only p non-zero entries, it is easily verified that

$$J^{(k)}V_2^{(k)} = 0 \; . \tag{14.25}$$

Thus expressing

$$\underline{\delta x}^{(k)} = V_1^{(k)}\underline{w}_1^{(k)} + V_2^{(k)}\underline{w}_2^{(k)} \; , \tag{14.26}$$

substituting into (14.14), and using (14.25) gives

$$J^{(k)T}J^{(k)}V_1^{(k)}\underline{w}_1^{(k)} + \lambda_k V_1^{(k)}\underline{w}_1^{(k)} + \lambda_k V_2^{(k)}\underline{w}_2^{(k)} = -J^{(k)T}\underline{f}^{(k)} .$$

After multiplying this on the left by $V_2^{(k)T}$ and using (14.25) again, we obtain

$$\lambda_k V_2^{(k)T}V_2^{(k)}\underline{w}_2^{(k)} = \underline{0} .$$

Since $V_2^{(k)}$ is orthogonal, it follows that $\underline{w}_2 = \underline{0}$ if $\lambda_k \neq 0$. In the GN case, since

$$J^{(k)+} = V^{(k)}\begin{bmatrix} \mathrm{diag}[\sigma_1^{-1} \ldots \sigma_p^{-1}] & 0 \\ 0 & 0 \end{bmatrix}U^{(k)T} ,$$

direct calculation of $J^{(k)+}\underline{f}^{(k)}$ shows that $\underline{\delta x}^{(k)}$ is again a linear combination of the first p columns of $V^{(k)}$.

4. Taking account of second derivative contributions

The methods discussed so far do not take into account the contribution of the second derivatives of $\underline{f}$ to the Hessian of F. In the large-residual case this must be done if superlinear convergence is to be achieved. Two algorithms aimed at remedying this omission will be described here and in the following section. First we describe a scheme of Gill and Murray (1978) which forms the basis of a suite of programs in the chapter on minimisation of functions in the NAG software library.

In view of equation (14.13), it is clear that the value of ρ^* is a measure of the importance of the second derivative terms. This is very closely related to the quantity

$$(\text{least singular value of } J^*)^2 \; (|| \sum_{i=1}^{m} f_i^* G_i^* ||) .$$

Thus we see that if we adopt a GN type of approach, then when the least singular value of J is small, some account should be taken of the second derivatives of $\underline{f}$. The algorithm of Gill and Murray (1978) does this; it groups the singular values of J into two sets according to their magnitudes, and takes action to enable steps to be taken along directions corresponding to the small singular values. Thus the effect of stepping only in the range of J which was raised in the previous section is considerably reduced.

If we let

$$B = \sum_{i=1}^{m} f_i(\underline{x})\, G_i(\underline{x}) \,, \tag{14.27}$$

then, using the singular value decomposition (14.23) with $p = n$, the equation for the Newton direction (omitting the superscripts for simplicity)

$$(J^T J + B)\underline{s} = -J^T \underline{f} \tag{14.28}$$

can be recast in the form

$$(S^2 + V^T BV)\underline{b} = -\Sigma\, U^T \underline{f} \tag{14.29}$$

with $\underline{s} = V\underline{b}$.

Gill and Murray suggest that $||B||$ is often small compared with $||J^T J||$ and that ill-conditioning of (14.29) is often due to $J^T J$. The notion of the <u>grade</u> of J is thus introduced. This is the integer r for which the singular values of J are such that $\sigma_i \,/\, \sigma_1$ is greater than some threshold for $i < r$ (and less than it for $i \geqslant r$). Then the following partitions of S are set up:

$$S_1 = \text{diag}[\sigma_1 \ldots \sigma_r] \,, \quad S_2 = \text{diag}[\sigma_{r+1} \ldots \sigma_n] \,.$$

Similarly V, $\underline{b}$ and $U\underline{f}$ are partitioned in like manner to obtain

$$\begin{bmatrix} S_1^2\, \underline{b}_1 + V_1^T\, BV_1\underline{b}_1 + V_1^T\, BV_2\underline{b}_2 \\ \\ S_2^2\, \underline{b}_2 + V_2^T\, BV_1\underline{b}_1 + V_2^T\, BV_2\underline{b}_2 \end{bmatrix} = \begin{bmatrix} -S\underline{f}_1 \\ \\ -S\underline{f}_2 \end{bmatrix} . \tag{14.30}$$

If $||B|| < ||S_1||^2$, and $||B|| < ||S_2||^2$ then we may set up an iterative scheme to determine $\underline{s}$ by

$$S_1^2 \underline{b}_1^{(i+1)} = -S_1\underline{f}_1 - V_1^T B\underline{s}^{(i)} \tag{14.31}$$

$$(S_2^2 + V_2^T BV_2)\underline{b}_2^{(i+1)} = -S_2\underline{f}_2 - V_2^T B\underline{b}_1^{(i)} . \tag{14.32}$$

Whenever $(J^TJ + B)$ is positive definite, so too are $S_1{}^2$ and $(S_2{}^2 + V_2^T BV_2)$ and equations (14.31) and (14.32) can be solved using Cholesky factorisation (see Chapter 1, section 2.4). The choice of grade is made to counteract the ill-conditioning in (14.28). If

$$||B||\ ||S_1^{-2}||\ \left(1 + ||B||\ ||(S_2^2 + V_2^T BV_2)^{-1}||\right) < 1 ,$$

then the iterative scheme converges. The value of r can be decreased if the convergence rate appears slow.

When $(J^TJ + B)$ is indefinite, the LDL^T factorisation implicit in the Cholesky decomposition (L lower triangular, D diagonal with positive elements) as applied to $(S_2^2 + V_2^T BV_2)$ is modified in the manner described in Gill and Murray (1974); the coefficient matrix of (14.32) is perturbed to $(S_2^2 + V_2^T BV_2 + E)$, so that it becomes positive definite. The resulting line search direction is checked to ensure that it is not close to orthogonality with the gradient of F; if the check test fails, the direction $\underline{s}$ is recomputed with the modified factorisation applied to the entire matrix.

If a stationary point is reached, the LDL^T factorisation is computed to determine whether the necessary conditions for a local minimum are satisfied. If not, a direction of negative curvature can be generated as in Gill and Murray (1974) and used as a search direction. While function reductions of greater than 1% are achieved, ordinary Gauss-Newton plus line search steps are taken. When the reduction falls below 1%, graded steps are taken until a 10% reduction has been made, when ordinary steps are resumed. The grade r is chosen as that which approximately balances the condition numbers of S_1 and S_2 - specifically, the value of r which minimises

$$\frac{\sigma_1}{\sigma_r} + \frac{\sigma_r}{\sigma_\ell} ,$$

σ_ℓ being the last non-zero singular value.

Gill and Murray tested several versions of their algorithm, the first in which B was explicitly evaluated, the second when B was approximated by differences and a third using Quasi-Newton type approximations to B. The best updating rule appeared to be that based on the BFGS formula (see Fletcher 1980). The performance of this last version was below expectation, and it was suggested that this was due to the problem of obtaining a good approximation to B. The Quasi-Newton condition could only be applied at the current iterate, whereas in more usual applications, n such conditions are required.

5. A trust region approach incorporating second derivative information

The trust region approach of the Levenberg-Marquardt method can easily be modified to include the second derivative terms B in the Hessian H of F. It is not necessary to calculate B precisely at every iteration, especially since it may not be the major contributor to H. The algorithm of Dennis, Gay and Welsch (1981) (abbreviated to DGW in what follows) adaptively selects either a GN, or GN plus second derivative estimate model. This selection is combined with a trust region method of the type used in the algorithm of Moré (outlined in section 3) moderated by a strategy due to Gay (1981). It is interesting to note that Gay takes particular action to deal with a direction of negative curvature, as do Gill and Murray in their algorithm.

There are many details to be attended to in the DGW method. In particular, the approximation to B and its updating can be treated in several ways, and a compromise between economy and accuracy is required. Since updating techniques do not generate arbitrarily good second derivative approximations, the algorithm may perform poorly on small-residual problems if the sequence of iterates $B^{(k)}$ does not tend to zero. The Oren-Luenberger scaling strategy (Oren 1974) is used to resolve this difficulty; its intention is to shift the spectrum of the approximation to overlap that of the true second derivative term. Dennis, Gay and Welsch refer to this as 'sizing'.

The second derivative term satisfies the approximate relation

$$B^{(k+1)}\underline{\delta x}^{(k)} \simeq ((J^{(k+1)})^T - (J^{(k)})^T)\underline{f}^{(k+1)} \qquad (14.33)$$

which follows from

$$\sum_{i=1}^{m} f_i^{(k+1)} G_i^{(k+1)} \underline{\delta x}^{(k)} \simeq \sum_{i=1}^{m} f_i^{(k+1)} (\underline{\nabla f}_i^{(k+1)} - \underline{\nabla f}_i^{(k)})$$

and the definition of J. The relation (14.33) does not determine $B^{(k+1)}$ uniquely; it is determined as the closest element to $B^{(k)}$ which is both symmetric and satisfies the requirement (14.33). The distance of $B^{(k+1)}$ from $B^{(k)}$ is measured in a norm which is intended to take into account the scaling near $\underline{x}^{(k)}$ and $\underline{x}^{(k+1)}$ as implied by the magnitudes of the derivatives at these points.

The sizing is done by updating $\tau_k B^{(k)}$, not $B^{(k)}$, where τ_k is a scalar not greater than unity. The sizes of $B^{(k)}$ and $B^{(k+1)}$ are compared by the ratio $((\underline{\delta x}^{(k)})^T B^{(k+1)} \underline{\delta x}^{(k)} / (\underline{\delta x}^{(k)})^T B^{(k)} \underline{\delta x}^{(k)})$, which can be computed before $B^{(k+1)}$ is calculated, using (14.33). Thus we set

$$\tau_k = \min \left\{ 1, \frac{(\underline{\delta x}^{(k)})^T((J^{(k+1)})^T - (J^{(k)})^T)\underline{f}^{(k+1)}}{(\underline{\delta x}^{(k)})^T \, B^{(k)} \, \underline{\delta x}^{(k)}} \right\}. \tag{14.34}$$

Initially the Gauss-Newton model is used and the Lagrange multiplier λ corresponding to an ellipsoidal trust region determined as in Moré's paper. A complete iteration contains several decisions on the model according to the predicted model decrease, compared with the decrease in the objective function. If the observed decrease is greater than 10% of the predicted decrease, and $\lambda_k > 0$, the condition

$$\underline{\delta f}^{(k)} \leqslant 0.75 \, (\underline{\nabla f}^{(k)})^T(\underline{x}^{(k+1)} - \underline{x}^{(k)}) \tag{14.35}$$

is checked. If it holds, then the trust region diameter is increased and a further candidate $\underline{x}^{(k+1)}$ is computed and tested in (14.35). If (14.35) is not satisfied, the previously calculated candidate is accepted. If a decrease of less than 10% of the predicted decrease is obtained, then the alternative model (i.e. one incorporating or removing the second derivative terms) is used. Provided the new predicted decrease is less, by a factor of 2/3, than that in the old model, and the step reduces $\underline{f}$ more than in the old model, the new model is accepted. If the prospective new model does not satisfy these requirements, then the trust region is shrunk and the current model unchanged. If the function reduction is less than 0.01% of the model prediction, $\underline{x}^{(k+1)}$ is taken equal to $\underline{x}^{(k)}$.

DGW contains a discussion of convergence criteria and a large range of tests on the algorithm is given, varying the sizing, model switching criteria, trust region scaling etc., and comparing the results with tests on a general purpose Quasi-Newton scheme. These points will be summarised in the next section.

6. Termination criteria and related matters

An important problem to address when the coding of a Mathematical Programming algorithm is planned is the decision of when to terminate. The routine must recognise and cater for convergence or the lack of progress towards an optimal point. These are not easy decisions to make; a premature exit should be avoided, but on the other hand we also want the return of an optimal point to be consistent and reliable. Good discussions of termination criteria are contained in Gill, Murray and Wright (1981), Hiebert (1981) and Dennis, Gay and Welsch (1981). GMW make the important point that not all exit criteria will be independent, for example function reduction, steplength and small derivative values must be interrelated. For precise details of exit criteria, the reader is referred to the source material; amongst the codes reviewed here a total of about eight tests are used, with most individual routines testing about five conditions. The essence of these tests is as follows. Upper bounds are placed on the number of iterations and function evaluations to prevent endless cycling; then function values, derivative values, steplength decrease, steplength size, and function reduction are evaluated in relative terms. Function values and derivative values are also usually safeguarded by testing their absolute values. Individual codes also include criteria which are relevant to that scheme only. Thus the Gill and Murray algorithm tests that the necessary second derivative conditions are satisfied, while DGW tests that the model function reduction is not greater than twice the observed function reduction at the final iteration.

The paper of Hiebert (1981) compares fourteen readily accessible codes on 72 examples, and Nazareth (1980) tests six algorithms on eleven large-residual test problems. Objective testing is a difficult task involving the selection of suitable test examples and the definition of fair and relevant performance indices. These difficulties are recognised by most authors and test results are usually presented without comment, leaving the task of assessment to the reader. In the case of nonlinear least squares, selection of problems presents little difficulty. A large variety of examples may be found in the literature involving trigonometric and exponential functions; linear, almost linear, rank-deficient problems; data fitting, large- and small-residual problems.

Hiebert (1981) evaluated the algorithms from a user's viewpoint, in that the purpose was to examine and recommend routines for inclusion in a software library. Emphasis was placed on robustness (the ability to find a solution), reliability (the ability to identify an exit correctly), and the communication interface, i.e. calling the subroutine and the documentation thereof. Nazareth's purpose was to compare algorithms, not codes, and in particular four schemes which take account of second derivative terms were compared with a general Quasi-Newton scheme and a Levenberg-Marquardt algorithm. Dennis et al. compared their algorithm with a Quasi-Newton routine.

Amongst Hiebert's conclusions, the following are perhaps the most striking -

(i) the Levenberg-Marquardt and augmented Gauss-Newton methods differ hardly at all in their performance;

(ii) algorithms using numerical estimates for derivatives performed as well as those using analytical derivatives;

(iii) the National Physical Laboratory codes LSQFDN, LSQFDQ and LSQNDN were the most reliable, all codes except one were robust on reasonably scaled problems, and the MINPACK codes LMDER and LMDIF were more robust on badly scaled problems.

Nazareth's main conclusion was that the Dennis et al. algorithm was consistently superior and that the general Quasi-Newton routine could not match the nonlinear least-squares routines. This last conclusion is supported by test results given by Dennis et al. of comparison between their augmented Gauss-Newton and a general Quasi-Newton method. These observations indicate the progress made with nonlinear least-squares, for McKeown (1975) tested large-residual problems and found Quasi-Newton methods better than special codes available at that time.

Once again we emphasise that these are very brief summaries of the articles referenced.

7. Special purpose algorithms

In a recent article on nonlinear fitting Dennis (1982) states, of nonlinear least-squares, that "there is not much left for us to do ..." . This statement might be accepted for general purpose unconstrained problems, but for constrained problems and those with a special structure, much remains to be done. Hiebert (1981) notes the need for algorithms in two problem areas: those in which the constraints are simple bounds, and those in which the constraints have a more general form.

Holt and Fletcher (1979) have developed an algorithm for simple bounds with a special structure and Wright and Holt (1985a) have developed an algorithm for general linear inequality constraints.

Separable nonlinear least-squares problems arise when the variables are grouped in ways which give the problems a block structure, or in which a subset of the variables appears linearly. In principle, all problems can be interpreted as separable; see for example Ruhe and Wedin (1980). Kaufman and Pereyra (1978) and Golub and Pereyra (1973) also contain descriptions of algorithms for separable least-squares problems.

Further structure can be identified in large sparse problems. These arise in earthquake ray path and travel time problems, for example. Until recently, only the Harwell code NS03A took account of sparsity; however Wright and Holt (1985b) have recently reported on a Levenberg-Marquardt method which exploits sparsity in the sequence of linear sub-problems arising at each iteration. In the same context, objective functions arise in which the component functions in the sum of squares may be non-differentiable at isolated points. An algorithm which handles such problems has been developed by Wright (1984).

8. Software

Some generally available routines for nonlinear least-squares problems are listed below. This list is essentially an updated version of the routines tested by Hiebert.

The MINPACK software from Argonne National Laboratory includes LMDER, LMDIF (Moré 1978). The National Physical Laboratory routines LSQFDN, LSQFDQ, LSQNDN (Gill and Murray 1978) are also available in Chapter E04 of the NAG Library. The algorithm NL2SOL of Dennis, Gay and Welsch (1981) is published in ACM TOMS as Algorithm 573.

Other routines are in the Harwell Library (NS03A; Fletcher 1971), and in IMSL (ZXSSQ).

If a prospective user is unable to acquire any of this software, a Fortran program is given by Osborne (1976), and Moré (1978) gives sufficient detail to enable his method to be coded with the aid of a QR factorisation routine.

Acknowledgments

Thanks are due to John Holt and Stephen Wright for their expert criticism and guidance.

CHAPTER 15

Problems in Time-Series Analysis

G. Tunnicliffe Wilson

1. Time-series modelling

We consider models of a series of observations $x_1, \ldots, x_n$ made at equally-spaced points in time - called a discrete time-series. The models are aimed at describing or representing the relationship between successive values. The main purpose is usually prediction of future values, but more general applications are in the extraction of various features from the time-series by filtering - e.g. seasonal adjustment.

It is rarely the case that the system from which the series arises is so well understood that a mechanistic (or structural) model may be confidently written down, and its free parameters estimated. More usually an empirical approach is employed, whereby a selection is made from a wide class of general models which have been proved by experience to be capable of fitting a variety of data, and this approach is followed here.

The usual assumptions of empirical modelling are:

(i) Stationarity. The data arises from a time homogeneous system which is in statistical equilibrium - this is in contrast to systems whose characteristics vary with time. The only relaxation of stationarity is to the borderline of stable systems which admits periodic and trend-like behaviour.

(ii) Linearity. The relationships between successive series values, and the operations applied in prediction and filtering are linear in the series values - though the model may contain free parameters which appear in a nonlinear manner.

(iii) Finiteness of state dimension. The system being modelled can be adequately represented using a finite set of variables in the model equations. This contrasts with the descriptive approach, in which an infinite set of autocorrelations (see section 2) or a continuous spectrum (see section 6) is used. In practice the amount of data available may prescribe whether one uses a simple approach which is prodigal in its use of parameters to approximate the system, or a sophisticated model which is parsimonious in its use of parameters.

(iv) Least-squares fitting criteria. This applies not only to model fitting, i.e. parameter estimation, but also to prediction and filtering. Together with linearity, it implies that only second-order statistics - involving products $x_t x_s$ of the data - are required in the modelling, and the stationarity assumption can be restricted to first and second-order moments.

There are several technical difficulties in time-series analysis which may be treated explicitly, or possibly 'glossed over' by some implicit procedure. They concern:

(a) End effects. These are due to the effect which values prior to the period of observation might have had on the values actually observed. Such effects would normally be transient, but even in long data sets are best not ignored. The assumption of stationarity can help.

(b) Resolution. The finiteness of the data may make it difficult or impossible to distinguish certain features in the model structure - or one feature may mask another if insufficient care is taken in modelling.

(c) Noise. In contrast with the statistical analysis of designed experiments, the precision with which parameters may be estimated is limited by the very nature of the data. This must be appreciated to avoid the detection of spurious features.

These comments will be illustrated by specific examples in the following sections.

2. Autocorrelation estimates and autoregressive modelling

The autocovariance at lag k, $\gamma_k = \text{Cov}(x_t , x_{t+k})$, is independent of t for a stationary series. The autocorrelation is $\rho_k = \gamma_k / \gamma_0$. Estimates of these over a range of lags $1 \leqslant k \leqslant K$ are important as descriptive quantities, and are used in fitting finite parameter models. The sequence ρ_k is known as the autocorrelation function or acf. Estimates C_k of γ_k are mostly of the form

$$C_k = \sum_{t=1}^{n-k} (x_t - \bar{x})(x_{t+k} - \bar{x}) / m \qquad (15.1)$$

where $\bar{x}$ is the sample average. The divisor most commonly used at present is $m = n$, which introduces a bias factor of approximately $(1 - k/n)$ but has advantages discussed in Jenkins and Watts (1968). The estimate r_k of ρ_k is usually c_k / c_0. Other divisors are recommended e.g. by Davies, Triggs and Newbold (1977) when constructing chi-squared tests for the presence of non-zero autocorrelation over a range of lags. The main criticism of the 'natural' divisor $m = n-k$ is that it can lead to a lack of positive definiteness, with values of $|r_k| > 1$. The end effect is revealed here in that use of the divisor $m = n$ means implicitly extending the observations outside the range $1 \leqslant t \leqslant n$ with values such that $x_t - \bar{x} = 0$. For a highly autocorrelated 'smooth' series, this leads to discontinuities with a severe distorting effect on any predictive application of r_k.

An autoregressive model has the form of a prediction equation

$$x_t = c + \varphi_1 x_{t-1} + \ldots + \varphi_p x_{t-p} + e_t \qquad (15.2)$$

where the sequence e_t is assumed to be an uncorrelated series with zero mean and constant variance. In many applications it is used as an approximation when the ARMA models of section 3 would be more suitable. In order that a sequence of values x_t, x_{t+1}, ... constructed recursively by (15.2) should reach statistical equilibrium, the characteristic polynomial $\varphi_p(z) = 1 - \varphi_1 z - \ldots - \varphi_p z^p$ must satisfy the classical stability condition, that for any factor $(1 - \alpha z)$ the constraint $|\alpha| < 1$ holds.

There is a simple relationship between $\varphi_1, \ldots, \varphi_p$ and $\rho_1, \ldots, \rho_p$ for the series model (15.2) - the Yule-Walker equations:

$$\rho_i = \sum_{j=1}^{p} \varphi_j \rho_{i-j}, \quad i = 1, \ldots, p . \qquad (15.3)$$

Estimates r_k of ρ_k can be substituted in this to obtain estimates $\hat{\varphi}_k$ of the autoregressive parameters. Provided the $(p+1)$ square matrix P with entries $P_{ij} = \rho_{i-j}$ is positive definite the stability condition will hold. Here we see the conflict between the desirable positive definiteness characteristic of using $m = n$ in (15.1) and the undesirable bias. Much algorithmic effort has been devoted to this problem. Solutions are obtained from the rapid recursive scheme for solving (15.3) for φ_k given ρ_k (and its reversal), given by Levinson (1947) and Durbin (1960). The dependence of any solution

$\varphi_1, \ldots, \varphi_p$ upon the choice of p is emphasised by inserting another suffix: thus, given $\varphi_{k,1}, \ldots, \varphi_{k,k}$, one may calculate

$$\left.\begin{aligned} \varphi_{k+1,k+1} &= (\rho_{k+1} - \sum_{j=1}^{k} \varphi_{k,j}\,\rho_{k+1-j}) / \tau_k \\ \varphi_{k+1,j} &= \varphi_{k,j} - \varphi_{k+1,k+1}\,\varphi_{k,k+1-j}\,, \quad j = 1,\ldots,k \\ \text{and} \quad \tau_{k+1} &= \tau_k\,(1 - \varphi^2_{k+1,k+1})\,, \quad \text{taking } \tau_0 = 1 \end{aligned}\right\} \quad k = 1,\ldots,p-1. \tag{15.4}$$

The sequence $\varphi_{k,k}$ for $k = 1,2,\ldots$ is called the partial autocorrelation function. The three sets $\rho_1, \ldots, \rho_p$; $\varphi_1, \ldots, \varphi_p$ ($\equiv \varphi_{p,1}, \ldots, \varphi_{p,p}$) and $\varphi_{1,1}, \ldots, \varphi_{p,p}$ are equivalent in that any one can be used to derive the others, so estimation of the acf may be considered by estimating one of the other sets. Moreover 'positivity' is characterised by $|\varphi_{k,k}| < 1$ for $k = 1,\ldots,p$, a condition which is readily checked, or constrained to be satisfied whilst avoiding bias. An excellent survey of autoregressive model estimation based on modifications of the above algorithm is given by Kay and Marple (1981), with a view to spectral analysis (see section 6); we shall note only one point here. Equation (15.2) may be fitted by ordinary least squares by letting t range from 1+p to n. The resulting normal equations have entries which are only slight variations on the Yule-Walker equations (15.3) with r_k replacing ρ_k. The stability condition on the estimates is not ensured, but the solution may again be obtained by a recursive scheme.

Finally it is possible to fit (15.2) under stationarity assumptions using maximum likelihood methods. The results are generally very similar to the ordinary least-squares estimates, but are considered in a more general context below.

3. Moving average and ARMA modelling

The moving average model supposes the observations to have arisen by a filtering process applied to an uncorrelated series e_t :

$$x_t = c + e_t - \theta_1 e_{t-1} - \dots - \theta_q e_{t-q} .$$

This model is appropriate when its characteristic property, that $\rho_k = 0$ for $k > q$, is inferred from small values of r_k for $k > q$, for some cut-off lag q. Thus $\rho_1 , \dots , \rho_q$ determine $\theta_1 , \dots , \theta_q$ subject only to an invertibility condition which is analogous to that for $\varphi_1 , \dots , \varphi_p$ in (15.2). The derivation of $\theta_1 , \dots , \theta_q$ from $\rho_1 , \dots , \rho_q$ is known as the Cramer-Wold factorisation, and a rapid (iterative) algorithm for this purpose, which is quite feasible for very large q, is given by Laurie (1982). It is a special case of spectral factorisation. The equations relating the coefficients are:

$$\rho_k = -(\theta_k - \theta_1\theta_{k+1} - \dots - \theta_{q-k}\theta_q) / (1 + \theta_1^2 + \dots + \theta_q^2). \quad (15.5)$$

One method of parameter estimation is to solve (15.5) for estimates $\hat{\theta}_k$, using r_k in place of ρ_k . This is not in general statistically efficient, and it is quite possible for no solution to exist. For example, consider the case $k = 1$, $\rho_1 = -\theta_1/(1 + \theta_1^2)$ which ranges over $[-\frac{1}{2} , \frac{1}{2}]$. A value of r_1 outside this range is quite possible, for which there is no corresponding θ_1 . Nevertheless, using better estimates of $\rho_1 , \dots , \rho_q$ which also incorporate information from $r_{q+1} , r_{q+2} , \dots ,$ Walker (1961) and Godolphin (1977) derive efficient and practical procedures based on these ideas.

An efficient two-stage estimation procedure proposed by Durbin (1959) is now receiving renewed attention. The first stage is to fit to the data an autoregressive model of high order p, using say the scheme of (15.4). A set of <u>inverse</u> autocorrelations ρi_k are then calculated using equation (15.5) but with p, φ_j replacing q, θ_j . These inverse autocorrelations are further subjected to the scheme (15.4) where they replace the ρ_k , and the recursions extended to order q (replacing p). The resulting coefficients $\varphi_{q,j}$ are taken as the estimates θ_j , $j = 1,\dots,q$. For example if $q = 1$,

$$\rho i_1 = -(\varphi_1 - \varphi_1\varphi_2 - \dots - \varphi_{p-1}\varphi_p)/(1 + \varphi_1^2 + \dots + \varphi_p^2)$$

whence $\hat{\theta}_1 = \rho i_1$. The efficiency of this scheme is proved by Bhansali (1980).

Maximum likelihood procedures may also be used for estimating the parameters, as discussed below.

The autoregressive-moving average (ARMA) model supposes that the observations arise from an (unobserved) uncorrelated series e_t according to the equation

$$x_t = c + \varphi_1 x_{t-1} + \ldots + \varphi_p x_{t-p} + e_t - \theta_1 e_{t-1} - \ldots - \theta_q e_{t-q} \,. \tag{15.6}$$

The justification for this model is that any single series x_t arising from a system which satisfies the assumptions (i) - (iv) in section 1, may be represented by such a model. The coefficients φ_j ,θ_j may in fact be (unknown) functions of a smaller set of parameters, but are usually taken as free. Exceptions are models which incorporate differencing and seasonality as expounded by Box and Jenkins (1970). The main algorithmic problems are well illustrated by confining attention to (15.6).

In theory a series following (15.6) can be approximated to arbitrary precision by an autoregressive or by a moving average model of sufficiently high order. In practice, with relatively small orders p, q, (15.6) is capable of representing a variety of series showing characteristics of cyclic or trend-like behaviour, which the 'pure' models (either $p = 0$ or $q = 0$) alone cannot adequately represent. Each 'mode' of the underlying system is represented by a linear or quadratic factor in the characteristic polynomial $\varphi_p(z) = 1 - \varphi_1 z - \ldots - \varphi_p z^p$, but will normally be balanced by moving average terms. For example, if the observations of x_t include a contaminating random error, it will usually be necessary to take $q = p$. Care taken in selecting and fitting an ARMA model can therefore reveal the main modes of the system, and lead to good extrapolations or forecasts. In particular, ARMA models probably provide the most sensitive means of resolving a pair of very similar modes (such as near-coincident cycles) in the data. There are however considerable difficulties in using them. Supposing that the orders p, q have been selected by some scheme, the estimation of parameters φ_j , θ_j is usually carried out by maximising a likelihood function. This is calculated by assuming that in (15.6), e_t is a random series from a Normal $(0,\sigma^2)$ distribution. The likelihood is then formally presented as

$$L \propto \sigma^{-n} M^{-\frac{1}{2}} \exp\left(-\tfrac{1}{2} Q/\sigma^2\right) ,$$

where $M = \det V$, $Q = \underline{x}^T V^{-1} \underline{x}$ and V has elements $V_{ij} = \gamma_{i-j} / \sigma^2$, which depend only on φ_j, θ_j. When maximised over σ, this yields a concentrated, or profile, likelihood $L \propto D^{-n/2}$ where $D = M^{1/n} Q$. The factor M does not depend on the data, but only on the parameters φ, θ. Moreover $M^{1/n} \to 1$ for all φ, θ as $n \to \infty$ so it is neglected in some procedures.

The quantity Q may be evaluated exactly or approximately by many devices, some based on the sample acf or sample spectrum, as in Hannan (1970). Most commmonly, the expression for Q is of the form of a sum of squares

$$Q = \sum_{t=1}^{n} e_t^2$$

where the e_t are regenerated from the data for any specified set of parameters φ, θ by rewriting (15.6) as

$$e_t = x_t - \varphi_1 x_{t-1} - \ldots - \varphi_p x_{t-p} + \theta_1 e_{t-1} + \ldots + \theta_q e_{t-q} - c \, .$$

Lack of knowledge of x_t for $t \leqslant 0$ (the end effect) prevents exact evaluation of $e_1, \ldots, e_n$, however, and a variety of algorithms are available which overcome this.

Box and Jenkins (1970) use the ingenious device of 'back forecasting' to estimate x_t for $t \leqslant 0$, whence $Q = \sum_{-L}^{n} e_t^2$, where L is a time origin prior to which the back forecasts are effectively zero. This does not yield an expression for M, which is neglected by them.

Newbold (1974) and Ljung and Box (1979) provide classical theory and algorithms for a finite procedure which estimates $x_0, \ldots, x_{1-p}$ and $e_0, \ldots, e_{1-q}$, yielding both Q and M exactly. A modification given by Tunnicliffe Wilson (1983) has some advantage in that it does not require calculation of theoretical autocorrelations of x_t, which all other schemes use as an intermediate step. The advantage is not so much a saving in direct computation, since the solution of least-squares equations for estimates $\hat{x}_0, \ldots, \hat{x}_{1-q}$ is still required. Rather, it allows sparsity of the coefficients, and the specification of $\varphi_p(z)$ as a product of factors, to be conveniently exploited, resulting in better numerical conditioning. Convenient computation of analytic derivatives also follows, and in fact the <u>whole</u> set of derivatives, and a good approximation to the <u>whole</u> Hessian of the likelihood, can each be computed with approximately the same effort as a single evaluation of the likelihood. Many published algorithms present calculation of the likelihood alone, and recommend the use of appropriate optimisation routines with numerical derivatives, and this should be borne in mind in any comparison. Close to the natural boundary of the parameter space, numerical derivatives can be inadequate. In these nearly singular regions, convergence of optimisation procedures may be slow at best, and analytic derivatives are valuable in avoiding any problems.

In recent years, methods have been used which calculate Q as a weighted sum of squares of innovations f_t based upon the finite data set:

$$f_t = x_t - E(x_t \mid x_{t-1}, \ldots, x_1)$$

so that

$$Q = \sum_1^n (f_t / h_t)^2, \qquad M = \prod_1^n h_t^2 .$$

Although $e_t - f_t \to 0$ and $h_t \to 1$ as t increases, the innovations f_t and the residuals e_t may differ for small values of t. The projection techniques for constructing f_t may be based on direct use of Cholesky factorisation as in Ansley (1979), or on the Kalman filter as in Gardner, Harvey and Phillips (1980). They are computationally efficient, and by exploiting the special structure of the ARMA model within the Kalman filter framework, Mélard (1984) has produced an algorithm which is extremely economical in its use of arithmetical operations and core storage.

The importance attached in recent years to the use of exact maximum likelihood estimation has arisen through appreciation of the bias inherent in other methods. The maximum likelihood equations are unbiased, and the solutions are constrained to satisfy the stability conditions - properties highly desirable in ARMA models. For example, for the model with p = 1, q = 0 and zero mean, the estimate of φ is the solution of the coupled equations:

$$\varphi = \left[\sum_1^{n-1} x_t x_{t+1} \right] \Big/ \left[\sum_2^{n-1} x_t^2 + \hat{\sigma}_x^2 \right] ;$$

$$\hat{\sigma}_x^2 = \{x_1^2 + \sum_2^n (x_t - \varphi x_{t-1})^2 / (1 - \varphi^2)\}/n .$$

There are however considerable problems arising from the fact that for a general ARMA (p,q) model, the likelihood can be a multi-modal function, and optimisation procedures may fail to locate the best model. Methods which are less than fully efficient statistically, but which can directly yield consistent parameter estimates, are therefore useful in providing initial parameter values for the optimisation. Many of these have a common basis, as explained by Piccolo and Tunnicliffe Wilson (1984). Currently, the method of Durbin (1960) has been reviewed by Hannan and Rissanen (1982). This again depends on a first stage of fitting a high-order autoregression to the data, but now with the aim of extracting the residuals e_t . The model (15.6) is then viewed as a linear regression for all the coefficients, φ_j , θ_j , with the entries e_{t-1} ,..., e_{t-q} being replaced by the known residuals. The canonical correlation analysis of Akaike (1976) uses a singular value decomposition to detect the pattern in the sample acf due to the autoregressive part of the model. Order selection criteria are generally part of these procedures. End effects are generally ignored in the first place.

4. Multiple time-series models

We consider first regression with ARMA error structures, i.e. a model relating y_t to explanatory variables $x_{j,t}$, for example

$$y_t = c + bx_t + n_t$$

where the error or noise n_t follows an ARMA process. The AR(1) process $n_t = \varphi n_{t-1} + e_t$ is common. A classical procedure due to Cochrane and Orcutt (1949) estimates the regression coefficient b and autoregressive coefficient φ alternately until convergence. More recent algorithms estimate all coefficients simultaneously (but iteratively).

Transfer function models allow a lagged response in the variable x_t . The model used by Box and Jenkins (1970) has the form

$$y_t = c + z_t + n_t ,$$

where $z_t = \delta z_{t-1} + \omega x_{t-b}$. This is formally equivalent to

$$y_t = c + \omega x_{t-b} + \omega\delta x_{t-b-1} + \omega\delta^2 x_{t-b-2} + \ldots + n_t .$$

Again, n_t follows an ARMA model. This form is known to economists as a distributed lag model and should not be confused with ARMAX models in which lagged values of y_t are introduced, e.g.

$$y_t = c + \delta y_{t-1} + \omega x_{t-b} + n_t .$$

In all cases the end effects now involve the explanatory variable x_t, besides the ARMA error n_t. The user of any algorithms should take careful note of how this is treated. One possibility is to introduce any unknown values x_t for $t \leqslant 0$ as nuisance parameters.

Multivariate ARMA models are appropriate when two or more series are interdependent. The model (15.6) is generalised to vector series $\underline{x}_t$ and $\underline{e}_t$, with matrix coefficients φ_j, θ_j. Many of the univariate considerations generalise, including estimation algorithms such as those presented by Nicholls and Hall (1979), Osborn (1977), Hillmer and Tiao (1979). There is an added complexity due to the topology of the parameter space. The work by Hannan and Kavalieris (1984) covers many important aspects of this problem.

5. Recursive algorithms

The Kalman filter algorithm for state estimation and prediction has been long used in time-series. Harrison and Stevens (1976) applied it in prediction. More recently it has been used for likelihood calculations - its use is reviewed by Harvey (1984). Recursive parameter estimation, by which parameters are updated as each new observation comes to hand, has been virtually the rule (and 'batch' estimation the exception) in control engineering literature, and the field is vast. One particularly successful scheme for transfer function model estimation exploits ideas of instrumental variables, and has been developed by Young and Jakeman (1979), and Jakeman and Young (1980). Models of nonlinear time-series phenomena have been researched extensively. Priestley (1980) reviews and generalises these models.

6. Spectral methods of analysis

Spectral methods of time-series analysis are basically motivated by the search for cycles in data, i.e. patterns of the form

$$R \cos(\omega t + \varphi) = A \cos \omega t + B \sin \omega t \tag{15.7}$$

where ω is the angular frequency. The frequency expressed as the number of cycles per unit time is $f = \omega/2\pi$, and the period of the cycle is $p = 1/f$. The amplitude is R, and the phase φ.

If the data $x_1, \ldots, x_n$ consists of a finite combination of pure cycles, the frequencies present may be detected by exploiting asymptotic orthogonality properties:

$$\lim (2/n) \sum_1^n \cos \omega t \cos \omega' t = \lim (2/n) \sum_1^n \sin \omega t \sin \omega' t = \begin{cases} 1, & \omega = \omega', \\ 0, & \omega \neq \omega', \end{cases}$$

$$\lim (2/n) \sum_1^n \cos \omega t \sin \omega' t = 0 .$$

Thus if

$$x_t = \sum_\omega R(f) \cos(\omega t + \varphi(f)), \tag{15.8}$$

and we let

$$\left.\begin{aligned} \hat{A}(f) &= (2/n) \sum x_t \cos \omega t, \quad \hat{B}(f) = (2/n) \sum x_t \sin \omega t, \\ \hat{R}(f)^2 &= | (2/n) \sum x_t e^{i\omega t} |^2 = \hat{A}(f)^2 + \hat{B}(f)^2, \end{aligned}\right\} \tag{15.9}$$

then as $n \to \infty$, $\hat{A}(f), \hat{B}(f), \hat{R}(f) \to A(f), B(f), R(f)$. The natural range of frequencies f is $[0,0.5]$, i.e. $p \geqslant 2$, because a cycle with any higher frequency cannot be distinguished by sampling at discrete times (integer t), from a cycle with frequency in this range. It is said to be aliased, and the highest frequency of one cycle per two sampling intervals is called the Nyquist frequency. We shall call the quantity

$$S^*(f) = (n/2)\,\hat{R}(f)^2 , \qquad 0 \leq f \leq 0.5 , \tag{15.10}$$

the sample spectrum of $x_1, \ldots, x_n$. The reason for the factor $(n/2)$ will become clear later. Various scaling factors are in use - the important feature is the shape of $S^*(f)$ as a function of f. Given data following (15.8), the limit of $(2/n)S^*(f)$ is thus $R(f)^2$ at any represented frequency f, and zero at any other frequency. For finite data sets, $S^*(f)$ is a continuous function with sharp peaks of height approximately $(n/2)R(f)^2$ and width $1/n$ at the represented frequencies.

If the only frequencies present in the model (15.8) are harmonic frequencies corresponding to a fundamental period of n, i.e. $f_j = j/n$, $0 \leq j \leq n/2$, then $\hat{R}(f)$ is exactly equal to $R(f)$ for these frequencies. Such a situation is rare, being appropriate for example when the data is known to be exactly periodic with $x_{t+n} = x_t$, or is spatial, being measured at n equally-spaced points on the perimeter of a circular object. More usually, the sample spectrum would be computed at a somewhat finer grid of frequencies, in order not to miss a peak half-way between harmonic frequencies, taking for example $f_j = j/(4n)$. The Fast (Finite) Fourier transform (see Chapter 8) can be used for this computation.

In trying to represent the cycles present in data, the sample spectrum is limited even when the data model (15.8) is valid. Difficulties arise due to

(i) frequencies f which are closer than $1/n$, i.e. which differ by less than one cycle in the data length. Their spectral peaks are not resolved in $S^*(f)$;

(ii) frequencies f which are separated by appreciably more than $(1/n)$ but which are highly disparate in magnitude. Each spectral peak in $S^*(f)$ is surrounded by sidelobe peaks at distances $\pm (j + \frac{1}{2})/n$ from f, which have approximate comparative heights $1/\{\pi^2 (j + \frac{1}{2})^2\}$, for $j = 1,2,\ldots$. A sidelobe due to a large component may mask the main peak of a small component at a frequency some distance away - a phenomenon known as leakage.

Note that these limitations of resolution are not necessarily inherent in the data. If it were known that the model (15.8) held exactly, with say two components, then fitting the model by least squares and searching over all possible pairs of frequencies would determine the components exactly. The limitation of the sample spectrum is due to the approximate nature of the orthogonality properties in finite samples. The sidelobes may be considered as arising from an end effect whereby series values outside the range

$t = 1,\ldots,n$ are implicitly taken to be zero.

Though motivated by the search for cycles, the sample spectrum can be viewed as a device for data transformation which has a valid interpretation in other contexts, in particular when the data is stochastic. In the simple case when x_t is a random sample from a Normal distribution, with zero mean and variance σ^2, the linear transformation to $\hat{A}(f)$ and $\hat{B}(f)$ in (15.9) is orthogonal for the harmonic frequencies f_j , whence $\hat{A}(f_j)$, $\hat{B}(f_j)$ are also independent and Normal, with zero means and variances $2\sigma^2/n$. It follows that $S^*(f_j)$ at harmonic frequencies f_j , form a random chi-squared distribution on two degrees of freedom, or equivalently, are exponentially distributed with expectation $2\sigma^2$. The mid-90% range of this distribution is [0.051,3.0] about its mean - a range of variation of the order 60, so that if $S^*(f)$ is graphed continuously, it fluctuates wildly from one harmonic frequency to the next, producing many peaks which the incautious analyst might falsely interpret as being due to cycles.

We note that

$$S^*(f) = 2\{C_0 + 2 \sum_{1}^{n-1} C_k \cos \omega k \} , \qquad (15.11)$$

where C_k is the autocovariance estimate given in (15.1), using the divisor $m = n$. For a stationary time-series the sample spectrum has therefore a natural interpretation as an 'estimate' of the theoretical spectrum, defined as

$$S(f) = 2\{\gamma_0 + 2 \sum_{1}^{\infty} \gamma_k \cos \omega k \} . \qquad (15.12)$$

However, although $E(S^*(f)) \simeq S(f)$, the distribution of $S^*(f)$ about $S(f)$ is (under wide assumptions) precisely that described for the simple case above (which is a special case of a stationary series).

We are assuming that (15.12) converges, in which case we say that the series has a <u>continuous</u> spectrum $S(f)$. In practice, the series may have a mixed spectrum, with a continuous part (often due to 'noise') and a <u>discrete</u> part (the signal) consisting of pure sinusoids as in (15.8). The problem is to get good estimates of both these parts. At any particular frequency the sample spectrum has expected amplitude $S(f) + (n/2)R(f)^2$ where $R(f)$ is the amplitude of the discrete component. Detection of the signal peak above the noise therefore depends on the signal to noise ratio $R(f)^2/S(f)$ at that frequency, and on the data length n.

The continuous spectrum may also have peaks (possibly narrow) which are due to resonant modes in the stochastic system generating the data. Furthermore, the cycles in data may not be pure, but may have irregularities or be subject to various modulations so that the 'power' $\frac{1}{2}R(f)^2$ associated with the sharp peak of height $(n/2)R(f)^2$ and width $(1/n)$ in the spectrum, becomes spread over a broader peak. Estimating a mixed spectrum is therefore not a clear cut procedure.

7. Spectrum smoothing

Assuming the spectrum to be continuous, the fluctuating sample spectrum may be smoothed to provide an estimate of the theoretical spectrum with, asymptotically, arbitrary precision (as $n \to \infty$).

Widely referenced works are by Blackman and Tukey (1959), Jenkins and Watts (1968) and more recently Bloomfield (1976). This last book probably represents the present state of the art, so far as 'traditional' direct methods of spectral estimation are concerned. The paper by Kay and Marple (1981) contrasts these methods with 'non-traditional' methods which are discussed in section 8 below. (See also the papers by Carter and Nuttall (1983) and Nuttall (1983).)

Three procedures are used.

(i) Sectioning the data into (possibly overlapping) subsets. The idea is to average the sample spectra of these subsets to reduce the variability of the estimate.

(ii) Tapering of the data (subsets) down to zero at both ends. This reduces the end effect discontinuity and reduces the sidelobe (leakage) problem. For an impressive example which also focuses on the problem and treatment of outliers, see Kleiner, Martin and Thomson (1979).

(iii) Windowing of the autocovariances. The spectrum estimate of the (possibly tapered) data is calculated from its sample autocovariances C_k as

$$\hat{S}(f) = 2\{C_0 + 2\sum_{1}^{M} \omega_k C_k \cos \omega k\}$$

where M is a truncation lag, and the lag window weights ω_k reduce to zero at lag M. This device smooths the spectrum.

Sectioning of the data is now uncommon outside some engineering and physical science circles, since windowing as in (iii) is designed to achieve the same end. Tapering as in (ii) has also been partly dropped in favour of pre-whitening. This is because the end effects are often associated with low frequencies in the data. Pre-whitening is a process of filtering the data by calculating, say $x'_t = x_t - \varphi x_{t-1}$ for some φ close to unity. This near-differencing reduces end effect discontinuities and sidelobes. The spectrum of x'_t is calculated, then divided by $|1 - \varphi e^{i\omega}|^2$ to estimate that of x_t. This technique is also considered by Kleiner et al. (1979).

For some years now, frequency window spectrum smoothing has also been employed. Possibly after tapering or pre-whitening, the sample spectrum $S^*(f)$ is calculated directly (using the FFT) as in (15.10). The smoothed spectrum estimate $\hat{S}(f')$ say, is obtained at each frequency f', by averaging the values of $S^*(f)$ over an interval (of chosen half-width δ) centred upon f'. The average need not be uniform, but may attach more weight to values towards the centre of the interval, according to a frequency window shape. For a uniform window the estimate is conveniently calculated by forming

$$T^*(f) = \int_0^f S(f')\,df' ,$$

and then

$$\hat{S}(f) = \{T^*(f+\delta) - T^*(f-\delta)\}/(2\delta) .$$

The integration and differencing are of course performed 'discretely' in practice, and are efficient operations. Moreover if the whole operation is repeated, possibly with a different value of δ, the net effect is to use a frequency window of trapezium shape, with somewhat smoother results.

Lag and frequency window smoothing are equivalent operations, the only difference being the finite cut-off of the lag window which implies an equivalent frequency window covering the whole range $[0, \frac{1}{2}]$, and the finite interval of the frequency window which implies a lag window extending to the maximum lag (n-1). The frequency window is currently more in fashion.

Whatever method is used, it is important that the statistical properties be clearly presented. A user of these methods should be able to specify the bandwidth of the smoothing, that is, the effective width of the frequency window implicitly or explicitly used. Estimates at frequencies separated by more than this bandwidth are effectively independent. The precision of the estimate depends on two considerations. Bias occurs due to leakage – flattening of peaks and filling in of dips – if the bandwidth is too large. A higher variance of the estimate results if the bandwidth is too small. Confidence intervals for the estimate which allow for the variance (but cannot allow for bias) should always be calculated.

The main interest is often the detection of peaks in the spectrum, but it is instructive to appreciate the natural variability of the spectrum even in large samples. With 1000 data points, and a modest bandwidth of 1/40, the confidence interval limits are in the ratio approximately 2:1.

The real challenge is to estimate mixed spectra. The difficulty with spectrum smoothing is that peaks due to the discrete part are smoothed out (into the shape of the frequency window). By comparing spectrum estimates with low resolution (wide bandwidth) and high resolution (narrow bandwidth) some progress can be made on this problem; see for example the comprehensive text by Priestley (1981). Kay and Marple (1981) present a testing example with a data set of 64 points containing three discrete sinusoids besides noise. The continuous spectrum due to the noise is theoretically low at the frequencies of the discrete components, so that, statistically, sufficient information is present to detect these cycles. It is in this context that the inadequacy of spectral estimation methods so far described, becomes evident.

8. Autoregressive spectrum estimation

Although discrete and continuous spectra are dissimilar in so far as they are associated with deterministic and stochastic processes respectively, they are similar in that they may both be represented, or well approximated, by autoregressive (-like) equations. Thus the cycle (15.7) satisfies $x_t = \varphi_1 x_{t-1} + \varphi_2 x_{t-2} + e_t$ where $\varphi_1 = 2\cos\omega$, $\varphi_2 = -1$ and $\mathrm{var}(e_t) = 0$. This is a limiting form of autoregression. Parzen (1968, 1969) proposed that a suitably high-order autoregressive model (as in (15.2)) should well represent any stationary series, and from this the spectrum may be derived as

$$\hat{S}(f) = 2\,\hat{\sigma}_e^2 \left| 1 - \hat{\varphi}_1 e^{i\omega} - \ldots - \hat{\varphi}_p e^{ip\omega} \right|^{-2} .$$

The choice of order p is comparable to the choice of bandwidth (which is approximately 1/(2p) in an asymptotic sense). It may be chosen automatically using various criteria, following Akaike (1969).

In so far as the autoregressive coefficients are determined from the first p acf values r_k, the method may be viewed as a device for extrapolating the acf beyond lag p, by using the first p values. The bias which derives from the lag window of spectrum smoothing as in section 7 is therefore avoided.

The estimation procedure for the autoregression can be critical if high resolution of the spectrum is important. If the usual estimate of r_k (as discussed in section 2, with the divisor m = n) is used together with the Levinson-Durbin algorithm, then resolution can be lost for the reason previously mentioned, i.e. the end effect of implicitly prefixing zeros to the data. Following from Burg (1978) various devices have been used to avoid this effect. Thus for example, ρ_1 may be estimated by a sample value of

$$r_1 = \left\{ \sum_1^{n-1} x_t x_{t+1} \right\} \Big/ \left[\left\{ \sum_1^{n-1} x_t^2 \right\} \left\{ \sum_1^{n-1} x_{t+1}^2 \right\} \right]^{\frac{1}{2}}$$

which satisfies $|r_1| < 1$ but avoids the bias factor of $(1 - 1/n)$ in the previous value. (We are assuming for convenience that x_t is mean-corrected.) The 'obvious' generalisation to higher lags does not however ensure positivity in the whole sample acf. Instead, the method focuses upon recursive estimates of the partial acf, which for example may be calculated at lag k+1 as

$$\hat{\varphi}_{k+k,k+1} = \left\{ \sum_{2+k}^{n} e_t f_{t-k-1} \right\} \Big/ \left[\left\{ \sum_{2+k}^{n} e_t^2 \right\} \left\{ \sum_{2+k}^{n} f_{t-k-1}^2 \right\} \right]^{\frac{1}{2}} .$$

Here e_t is the forward prediction error, and f_t the backward prediction error, calculated for example as

$$e_t = x_t - \varphi_{k,1}\, x_{t-1} - \dots - \varphi_{k,k}\, x_{t-k} ,$$

$$f_t = x_t - \varphi_{k,1}\, x_{t+1} - \dots - \varphi_{k,k}\, x_{t+k} ,$$

although they may also be updated by simple recursions. Positivity follows because $|\hat{\varphi}_{k,k}| < 1$.

Much higher resolution may be obtained in the spectrum estimate by this means. The ilustrations given by Kay and Marple (1981, p. 1409) are very instructive in this context. Several new algorithms are compared by Goutis and Ibrahim (1983) who claim to achieve further improvement for the particular case of sine wave signals in noise.

9. Cross-spectrum estimation

An early, and still important, reason for cross-spectrum analysis was the estimation of the transfer function relationship between two series y_t and x_t . In the (discrete) time domain, the basic model is

$$y_t = v_0 x_t + v_1 x_{t-1} + v_2 x_{t-2} + \ldots + n_t , \qquad (15.13)$$

where n_t is the noise, and all of y_t , x_t , n_t are assumed to be jointly stationary. In the frequency domain the convolution of the impulse response sequence v_k with the input x_t , becomes a simple product:

$$y(f) = V(f)\, x(f) + n(f) \qquad (15.14)$$

where the quantities in this equation are the Fourier transforms of those in (15.13), ignoring the end effects. Given that these transforms are complex-valued, (15.14) provides two regression equations for the two unknown parts of the transfer function V(f) at any frequency f (the values of y(f) and x(f) being computed from the data). To obtain sufficient degrees of freedom to estimate both V(f) and the noise spectrum, the equations (15.14) are grouped for frequencies within a set of chosen bandwidths, over each of which V(f) is assumed to be relatively constant. The estimates are in practice derived from the Var-Covariance matrix of y_t , x_t in the frequency domain, with entries which are, for example, estimates of the cross-spectrum

$$S_{yx}(f) = \sum_{-\infty}^{\infty} \gamma_{yx,k}\, e^{i\omega k} ,$$

where $\gamma_{yx,k}$ is the lagged cross covariance, $\mathrm{Cov}(y_t , x_{t+k})$. In a similar notation $S_{yy}(f)$, $S_{xx}(f)$ are the auto spectra of y_t , x_t . These may all be estimated for any chosen bandwidth using lag or frequency domain smoothing windows with the sample auto and cross covariances. The transfer function and noise spectrum estimates are then

$$\hat{V}(f) = \hat{S}_{yx}(f) \,/\, \hat{S}_{xx}(f)\ , \quad \hat{S}_{n}(f) = \hat{S}_{yy}(f) - |\hat{S}_{yx}(f)|^2 \,/\, \hat{S}_{xx}(f)\ .$$

The squared coherency W(f) between the series measures the strength of their relationship at each frequency, and is estimated by

$$\hat{W}(f) = |\hat{S}_{yx}(f)|^2 \,/\, \{\hat{S}_{xx}(f)\ \hat{S}_{yy}(f)\}\ .$$

It is difficult to find a better text than that of Jenkins and Watts (1968) for spelling out the practical concerns in this estimation procedure. Parzen (1967) sets out the basic statistics of the problem. A new concern is the reduction of bias by proper alignment shifts of the two series. The calculation should also give standard errors of the estimates, which are essential if correct application is to be made. For the transfer function, the information is usually presented as a gain estimate $|\hat{V}(f)|$ and a phase estimate $\arg(\hat{V}(f))$ which are more readily interpreted.

Estimates of the impulse response $\hat{v}_k$ and the noise acf $\hat{r}_{n,k}$ may be derived for use in parametric modelling.

Advances in this area have not been major, except in specific applications, where consideration is given to design of the input signal x_t. The use of multivariate autoregression for cross-spectrum estimation has however received a fair amount of attention, for example by Marple and Nuttall (1983). An important point made in their conclusion is that, <u>unlike</u> the lag or frequency window techniques, the resulting auto spectrum estimate $\hat{S}_{xx}(f)$ say, depends on whether the single series x_t, or the pair of series x_t and y_t, is used, to fit a univariate or bivariate autoregression respectively.

10. Filtering

At one time the province of electrical circuit designers, filtering is now very much a digital technique. The fundamental equation is the calculation of one series y_t say from another, x_t, according to

$$y_t = \sum_k \alpha_k\, x_{t-k}$$

where k may range over positive and/or negative integer lags, and may in theory have infinite range. The spectrum of y_t is related to that of x_t by $S_y(f) = G(f)^2 S_x(f)$ where $G(f) = |\alpha(f)|$ and $\alpha(f) = \sum_k \alpha_k e^{ik\omega}$.

By design of the coefficients α_k, the filter can enhance some frequency components and remove others. Users should be aware that although the gain $G(f)$ is sometimes all that is specified for a filter, it is possible for many filters to have the same gain. Full specification may require a further assumption, for example that the filter is two-sided and symmetric ($\alpha_k = \alpha_{-k}$), i.e. zero-phase, or is physically realizable and invertible ($\alpha_k = 0$ for $k < 0$ and $\sum \alpha_k z^k \neq 0$ for $|z| < 1$).

Symmetric filters of the form $\alpha_k = \omega_k \cos \omega' k$, where ω_k is a weight function similar to that used for lag windows, are useful in selecting frequency components in a band centred on ω'. Realizable filters which achieve similar aims can be based on ARMA (or transfer function model) type recursive equations, such as

$$y_t = \theta_0 x_t - \theta_1 x_{t-1} + \varphi_1 y_{t-1} + \varphi_2 y_{t-2} .$$

A combination of forward and backward filtering of this type can be used to obtain symmetry if that is desired.

As a technique of signal processing, filtering is attractive to the scientist searching for data cycles - see Currie (1981). It is open to abuse if (as is commonly the situation) no error limits are provided for the extracted signal. Ideas of optimal linear filtering based upon signal + noise models are necessary for statistical rigour. The Wiener (1949) filter provides this rigour, and is presented for example by Whittle (1983). Recent applications to economic signal processing (seasonal adjustment) are presented by Burman (1980). In this context the proper treatment of end effects is particularly important. The use of forecasting and back forecasting, based upon an ARMA model, to extend the data, is particularly appropriate for Wiener filtering. The Kalman filter, originally conceived as an extension of the Wiener filter to models with time-dependent coefficients, is also particularly appropriate for handling end effects in the stationary situation. The filter may be augmented by the backward smoothing algorithm to provide optimal two-sided filtering (with identical results to the Wiener filter). Sage and Melsa (1971) provide the basic algorithms.

11. Availability of software

Many statistical packages such as SPSS, BMD, MINITAB contain time-series facilities. The user should check the estimation criterion used, and whether parameter constraints might also be considered useful. Many packages cover univariate ARMA modelling, but not Transfer Function modelling. Facilities for spectral analysis are included in SPSS and BMD.

The NAG Library contains a range of routines including ARMA and Transfer Function model estimation, and univariate and cross-spectral analysis. The GENSTAT package distributed by NAG covers basically the same features, with rather more options. It contains a basic Fourier transformation directive, which allows short programs or macros, using say ten directives, to be written for spectral analysis, including autoregressive spectrum estimation.

There are also special-purpose packages written for 'signal processing'. NAG distributes TSA, which besides spectral analysis contains facilities for data manipulation, regression, ARIMA modelling and filtering such as seasonal adjustment. Graphical output facilities are available.

Software originally developed at Lancaster under G.M. Jenkins is available from Gwilym Jenkins and Partners, and that developed at Madison under G.E.P. Box is available through Automatic Forecasting Systems. A program called CAPTAIN for recursive modelling, whereby parameters are updated as each new observation comes to hand, is available from Professor P.C. Young at Lancaster University.

The Signal Processing and Applications Group (SPAG) at Cranfield Institute of Technology supply both hardware and software for processes including signal digitisation, data editing and filtering, spectral analysis, and graphical presentation. The Institute of Sound and Vibration Research (ISVR) at Southampton University has a similar product.

CHAPTER 16

Advances in Algorithms for Surface Fitting

J.G. Hayes

1. Introduction

As in Chapter 10 on curve fitting, I shall be presenting a variety of algorithms from the past few years. They are mostly for fitting bivariate splines by least squares or by interpolation, but there are some departures. For the most part, they are developments from the curve-fitting algorithms and I shall not usually repeat the earlier details. Thus after covering, in the next section, some additional notation and definitions relating to splines, the following three sections will concern the spline surface routines in DASL. This is our new Data Approximation Subroutine Library (Anthony, Cox and Hayes 1982), which is available from NAG. It represents the bulk of data-fitting software produced at NPL since the routines in the NAG Library, discussed (some of them in prospect) in Hayes (1978).

An extra factor arises in surface problems: the arrangement of the data points in the plane of the two independent variables. We shall encounter two cases: that where the data points are arbitrarily disposed (often called scattered data) and that where the points lie at all the nodes of a rectangular mesh. This latter case is amenable to the development of much faster computational methods. The two cases will be considered separately in sections 3 and 4, and together, for the periodic situation, briefly in section 5. In sections 6 and 7, we shall be concerned with further developments from work discussed for curve fitting: respectively, automatic fitting algorithms for both the above data arrangements and a monotonic interpolation algorithm for the mesh case.

In the final section 8, we shall look at algorithms for scattered data which really have no counterpart in one variable. The bivariate spline method of section 3 is especially appropriate to smooth surfaces, and probably not more than a few hundred data points. These other methods, in contrast, have been developed mainly for surfaces with many peaks and troughs and with data ordinates that often look random except in relation to near neighbours. Typically, the data are measurements of some geophysical quantity at points scattered over a bit of the earth's surface. There are many data points.

Again I should comment that my omission of polynomials from the discussion is not because they are no longer important (indeed for surfaces they are probably just as important as splines) but because the methods were established long ago. The still-current algorithms are discussed in Hayes (1978).

2. Notation and definitions

We now have two independent variables x and y, the dependent variable being f still. We shall be concerned with the rectangle in the (x, y) plane given by

$$x_{min} \leqslant x \leqslant x_{max} \quad \text{and} \quad y_{min} \leqslant y \leqslant y_{max}, \tag{16.1}$$

which contains our data. Commonly x_{min} and x_{max} will be respectively set equal to the smallest and largest of the data values of x, and correspondingly for y. Again we have a set of interior knots λ_i, $i = 1,\ldots,N_x$, defined as before for the variable x, and now we also have a corresponding set μ_j, $j = 1,\ldots,N_y$, for the variable y. The lines $x = \lambda_i$ and the lines $y = \mu_j$ divide the rectangle defined by (16.1) into rectangular panels.

A <u>bivariate (polynomial) spline</u>, with the two knot sets $\{\lambda_i\}$ and $\{\mu_j\}$, and of orders n_x and n_y, is defined as having the form

$$s(x, y) = \sum\sum \alpha_{u,v}\, x^u y^v$$

in each of the above panels (where the summations run from $u = 0$ to n_x-1 and from $v = 0$ to n_y-1) with the property that

$$\frac{\partial^{p+q} s(x, y)}{\partial x^p \partial y^q}, \quad \text{for} \quad 0 \leqslant p \leqslant n_x-2 \quad \text{and} \quad 0 \leqslant q \leqslant n_y-2 ,$$

is continuous throughout the rectangle (16.1). Adding knots appropriately at or outside each end of each of the intervals in (16.1), we can in the same way as before define two sets of (univariate) B-splines, $N_{n_x,i}(x)$ and $N^*_{n_y,j}(y)$. Any bivariate spline as defined above then has a unique B-spline representation

$$s(x, y) = \sum \sum c_{i,j} N_{n_x,i}(x) N^*_{n_y,j}(y) \tag{16.2}$$

in the rectangle (16.1), where the summations run from $i = 1$ to q_x and from $j = 1$ to q_y . Here $q_x = N_x + n_x$ and $q_y = N_y + n_y$.

Scattered data points will be denoted by (x_h , y_h , f_h) with weights w_h , $h = 1,2,\ldots,m$. Then the observation equations are

$$s(x_h , y_h) = f_h , \quad h = 1,2,\ldots,m , \tag{16.3}$$

with $s(x, y)$ as in (16.2), and for the least-squares fitting problem we minimize the weighted residual sum of squares

$$\sum_{h=1}^{m} e_h^2 , \tag{16.4}$$

where

$$e_h = w_h [f_h - s(x_h , y_h)] . \tag{16.5}$$

(The interpolation problem with scattered data cannot sensibly be attacked by bivariate splines, and alternative ways are in section 8.)

Data points on a rectangular mesh are denoted by $(x_h , y_k , f_{h,k})$, $h = 1,2,\ldots,m_x$ and $k = 1,2,\ldots,m_y$. Only the case of unit weights is considered: general weights prevent the use of the faster algorithms and the points would have to be treated as scattered points. The observation equations are now

$$s(x_h, y_k) = f_{h,k}, \quad h = 1,\ldots,m_x, \quad k = 1,\ldots,m_y, \tag{16.6}$$

with s(x, y) as in (16.2). In the interpolation problem (which is satisfactory with this data), these equations are satisfied exactly. In the least-squares problem, the residual sum of squares is

$$\sum\sum e_{h,k}^2, \tag{16.7}$$

where

$$e_{h,k} = f_{h,k} - s(x_h, y_k), \tag{16.8}$$

and the summations run from h = 1 to m_x and from k = 1 to m_y.

3. Bivariate splines with scattered data

For this case, the observation equations can be written in matrix form as

$$A\,\underline{c} = \underline{f}, \tag{16.9}$$

where $\underline{c}$ denotes $[c_{1,1}\ c_{2,1} \ldots c_{q_x,q_y}]^T$, A is an $m \times (q_x q_y)$ matrix with its columns correspondingly ordered and with row h containing the values of the B-spline products at (x_h, y_h), and $\underline{f}$ is the vector of values of the f_h. The matrix A is a band matrix, with bandwidth $(n_y-1)q_x + n_x$, provided the data points are appropriately ordered. This ordering is according to the particular panel $(\lambda_i \leq x < \lambda_{i+1}, \mu_j \leq y < \mu_{j+1})$ into which each data point falls.

NAG routine E02DAF, for the special case of bicubic splines $(n_x = n_y = 4)$, computes a least-squares solution to the weighted form of (16.9) by first reducing WA to triangular form using Givens rotations. Here $W = \text{diag}\,[w_h]$. To get full advantage from the band form, an auxiliary routine E02ZAF was provided to arrange the data in panel order (using a linked list, not actually changing the actual order). Since it is common in the bivariate case for A to be rank-deficient (and there is nothing corresponding to the

Schoenberg-Whitney conditions of the univariate case), E02DAF was designed to give a solution in the non-unique situation. The solution computed in this situation is the minimum-norm solution.

The new routine in DASL deals with general order, different in the two variables if required. It still obtains the solution via Givens rotations, but applied as described by Cox (1981) allowing the data points to be taken in any order. This is more beneficial than for one variable, as the points in the bivariate case would have to be ordered after every change in the choice of knots to restore the band form. Of course, some penalty in efficiency is involved in the Givens reduction when the points are out of order, and that penalty could be significant. In the rank-deficient case, instead of computing the minimum-norm solution, the new routine introduces resolving constraints based on those described for the curve case. If $c_{i,j}$ is an indeterminate coefficient, the resolving constraint expresses $c_{i,j}$ as a weighted average of $c_{i-1,j}$, $c_{i+1,j}$, $c_{i,j-1}$ and $c_{i,j+1}$ (if they exist). Thus again the 'missing' $c_{i,j}$ are filled in smoothly so that any inadequacies of the data will tend to be made good in a smooth way. Related routines in DASL are an evaluation routine and two routines providing first partial derivatives with respect to either x or y, in the form of bivariate splines. Higher derivatives can be obtained by repeated use of the routines and cross derivatives by using both routines.

4. Bivariate splines with data on a mesh

The gain in computational speed that is possible for this case stems from the fact that it can be treated as a series of curve-fitting problems, first fitting the data on each line in one direction and then fitting the coefficients so produced, by a set of splines. For each value of k in turn, a spline curve in x is fitted to the data $(x_h\ ,\ f_{h,k})$, $h = 1,\ldots,m_x$, yielding B-spline coefficients $d_{i,k}$, say. Then for each value of i in turn $(i = 1,\ldots,q_x)$, a spline curve in y is fitted to the data $(y_k\ ,\ d_{i,k})$, $k = 1,\ldots,m_y$. The coefficients of these splines are the required $c_{i,j}$ in (16.2).

Algebraically, the process is first of all the solution of the system

$$A_x D = F\ , \qquad (16.10)$$

where A_x is a band matrix of dimensions $m_x \times q_x$ and bandwidth n_x containing in row h the values at x_h of the B-splines in x, F is the $m_x \times m_y$ rectangular matrix of values of $f_{h,k}$, and D is a $q_x \times m_y$ rectangular matrix of the 'intermediate coefficients'. Thus (16.10) can be seen as several sets of linear equations with different right-hand sides (the columns of F) but the same left-hand matrix A_x . The solution of (16.10) can be obtained just as before, first reducing A_x to upper triangular form by means of Givens rotations. In the corresponding manner, we then solve

$$A_y C^T = D^T , \tag{16.11}$$

where A_y is the counterpart of A_x for the y-variable and C denotes the $q_x \times q_y$ rectangular matrix of the required values of $c_{i,j}$.

The extent of the gain in speed is greater than might be imagined. With a 30×30 mesh of data points and 10 interior knots in each direction, using a bicubic spline, the reduction in the number of multiplications over employing the routine of the previous section, designed for general data, is a factor of about 400.

The routine in DASL requires the x_h to be ordered and the y_k also. A unique solution is (in theory) guaranteed if the Schoenberg-Whitney conditions are satisfied in each variable separately, and the routine checks for this. If numerical rank deficiency is encountered in solving either (16.10) or (16.11), resolving constraints are introduced as in the univariate case. The routine also requires the two knot sets to be provided, but the simple routine provided for the univariate case can be used to give at least a start.

The interpolation problem is solved also through (16.10) and (16.11), with $m_x = q_x$ and $m_y = q_y$. The DASL routine for this problem selects its own interior knots according to the rule $\lambda_i = x_{r+1}$ if $n_x = 2r$ and $\lambda_i = \frac{1}{2}(x_{i+r-1} + x_{i+r})$ if $n_x = 2r-1$, for $i = 1,2,\ldots,N_x$, and similarly for the y-variable. This is known to be a good set in most circumstances. The equations (16.10) and (16.11) are solved by Gaussian elimination, without pivoting, so that the band form is not destroyed in the computations. This is quite safe, because the (now square) matrices A_x and A_y are totally positive. Thus the advantage of the band form is maintained in the basically faster computation of Gaussian elimination (vis-a-vis Givens rotations).

The routines mentioned in the previous section for evaluation and differentiation apply equally here.

5. Periodic spline surfaces

Corresponding to each of the three fitting routines of the previous two sections (a least-squares routine for scattered data and one for mesh data, and an interpolation routine for the latter case), there is a DASL routine in which the bivariate spline is periodic in one variable. These are appropriate, for example, to problems in which we are given data in cylindrical coordinates (r, z, θ) and wish to fit r as a function of z and θ. This involves defining one set of the B-splines in (16.2) with respect to a periodic knot-set, and then solving the observation equations with an algorithm that takes account of the particular structure. For mesh data, the algorithm follows directly from the curve case. For scattered data, all the non-zero elements of the matrix A in (16.9) which are outside the band structure can be arranged to lie in a number of the right-hand columns of the matrix, as in the univariate case. The algorithm given by Cox (1981) can again be applied, therefore.

There are also routines for evaluation and differentiation, corresponding to those above for non-periodic surfaces.

6. Almost automatic algorithms

In section 5 of Chapter 10, I described an algorithm which requires the user to choose only one parameter (other than the order of the spline) in fitting a spline curve. It chooses the knots by a simple ad-hoc system involving the results from least-squares fits as more knots are gradually added. Then with the knot set found, a kind of smoothing spline is determined which minimizes an intuitively acceptable measure of smoothness subject to the residual sum of squares for the given data being less than a constant S. It is the value of S, the smoothing factor, which the user has to choose.

Dierckx (1981b, 1982) has extended this work of his to surface fitting. For the case of scattered data, starting with no knots, just one knot at a time is added (on the grounds that the computation in this case can become substantial and can increase rapidly with the number of knots). At each addition, a spline is fitted by least squares and the residual sum of squares in each knot interval is computed. That is, for each i, the value of e_h^2, with e_h as in (16.5), is summed over all data points in the interval

$\lambda_i \leq x < \lambda_{i+1}$; similarly, for each j, e_h^2 is summed over all data points in $\mu_j \leq y < \mu_{j+1}$. Then a new knot is added in the interval having the largest sum, whether it be an x-interval or a y-interval (with the x-interval having preference in the case of equality). In fact, the knot is positioned at the weighted mean of the data values in the chosen interval (the x-values if it is an x-interval and correspondingly for y), the weights being proportional to the values of e_h .

The smoothing measure used in the univariate case is the sum of squares of the discontinuities at the knots of the (n-1)th derivative of the spline. To extend this to the bivariate case, Dierckx first derives conditions, linear in the B-spline coefficients, for all the possible derivative discontinuities of the bivariate spline to be zero. These conditions are of the form

$$\Sigma\ u_{t,i}\, c_{t,r} = 0, \quad i = 1,\ldots,N_x\ ; \quad r = 1,\ldots,q_y \tag{16.12}$$

and

$$\Sigma\ v_{r,j}\, c_{t,r} = 0, \quad j = 1,\ldots,N_y\ ; \quad t = 1,\ldots,q_x\ , \tag{16.13}$$

where the summations run from t = 1 to q_x and from r = 1 to q_y respectively. In (16.12), $u_{t,i}$ is the value of the discontinuity at $x = \lambda_i$ of the derivative of order n_x-1 of the B-spline $N_{n_x,t}(x)$. A corresponding definition applies to the $v_{r,j}$ in (16.13). Both sets of equations have a band structure. The fitting criterion is then

$$\text{minimize } D, \quad \text{subject to} \quad \delta \leq S\ , \tag{16.14}$$

where D is the smoothing measure defined as the sum of squares of the left-hand sides of (16.12) and (16.13), and δ is the weighted residual sum of squares. The minimization is with respect to the B-spline coefficients. The numerical solution of this problem is shown to contain the sub-problem:

$$\text{find the spline } s_p(x,\ y) \text{ to minimize } D + p\delta\ , \text{ for given p.} \tag{16.15}$$

The solution to (16.15) is the solution of the set of equations, linear in the B-spline coefficients, comprising the observation equations (16.9) and the two sets (16.12) and (16.13). The observation equations have the given weights w_h and each equation in (16.12) and (16.13) has a weight $1/\sqrt{p}$. Defining

$$F(p) = \sum_{h=1}^{m} w_h^2 [f_h - s_p(x_h , y_h)]^2 , \qquad (16.16)$$

the final, outer part of the computation to satisfy the chosen fitting criterion is to find the value of p such that

$$F(p) = S . \qquad (16.17)$$

The spline $s_p(x, y)$ corresponding to that value of p is the required solution.

Dierckx (1982) provides an algorithm in the same vein for the case of data on a rectangular mesh. A modification to the above smoothness measure is necessary in order that full advantage may be taken of this data arrangement: the main computation becomes equivalent to repeated curve fitting.

7. Monotonic interpolation

Carlson and Fritsch (1985) have extended to surfaces their univariate work on monotonic interpolation, in the case that the data points lie on a rectangular mesh. The meaning of monotonic is not quite so clear for surfaces, but the authors take what seems to be the most natural definition: that a surface is monotonic if all its sections at constant x are monotonic in the same direction (i.e. all increasing or all decreasing) and the sections for constant y also (though not necessarily in the same direction as the sections at constant x). The interpolating function is taken to be a bicubic polynomial on each panel of the data mesh, with first-derivative continuity across the panel boundaries. The function is described by its function value, first partial derivative parallel to each axis and the first cross-derivative, at each point of the mesh. The problem is thus one of choosing the other three parameters, given the first. The authors derive necessary and sufficient conditions on the parameters to ensure monotonicity, but simplify them to a set of sufficient conditions in order to develop a computational algorithm. This is clearly much more difficult than in the univariate case, but the above paper outlines an iterative algorithm, too involved to give here, and preliminary documentation of the code is in Fritsch and Carlson (1983).

8. Algorithms for large sets of scattered data

As stated in the introduction, we shall be concerned here with algorithms designed for surfaces with many peaks and troughs and many data points. A very large number of algorithms have been developed. Curiously (since the scattered data suggest measurements, and measurement errors that one usually wants to smooth out) nearly all of them are for interpolation. I shall confine myself mainly to these interpolation algorithms, their programs being more readily available.

The survey by Schumaker (1976) gives a very good introduction into the whole topic, approximation as well as interpolation. Barnhill (1977) surveys interpolation methods from a computer-aided design viewpoint and Sabin (1980) from the viewpoint of contour plotting (which indeed provided the motivation for many of the methods). The results of an impressive numerical comparison of interpolation methods are given by Franke (1979) and, updated a little and shorn of close on 300 pages of detailed results, in Franke (1982a). Franke tested 29 different programs (though paying most attention to 13 of them), which he classified into seven different categories. He compared them for such qualities as storage required, accuracy of approximation (when fitted to data from a known function), appearance of surface, sensitivity to choice of a parameter (if any), and timing for both preprocessing (which would include fitting, if any) and evaluation of the interpolant. Not surprisingly perhaps, no method is found to be universally best in all these qualities or over all data sets (and one must appreciate that, even in extensive tests such as these, the range of problems is necessarily severely limited). For any particular data set, Franke (1984) simply advises that as many algorithms as possible should be tried.

Here, I shall outline just two algorithms whose details have become available quite recently, since they illustrate the current tendency towards a 2-stage or 3-stage approach and have basic ideas which seem equally applicable to the approximation problem. They make use of two earlier methods so we shall need to look at those first, and in passing I will mention the algorithm which, in Franke's tests, was overall best of all the 'local' algorithms. The dichotomy between 'local' and 'global' algorithms is an important quality of an algorithm, 'global' meaning that the interpolated value at any point depends on all the data and 'local' that it depends only on data in the neighbourhood of the point. Global methods tend to give more acceptable results for small data sets, but for larger sets (Franke says 100 to 200 points) global methods tend to become computationally too expensive – though there are exceptions.

In the commonest form of the early methods, due to Shepard (1968), the interpolating function, $F(x, y)$, is given by

$$F(x, y) = \sum_{h=1}^{m} w_h(x, y) f_h \Big/ \sum_{h=1}^{m} w_h(x, y) , \qquad (16.18)$$

where

$$w_h(x, y) = d_h^{-2} , \quad \text{and} \quad d_h^2 = (x - x_h)^2 + (y - y_h)^2 . \qquad (16.19)$$

We see that $F(x, y)$ here is a weighted mean of the data ordinates f_h , the weights being functions of x and y. There is no representation of $F(x, y)$ in the sense used in earlier sections, which implies computationally that there is no preprocessing stage: the computation is wholly evaluation at the required values of x and y. This, and its global nature, makes the method potentially expensive. Also the first partial derivatives of $F(x, y)$ are zero at all the data points, limiting its smoothness. On the other hand, the fact that the interpolant lies within the range of the f_h is an advantage for the most part, avoiding the wild fluctuations which can arise with some methods – the more so when the method is made local, as below.

This basic method has been extended and modified in various alternative ways to improve its performance. It was one of these modifications (Franke and Nielson 1980) that turned out best overall of the local algorithms in Franke's tests.

The method is made local by modifying each $w_h(x, y)$ so that, while it still goes to infinity in the same manner as (x, y) approaches (x_h , y_h), it is zero outside a circle of some suitable radius centred on (x_h , y_h) and has zero slope everywhere on the circumference of the circle. This latter provision ensures that $F(x, y)$ has first-derivative continuity. This is the degree of continuity usually sought in these methods (of course, global Shepard has all its derivatives continuous). The problem of $F(x, y)$ having zero derivatives at the data points is overcome by replacing each f_h in (16.18) by a function $f_h(x, y)$ whose behaviour at (x_h , y_h) is more like that expected of $F(x, y)$ - since $F(x, y)$ will have the same tangent plane at (x_h, y_h) as does $f_h(x, y)$. In the Franke-Nielson modification (called the Modified Quadratic Shepard's Method in Franke's tests), each $f_h(x, y)$ is obtained as a weighted least-squares fit of a quadratic in x and y to data points in the neighbourhood of (x_h, y_h). The fit is constrained to pass through (x_h, y_h, f_h), which it must do to make F an interpolant. Computation of the coefficients in the m least-squares fits constitutes the pre-computation phase.

The other early method (Harder and Desmairis 1972) is based on the solution to the problem of a thin plate forced to pass through the data by point loads. The form of the interpolant is

$$Q(x, y) = \sum_{h=1}^{m} A_h d_h^2 \ln d_h + a + bx + cy , \qquad (16.20)$$

where d_h is as in (16.19). The values of A_h and of a, b and c are determined as the solution of the m linear equations formed by substituting the data points (x_i, y_i, f_i) in turn into (16.20), with f_i replacing $Q(x, y)$, together with the three linear equations $\sum_{h=1}^{m} A_h = 0$, $\sum_{h=1}^{m} A_h x_h = 0$ and $\sum_{h=1}^{m} A_h y_h = 0$. (This solution comprises the pre-computation stage.) It is one of the best global algorithms in Franke's tests.

Franke (1982b) has used this algorithm as part of a local algorithm. Consider a rectangle R in the (x, y) plane containing the data and divided into sub-rectangles with boundaries $x = \lambda_i$, $i = 0,\ldots,N_x+1$ and $y = \mu_j$, $j = 0,\ldots,N_y+1$ (using the same notation as for knots in the similar situation of section 2). Then, for each $i = 1,\ldots,N_x$ and $j = 1,\ldots,N_y$, an interpolant $Q_{i,j}(x, y)$ of the form (16.20) is derived from the data lying in the rectangle somewhat larger than the rectangle $R_{i,j}$, defined by $\lambda_{i-1} \leqslant x \leqslant \lambda_{i+1}$, $\mu_{j-1} \leqslant y \leqslant \mu_{j+1}$ (somewhat larger since this is found to improve results). Then nonnegative weight functions $W_{i,j}(x, y)$ are defined such that (with modifications near the boundary of R) each $W_{i,j}$ has a peak value of unity at (λ_i, μ_j), goes to zero with zero derivative all around the boundary of $R_{i,j}$ and is zero outside $R_{i,j}$. Including the modifications near the boundary, we also have that the sum of the $W_{i,j}$ over all i and j is identically equal to unity. The interpolant is then given by

$$F(x, y) = \sum \sum W_{i,j}(x, y)\, Q_{i,j}(x, y)\,, \qquad (16.21)$$

where the summations run from $i = 1$ to N_x and from $j = 1$ to N_y. The pre-computation phase is to derive the $Q_{i,j}$. Evaluation at any (x, y) involves a weighted mean with only four non-zero weights (less on a mesh boundary). The form of the $W_{i,j}$ is designed to provide a smooth transition from one set of local interpolants to the next, and provides first-derivative continuity. The choice of the λ_i and μ_j can be made by the user or left to the program.

The final method, due to Foley (1983), also uses a rectangular mesh but proceeds quite differently, in three stages. The first stage is to provide a value at each node of the mesh, the value being that of a quadratic in x and y fitted by least squares to the data nearest the node. The second stage is to compute a bicubic spline interpolant to these mesh values. Of course, this spline does not interpolate the original data, and the final stage is to apply a modified Shepard's algorithm to the differences between the data and the spline. This interpolant added to the spline gives the final interpolant. Again the mesh can be chosen by the user or by the program. The consideration behind this 3-stage algorithm is that many basic interpolation algorithms, such as Shepard's, tend to give not very smooth results, whereas a bicubic spline is inherently smooth, but cannot interpolate scattered data. The algorithm reduces the lack of smoothness in Shepard's method by applying the method only to the hopefully small correction surface. The method is, of course, global because of the bicubic spline, but we have already commented on the great speed advantage when interpolating mesh data, and also the evaluation of the bicubic spline is very fast. The method is also global, however, because a global form of Shepard is used (resulting in the second-derivative continuity of the spline being carried through to the final interpolant). Even so, Foley found satisfactory speed in tests with up to 800 data points.

APPENDIX

Addresses of Organisations Supplying Software

ACM Transactions on Mathematical Software
ACM Algorithms Distribution Service,
IMSL Inc., Sixth Floor, NBC Building,
7500 Bellaire Boulevard, Houston, Texas 77036, USA.

BMD
BMDP, 1964 Westwood Boulevard, Suite 202,
Los Angeles, California 90025, USA.

Gwilym Jenkins and Partners Ltd.
Parkfield, Greaves Road, Lancaster.

HARWELL Library
Computer Science and Systems Division, A.E.R.E. Harwell,
Didcot, Oxfordshire, OX11 ORA, UK.

IMSL
IMSL Inc., Sixth Floor, NBC Building,
7500 Bellaire Boulevard, Houston, Texas 77036, USA.

LINPACK (and MINPACK)
National Energy Software Center, Argonne National Laboratory,
9700 South Cass Avenue, Argonne, Illinois 60439, USA.

MINITAB
Pennsylvania State University, 1600 Woodland Road,
Abington, Pennsylvania 19001, USA.

NAG Library (also GENSTAT)
NAG Central Office, Mayfield House, 256 Banbury Road,
Oxford, OX2 7DE, UK.

SLATEC Library
National Energy Software Center, Argonne National Laboratory,
9700 South Cass Avenue, Argonne, Illinois 60439, USA.

SPSS
SPSS Inc., Suite 3000, 444 N. Michigan Avenue,
Chicago, Illinois 60611, USA.

REFERENCES

Abd-Elal, L.F. and Delves, L.M. (1976). A regularisation technique for a class of singular integral problems. *J. Inst. Math. Appl.* 18, pp. 37-47.

Abd-Elal, L.F. and Delves, L.M. (1984). Fredholm integral equations with singular convolution kernels. Preprint, Dept. of Statistics and Computational Mathematics, University of Liverpool.

Abramowitz, M. and Stegun, I.A. (eds.) (1964). *Handbook of Mathematical Functions with Formulas, Graphs and Mathematical Tables.* Nat. Bur. Standards. *Appl. Math. Series* 55, U.S. Govt. Printing Office, Washington, D.C.. Dover reprint (1968).

Adams, J., Swartztrauber, P. and Sweet, R. (1978). FISHPAK: Efficient FORTRAN subprograms for the solution of separable elliptic partial differential equations. National Center for Atmospheric Research, Boulder, Colorado.

Adams, L. (1983). M-step preconditioned conjugate gradient methods. NASA Contractor Report No. 172130.

Akaike, H. (1969). Power spectrum estimation through autoregression model fitting. *Ann. Inst. Statist. Math.* 21, pp. 407-419.

Akaike, H. (1976). Canonical correlation analysis of time series and the use of an information criterion. In *Systems Identification : Advances and Case Studies* (eds. R.K. Mehra and D.G. Lainiotis) pp. 27-96. Academic Press, New York.

Akin, J.E. (1976). The generation of elements with singularities. *Int. J. Numer. Methods Engng.* 10, pp. 1249-1259.

Akin, J.E. (1982). *Application and Implementation of the Finite Element Method.* Academic Press, New York.

Aktas, Z. and Stetter, H.J. (1977). A classification and survey of numerical methods for boundary value problems in ordinary differential equations. *Int. J. Numer. Methods Engng.* 11, pp. 771-796.

Albasiny, E.L. (1979). A subroutine for solving a system of differential equations in Chebyshev series. In *Codes for Boundary-Value Problems in Ordinary Differential Equations* (eds. B. Childs, M. Scott, J.W. Daniel, E. Denman and P. Nelson) *Lecture Notes in Computer Science* 76, pp. 280-286. Springer-Verlag, Berlin.

Anderssen, R.S. and Prenter, P.M. (1981). A formal comparison of methods proposed for the numerical solution of first kind Fredholm equations. *J. Aust. Math. Soc. B* 22, pp. 488-500.

Anselone, P.M. (1981). Singularity subtraction in the numerical solution of integral equations. *J. Aust. Math. Soc. B* 22, pp. 408-418.

Ansley, C.F. (1979). An algorithm for the exact likelihood of a mixed autoregressive moving average process. *Biometrika* 66, pp. 59-65.

Anthony, G.T., Cox, M.G. and Hayes, J.G. (1982). DASLDOC1 - An outline of the NPL Data Approximation Subroutine Library. National Physical Laboratory, Teddington, Middlesex.

Ascher, U., Christiansen, J. and Russell, R.D. (1979). COLSYS - A collocation code for boundary-value problems. In *Codes for Boundary-Value Problems in Ordinary Differential Equations* (eds. B. Childs, M. Scott, J.W. Daniel, E. Denman and P. Nelson) *Lecture Notes in Computer Science* 76, pp. 164-185. Springer-Verlag, Berlin.

Ascher, U., Christiansen, J. and Russell, R.D. (1981). Collocation software for boundary-value ODEs. *ACM Trans. Math. Software* 7, pp. 209-229.

Atkinson, K.E. (1976). *A Survey of Numerical Methods for the Solution of Fredholm Integral Equations of the Second Kind.* SIAM, Philadelphia.

Axelsson, O. (1979). On the preconditioned conjugate gradient method. In *Conjugate Gradient Methods and Similar Techniques* (ed. I.S. Duff). Harwell Report AERE-R9636. HMSO, London.

Babolian, E. and Delves, L.M. (1979). An extended Galerkin method for first kind Fredholm equations. *J. Inst. Math. Appl.* **24**, pp. 157-174.

Babuska, I. and Rheinboldt, W. (1978). Error estimates for adaptive finite-element computations. *SIAM J. Numer. Anal.* **15**, pp. 736-754.

Bader, G. and Deuflhard, P. (1983). A semi-implicit mid point rule for stiff systems of ordinary differential equations. *Numer. Math.* **41**, pp. 373-398.

Bailey, P.B. (1966). Sturm-Liouville eigenvalues via a phase function. *SIAM J. Appl. Math.* **14**, pp. 242-249.

Bailey, P.B. and Shampine, L.F. (1979). Automatic solution of Sturm-Liouville eigenvalue problems. In *Codes for Boundary-Value Problems in Ordinary Differential Equations* (eds. B. Childs, M. Scott, J.W. Daniel, E. Denman and P. Nelson) *Lecture Notes in Computer Science* **76**, pp. 274-279. Springer-Verlag, Berlin.

Bailey, P.B., Gordon, M.K. and Shampine, L.F. (1978). Automatic solution of the Sturm-Liouville problem. *ACM Trans. Math. Software* **4**, pp. 193-208.

Baker, C.T.H. (1968). On the nature of certain quadrature formulas and their errors. *SIAM J. Numer. Anal.* **5**, pp. 783-804.

Baker, C.T.H. (1977). *The Numerical Treatment of Integral Equations.* Oxford University Press. (2nd printing 1978.)

Baker, C.T.H. (1982). An introduction to the numerical treatment of Volterra and Abel-type integral equations. In *Topics in Numerical Analysis* (ed. P.R. Turner) *Lecture Notes in Mathematics* **965**, pp. 1-35. Springer-Verlag, Berlin.

Baker, C.T.H. and Hodgson, G.S. (1971). Asymptotic expansions for integration formulas in one or more dimensions. *SIAM J. Numer. Anal.* **8**, pp. 473-480.

Baker, C.T.H. and Keech, M.S. (1978). Stability regions in the numerical treatment of Volterra integral equations. *SIAM J. Numer. Anal.* **15**, pp. 394-417.

Baker, C.T.H. and Miller, G.F. (eds.) (1982). *Treatment of Integral Equations by Numerical Methods.* Academic Press, London.

Baker, C.T.H. and Phillips, C. (eds.) (1981). *The Numerical Solution of Nonlinear Problems.* Clarendon Press, Oxford.

Banfield, C.F. (1978). Singular value decomposition in multivariate analysis. In *Numerical Software - Needs and Availability.* (ed. D.A.H. Jacobs) pp. 137-149. Academic Press, London.

Bank, R.E. and Sherman, A.H. (1981). An adaptive multi-level method for elliptic boundary-value problems. Computing **26**, pp. 91-105.

Barker, V. (ed.) (1976). *Sparse Matrix Techniques. Lecture Notes in Mathematics* **572**. Springer-Verlag, Berlin.

Barnhill, R.E. (1977). Representation and approximation of surfaces. In *Mathematical Software III* (ed. J.R. Rice) pp. 69-120. Academic Press, New York.

Barrodale, I. (1978). Best approximation of complex-valued data. In *Numerical Analysis. Proceedings, Biennial Conference. Dundee 1977* (ed. G.A. Watson) *Lecture Notes in Mathematics* **630**, pp. 14-22. Springer-Verlag, Berlin.

Barrodale, I. and Chow, K. (1982). Least absolute values and least squares methods for identifying and trimming outliers. Report #DCS-22-IR, Dept. of Computer Science, University of Victoria, B.C.

Barrodale, I. and Phillips, C. (1974). An improved algorithm for discrete Chebyshev linear approximation. In *Proceedings of the Fourth Manitoba Conference on Numerical Mathematics* (ed. R. Stanton) pp. 177-190. University of Manitoba Press, Winnipeg.

Barrodale, I. and Phillips, C. (1975). Algorithm 495: Solution of an overdetermined system of linear equations in the Chebyshev norm [F4]. *ACM Trans. Math. Software* 1, pp. 264-270.

Barrodale, I. and Roberts, F.D.K. (1973). An improved algorithm for discrete L_1 linear approximation. *SIAM J. Numer. Anal.* 10, pp. 839-848.

Barrodale, I. and Roberts, F.D.K. (1974). Algorithm 478: Solution of an overdetermined system of equations in the L_1 norm [F4]. *Comm. ACM* 17, pp. 319-320.

Barrodale, I. and Roberts, F.D.K. (1978). An efficient algorithm for discrete L_1 linear approximation with linear constraints. *SIAM J. Numer. Anal.* 15, pp. 603-611.

Barrodale, I. and Roberts, F.D.K. (1980). Algorithm 552: Solution of the constrained L_1 linear approximation problem [F4]. *ACM Trans. Math. Software* 6, pp. 231-235.

Barrodale, I., Delves, L.M. and Mason, J.C. (1978). Linear Chebyshev approximation of complex-valued functions. *Math. Comp.* 32, pp. 853-863.

Barrodale, I., Zala, C.A. and Chapman, N.R. (1984). Comparison of the L_1 and L_2 norms applied to one-at-a-time spike extraction from seismic traces. *Geophys.* 49, pp. 2048-2052.

Bartels, R.H. and Stewart, G.W. (1972). Solution of the equation AX+XB=C. *Comm. ACM* 15, pp. 820-826.

Bartels, R.H., Conn, A.R. and Sinclair, J.W. (1978). Minimization techniques for piecewise differentiable functions - the L_1 solution to an overdetermined linear system. *SIAM J. Numer. Anal.* 15, pp. 224-241.

Barwell, V. and George, J.A. (1976). A comparison of algorithms for solving symmetric indefinite systems of linear equations. *ACM Trans. Math. Software* 2, pp. 242-251.

Bathe, K.J. and Wilson, E.L. (1976). *Numerical Methods in Finite Element Analysis*. Prentice-Hall, Englewood Cliffs, N.J.

Bauer, F.L., Rutishauser, H. and Stiefel, E. (1963). New aspects in numerical quadrature. In *Proc. Symp. Appl. Math.* 15, pp. 199-218. American Mathematical Society, Providence, R.I.

Bavely, C. and Stewart, G.W. (1979). An algorithm for computing reducing subspaces by block diagonalization. *SIAM J. Numer. Anal.* 16, pp. 359-367.

Beltyukov, B.A. (1965). An analogue of the Runge-Kutta method for the solution of a nonlinear integral equation of Volterra type. *Diff. Equa.* 1, pp. 417-433.

Berzins, M. and Dew, P.M. (1981). A generalised Chebyshev method for nonlinear parabolic PDEs in one space variable. *IMA J. Numer. Anal.* 1, pp. 469-487.

Berzins, M., Dew, P.M. and Furzeland, R.M. (1984). Software for time-dependent problems. In *PDE Software: Modules, Interfaces and Systems* (eds. B. Engquist and T. Smedsaas) pp. 309-322. North-Holland, Amsterdam.

Betts, P.L. (1979). A variational principle in terms of stream function for free surface flows. *Comput. Fluids* 7, pp. 145-153.

Bhansali, R.J. (1980). Autoregressive and window estimates of the inverse correlation function. *Biometrika* 67, pp. 551-566.

Bjorck, A. (1967). Iterative refinement of linear least squares solution I. *BIT* 7, pp. 257-278.

Bjorck, A. (1968). Iterative refinement of linear least squares solution II. *BIT* 8, pp. 8-30.

Bjorck, A. (1982). A block QR algorithm for partitioning stiff differential systems. Report LITH-MAT-R-82-44, Dept. of Mathematics, Linkoping University.

Bjorck, A. and Duff, I.S. (1980). A direct method for the solution of sparse linear least squares problems. Harwell Report CSS-79, A.E.R.E. Harwell, Oxfordshire.

Blackman, R.B. and Tukey, J.W. (1959). *The Measurement of Power Spectra from the Viewpoint of Communications Engineering*. Dover, New York.

Blommers, J.P. (1978). *Design Algorithms for Finite Impulse Response Digital Filters*. M.Sc. Thesis, Dept. of Mathematics, University of Victoria, B.C.

Bloomfield, P. (1976). *Fourier Analysis of Time Series*. Wiley, New York.

Box, G.E.P. and Jenkins, G.M. (1970). *Time Series Analysis : Forecasting and Control*. Holden-Day, San Francisco. Revised edition 1976.

Brandt, A. (1982). Guide to multigrid development. In *Multigrid Methods* (eds. W. Hackbusch and U. Trottenberg) *Lecture Notes in Mathematics* 960, pp. 220-312. Springer-Verlag, Berlin.

Brent, R.P. and Luk, F.T. (1985). The solution of singular value and symmetric eigenvalue problems on multiprocessor arrays. *SIAM J. Sci. Stat. Comp.* 6, pp. 69-84.

Brezinski, C. (1980). A general extrapolation algorithm. *Numer. Math.* 35, pp. 175-187.

Brodlie, K.W. (1977). Unconstrained minimization. In *The State of the Art in Numerical Analysis* (ed. D.A.H. Jacobs) pp. 229-268. Academic Press, London.

Brodlie, K.W. (1980). *Mathematical Methods in Computer Graphics and Design*. Academic Press, London.

Brunner, H. (1982). A survey of recent advances in the numerical treatment of Volterra integral and integro-differential equations. *J. Comp. Appl. Math.* 8, pp. 213-229.

Brunner, H. and te Riele, H.J.J. (1982). Volterra-type integral equations of the second kind with non-smooth solutions: high order methods based on collocation techniques. Report 118/82, Mathematisch Centrum, Amsterdam.

Brunner, H., Hairer, E. and Nørsett, S.P. (1982). Runge-Kutta theory for Volterra integral equations of the second kind. *Math. Comp.* 39, pp. 147-163.

Bulirsch, R. (1964). Bemerkungen zur Romberg-Integration. *Numer. Math.* 6, pp. 6-16.

Bunch, J.R. (1971). Analysis of the diagonal pivoting method. *SIAM J. Numer. Anal.* 8, pp. 656-680.

Bunch, J.R. and Parlett, B.N. (1971). Direct methods for solving symmetric indefinite systems of linear equations. *SIAM J. Numer. Anal.* 8, pp. 639-655.

Bunse-Gerstner, A. (1984). An algorithm for the symmetric generalized eigenvalue problem. *Lin. Alg. App.* 58, pp. 43-68.

Burg, J.P. (1978). Maximum entropy spectral analysis. In *Modern Spectrum Analysis* (ed. D.G. Childers) pp. 34-41. IEEE Press Selected Reprint Series, New York.

Burman, J.P. (1980). Seasonal adjustment by signal extraction. *J. Roy. Stat. Soc. A* 143, pp. 321-337.

Burton, T.A. (1983). *Volterra Integral and Differential Equations*. Academic Press, New York.

Butcher, J.C. (1975). A stability property of implicit Runge-Kutta methods. *BIT* 15, pp. 358-361.

Butcher, J.C. (1976). Implicit Runge-Kutta and related methods. In *Modern Numerical Methods for Ordinary Differential Equations* (eds. G. Hall and J.M. Watt) pp. 136-151. Clarendon Press, Oxford.

Byers, G. (1984). A LINPACK-style condition estimator for the equation $AX-XB^T=C$. *IEEE Trans. Auto. Cont.* AC-29, pp. 926-928.

Byrd, R.H. (1982). Algorithms for robust regression. In *Nonlinear Optimization 1981* (ed. M.J.D. Powell) pp. 79-84. Academic Press, London.

Byrne, G.D. and Hindmarsh, A.C. (1975). A polyalgorithm for the numerical solution of ordinary differential equations. *ACM Trans. Math. Software* 1, pp. 71-96.

Carey, G.F. and Oden, J.T. (1983). *Finite Elements, A Second Course.* Prentice-Hall, Englewood Cliffs, N.J.

Carlson, R.E. and Fritsch, F.N. (1985). Monotone piecewise bicubic interpolation. *SIAM J. Numer. Anal.* 22, pp. 386-400.

Carter, G.C. and Nuttall, A.H. (1983). Analysis of a generalised framework for spectral estimation, Part 1. *IEE Proc.* 130, Pt F, pp. 239-241.

Carver, M.B. and Boyd, A.W. (1979). A program package using stiff, sparse integration methods for the automatic solution of mass action chemical kinetics. *Inst. J. Chem. Kinet.* 11, pp. 1097-1108.

Carver, M.B. and MacEwan, S.R. (1981). On the use of sparse matrix approximation to the Jacobian in integrating large sets of ordinary differential equations. *SIAM J. Sci. Stat. Comp.* 2, pp. 51-64.

Chambers, J.M. (1977). *Computational Methods for Data Analysis.* Wiley, New York.

Chan, S.P., Feldman, R. and Parlett, B.N. (1977). Algorithm 517: A program for computing the condition numbers of matrix eigenvalues without computing eigenvectors. *ACM Trans. Math. Software* 3, pp. 186-203.

Chan, T.F. (1982a). An improved algorithm for computing the singular value decomposition. *ACM Trans. Math. Software* 8, pp. 72-83.

Chan, T.F. (1982b). Algorithm 581: An improved algorithm for computing the singular value decomposition. *ACM Trans. Math. Software* 8, pp. 84-88.

Chan, T.F. and Jackson, K.R. (1984). The use of iterative linear-equation solvers in codes for large systems of stiff IVPs for ODEs. Technical Report No. 170/84, Dept. of Computer Science, University of Toronto.

Chandler, G.A. (1980). Product integration methods for weakly singular second kind integral equations. Preprint, Australian National University, Canberra.

Chapman, N.R. and Barrodale, I. (1983). Deconvolution of marine seismic data using the L_1 norm. *Geophys. J. R. Astr. Soc.* 72, pp. 93-100.

Chapman, N.R., Barrodale, I. and Zala, C.A. (1984). Measurement of sound-speed gradients in deep-ocean sediments using L_1 deconvolution techniques. *IEEE J. Ocean. Eng.* OE-9, pp. 26-30.

Childs, B., Scott, M., Daniel, J.W., Denman, E. and Nelson, P. (eds.) (1979). *Codes for Boundary-Value Problems in Ordinary Differential Equations. Lecture Notes in Computer Science* 76. Springer-Verlag, Berlin.

Ciarlet, P.G. (1976). *The Finite Element Method for Elliptic Problems.* North Holland, Amsterdam.

Claerbout, J.R. and Muir, F. (1973). Robust modelling with erratic data. *Geophys.* 38, pp. 826-844.

Cline, A.K., Conn, A.R. and Van Loan, C.F. (1982). Generalising the LINPACK condition estimator. In *Numerical Analysis* (ed. J.P. Hennart) *Lecture Notes in Mathematics* 909, pp. 73-83. Springer-Verlag, New York.

Cochrane, D. and Orcutt, G.H. (1949). Application of least-squares regression to relationships containing autocorrelated error terms. *J. Amer. Statist. Assoc.* **44**, pp. 32-61.

Concus, P. and Golub, G.H. (1976). A generalised conjugate gradient method for non-symmetric systems of linear equations. In *Lecture Notes in Economics and Mathematical Systems* **134**, pp. 56-65. Springer-Verlag, Berlin.

Connor, J.J. and Brebbia, C.A. (1978). *Finite Element Techniques for Fluid Flow*. Newnes Butterworths, London.

Cooley, J.W. and Tukey, J.W. (1965). An algorithm for the machine calculation of complex Fourier series. *Math. Comp.* **19**, pp. 297-301.

Corduneanu, C. (1971). *Principles of Differential and Integral Equations*. Allyn and Bacon, Rockledge, N.J.

Courant, R. (1943). Variational methods for the solution of problems of equilibrium and vibrations. *Bull. Amer. Math. Soc.* **49**, pp. 1-23.

Cox, M.G. (1972). The numerical evaluation of B-splines. *J. Inst. Math. Appl.* **10**, pp. 134-149.

Cox, M.G. (1981). The least squares solution of overdetermined linear equations having band or augmented band structure. *IMA J. Numer. Anal.* **1**, pp. 3-22.

Cox, M.G. (1982). Practical spline approximation. In *Topics in Numerical Analysis* (ed. P.R. Turner) *Lecture Notes in Mathematics* **965**, pp. 79-112. Springer-Verlag, Berlin.

Crane, H.L., Gibbs, N.E., Poole, W.G. and Stockmeyer, P.K. (1976). Algorithm 508: Matrix bandwidth and profile reduction. *ACM Trans. Math. Software* **2**, pp. 375-377.

Crawford, C.R. (1973). Reduction of a band-symmetric generalized eigenvalue problem. *Comm. ACM* **16**, pp. 41-44.

Crisfield, M.A. (1979). Iterative solution procedures for linear and nonlinear structural analysis. Report TRRL LR-900. HMSO, London.

Crisfield, M.A. (1984). New solution procedures for linear and finite element analysis. In *The Mathematics of Finite Elements and its Applications V* (ed. J.R. Whiteman). Academic Press, London.

Currie, R.G. (1981). Solar cycle signal in earth rotation : non-stationary behaviour. *Science* **211**, pp. 386-389.

Curry, H.B. and Schoenberg, I.J. (1966). On Pólya frequency functions IV: the fundamental spline functions and their limits. *J. Anal. Math.* **17**, pp. 71-107.

Curtis, A.R. (1980). The Facsimile numerical integrator for stiff initial value equations. In *Computational Techniques for Ordinary Differential Equations* (eds. I. Gladwell and D.K. Sayers) pp. 47-82. Academic Press, London.

Curtis, A.R. (1983). Jacobian matrix properties and their impact on choice of software for stiff ODE systems. *IMA J. Numer. Anal.* **3**, pp. 397-415.

Curtis, A.R., Powell, M.J.D. and Reid, J.K. (1974). The estimation of sparse Jacobian matrices. *J. Inst. Math. Appl.* **13**, pp. 117-119.

Dahlquist, G. (1963). A special stability problem for linear multistep methods. *BIT* **3**, pp. 27-43.

Dahlquist, G. (1976). Error analysis for a class of methods for stiff nonlinear initial value problems. In *Numerical Analysis. Dundee 1975* (eds. A. Dold and B. Eckmann) *Lecture Notes in Mathematics* **506**, pp. 60-74. Springer-Verlag, Berlin.

Davidon, W.C. (1959). Variable metric methods for minimization. A.E.C. Res. and Develop. Report ANL-5990, Argonne National Laboratory, Argonne, Illinois.

Davies, A.J. (1980). *The Finite Element Method*. Clarendon Press, Oxford.

Davies, N., Triggs, C.M. and Newbold, P. (1977). Significance of the Box-Pierce Portmanteau statistics in finite samples. *Biometrika* 64, pp. 517–522.

Davis, P.J. and Rabinowitz, P. (1984). *Methods of Numerical Integration.* Academic Press, New York. 2nd edition.

de Boor, C. (1971a). CADRE: An algorithm for numerical software. In *Mathematical Software* (ed. J.R. Rice) pp. 417–449. Academic Press, New York.

de Boor, C. (1971b). On writing an automatic integration routine. *ibid*. pp. 201–209.

de Boor, C. (1972). On calculating with B-splines. *J. Approx. Th.* 6, pp. 50–62.

de Boor, C. (1977). Asymptotic error expansions for composite quadrature rules in several variables. Private communication.

de Boor, C. (1980). FFT as nested multiplication with a twist. *SIAM J. Sci. Stat. Comp.* 1, pp. 173–178.

de Doncker, E. (1978). An adaptive extrapolation algorithm for automatic integration. *SIGNUM Newsletter* 13, No.2, pp. 12–18.

de Doncker, E. (1979). New Euler Maclaurin expansions and their application to quadrature over the s-dimensional simplex. *Math. Comp.* 33, pp. 1008–1018.

de Doncker, E. and Piessens, R. (1976). A bibliography on automatic integration. *J. Comp. Appl. Math.* 2, pp. 273–280.

de Hoog, F.R. and Weiss, R. (1973). Asymptotic expansions for product integration. *Math. Comp.* 27, pp. 295–306.

Delbourgo, R. and Gregory, J.A. (1983). C^2 rational quadratic spline interpolation to monotonic data. *IMA J. Numer. Anal.* 3, pp. 141–152.

Delves, L.M. (1977). A fast method for the solution of integral equations. *J. Inst. Math. Appl.* 20, pp. 173–182.

Delves, L.M. and Abd-Elal, L.F. (1977). Algorithm 97. The Fast Galerkin algorithm for the solution of linear Fredholm equations. *Comput. J.* 20, pp. 374–376.

Delves, L.M. and Mohamed, J.L. (1985). *Computational Methods for Integral Equations.* Cambridge University Press.

Delves, L.M., Abd-Elal, L.F. and Hendry, J.A. (1979). A Fast Galerkin algorithm for singular integral equations. *J. Inst. Math. Appl.* 23, pp. 139–166.

Delves, L.M., Abd-Elal, L.F. and Hendry, J.A. (1981). A set of modules for the solution of integral equations. *Comput. J.* 24, pp. 184–190.

Dennis, J.E. (1977). Nonlinear least squares and equations. In *The State of the Art in Numerical Analysis* (ed. D.A.H. Jacobs) pp. 269–312. Academic Press, London.

Dennis, J.E. (1982). Algorithms for nonlinear fitting. In *Nonlinear Optimization 1981* (ed. M.J.D. Powell) pp. 67–78. Academic Press, London.

Dennis, J.E., Gay, D.M. and Welsch, R.E. (1981). An adaptive nonlinear least-squares algorithm. *ACM Trans. Math. Software* 7, pp. 348–368.

Deuflhard, P. (1980). Recent advances in multiple shooting techniques. In *Computational Techniques for Ordinary Differential Equations* (eds. I. Gladwell and D.K. Sayers) pp. 217–272. Academic Press, London.

Deuflhard, P. (1983). Order and stepsize control in extrapolation methods. *Numer. Math.* 41, pp. 399–422.

Deuflhard, P. and Bauer, H.J. (1982). A note on Romberg quadrature. Report No. 169, Sonderforschungs-bereich 123, University of Heidelberg.

Dew, P.M. and Walsh, J. (1981). A set of library routines for solving parabolic equations in one space variable. *ACM Trans. Math. Software* 7, pp. 295–314.

Dierckx, P. (1981a). An improved algorithm for curve fitting with spline functions. Report TW54, Dept. of Computer Science, Katholieke Universiteit Leuven.

Dierckx, P. (1981b). An algorithm for surface fitting with spline functions. *IMA J. Numer. Anal.* 1, pp. 267-283.

Dierckx, P. (1982). A fast algorithm for smoothing data on a rectangular grid while using spline functions. *SIAM J. Numer. Anal.* 19, pp. 1286-1304.

Dixon, L.C.W. (1972a). Quasi-Newton algorithms generate identical points. *Math. Progr.* 2, pp. 383-387.

Dixon, L.C.W. (1972b). Quasi-Newton algorithms generate identical points II: The proof of four new theorems. *Math. Progr.* 3, pp. 345-358.

Dixon, V.A. (1974). Numerical quadrature: a survey of available algorithms. In *Software for Numerical Mathematics* (ed. D.J. Evans) pp. 105-137. Academic Press, New York.

Dongarra, J.J. (1982). Algorithm 589. SICEDR: A Fortran subroutine for improving the accuracy of computed matrix eigenvalues. *ACM Trans. Math. Software* 8, pp. 371-375.

Dongarra, J.J. (1983). Improving the accuracy of computed singular values. *SIAM J. Sci. Stat. Comp.* 4, pp. 712-719.

Dongarra, J.J. and Moler, C.B. (1983). EISPACK - a package for solving matrix eigenvalue problems. Technical Memorandum ANL/MCS-TM-12, Argonne National Laboratory, Argonne, Illinois.

Dongarra, J.J., Gustavson, F.G. and Karp, A. (1984). Implementing linear algebra algorithms for dense matrices on a vector pipeline machine. *SIAM Rev.* 26, pp. 91-112.

Dongarra, J.J., Kaufman, L. and Hammarling, S.J. (1985). Squeezing the most out of eigenvalue solvers on high performance computers. Technical Memorandum ANL/MCS-TM-46, Argonne National Laboratory, Argonne, Illinois.

Dongarra, J.J., Moler, C.B. and Wilkinson, J.H. (1983). Improving the accuracy of computed eigenvalues and eigenvectors. *SIAM J. Numer. Anal.* 20, pp. 23-45.

Dongarra, J.J., Bunch, J.R., Moler, C.B. and Stewart, G.W. (1979). *LINPACK User's Guide*. SIAM Publications, Philadelphia.

Dormand, J.R. and Prince, P.J. (1978). New Runge-Kutta algorithms for numerical simulation in dynamical astronomy. *Celestial Mechanics* 18, pp. 223-232.

Dormand, J.R. and Prince P.J. (1980). A family of embedded Runge-Kutta formulae. *J. Comp. Appl. Math.* 6, pp. 19-26.

Draper, N.R. and Smith, H. (1966). *Applied Regression Analysis*. Wiley, New York.

Dubois, P.F., Greenbaum, A. and Rodrigue, G.H. (1979). Approximating the inverse of a matrix for use in iterative algorithms on vector processors. *Computing* 22, pp. 257-268.

Du Croz, J.J. (1983). Adapting the NAG Library to vector-processing machines. *NAG Newsletter 2/83*, pp. 23-33.

Duff, I.S. (1977). MA28 - a set of FORTRAN subroutines for sparse unsymmetric linear equations. Harwell Report AERE-R8730. HMSO, London.

Duff, I.S. (ed.) (1982a). *Sparse Matrices and their Uses*. Academic Press, London.

Duff, I.S. (1982b). A survey of sparse matrix software. Harwell Report AERE-R10512. HMSO, London.

Duff I.S. and Reid, J.K. (1976). A comparison of some methods for sparse least squares problems. *J. Inst. Math. Appl.* 17, pp. 267-280.

Duff, I.S. and Reid, J.K. (1978). An implementation of Tarjan's algorithm for block triangularisation of a matrix. *ACM Trans. Math. Software* 4, pp. 137-147.

Duff, I.S. and Reid, J.K. (1983). The multifrontal solution of indefinite sparse symmetric linear equations. *ACM Trans. Math. Software* 9, pp. 302–325.

Duff, I.S. and Stewart, G.W. (eds.) (1979). *Sparse Matrix Proceedings 1978*. SIAM, Philadelphia.

Durbin, J. (1959). Efficient estimation of parameters in moving-average models. *Biometrika* **46**, pp. 306–316.

Durbin, J. (1960). The fitting of time series models. *Rev. Int. Inst. Stat.* **28**, pp. 233–244.

Earnshaw, J.L. and Yuille, I.M. (1971). A method of fitting parametric equations for curves and surfaces to sets of points defining them approximately. *Comput. Aided Des.* 3, pp. 19–22.

Eisenstat, S.C., Elman, H.C. and Schultz, M.H. (1983). Variational iterative methods for nonsymmetric systems of linear equations. *SIAM J. Numer. Anal.* **20**, pp. 345–357.

Eisenstat, S.C., Gursky, M.C., Schultz, M.H. and Sherman, A.H. (1982). Yale Sparse Matrix Package 1: Symmetric codes. *Int. J. Numer. Methods Engng.* **18**, pp. 1145–1151.

El-gendi, S.E. (1969). Chebyshev solution of differential, integral and integro-differential equations. *Comput. J.* **12**, pp. 282–287.

Elman, H.C. (1981). Preconditioned conjugate-gradient methods for nonsymmetric systems of linear equations. Technical Report #203. Dept. of Computer Science, Yale University.

Engels, H. (1980). *Numerical Quadrature and Cubature*. Academic Press, New York.

Ericsson, T. (1983a). Implementation and application of the spectral transformation Lanczos algorithm. In *Matrix Pencils*. (eds. B. Kågström and A. Ruhe) *Lecture Notes in Mathematics* 973, pp. 177–188. Springer-Verlag, Berlin.

Ericsson, T. (1983b). Algorithms for large sparse symmetric generalized eigenvalue problems. Report UMINF-108.83, Dept. of Information Processing, University of Umeå.

Ericsson, T. (1983c). Notes on indefinite symmetric matrix pencils. Report UMINF-107.83. Dept. of Information Processing, University of Umeå.

Ericsson, T. and Ruhe, A. (1982). STLM - a software package for the spectral transformation Lanczos algorithm. Report UMINF - 101.8, Dept. of Information Processing, University of Umeå.

Essah, W., Delves, L.M. and Belward, J. (1986). A cross validation Galerkin algorithm for first kind integral equations. Preprint, Dept. of Statistics and Computational Mathematics, University of Liverpool.

Evans, G.A., Forbes, R.C. and Hyslop, J. (1983). Polynomial transformations for singular integrals. *Int. J. Computer Math.* **14**, pp. 157–170.

Fehlberg, E. (1970). Klassische Runge-Kutta Formeln vierter und niedrigerer Ordnung mit Schrittweiten-Kontrolle und ihre Anwendung auf Wärmelertungsprobleme. *Computing* 6, pp. 61–71.

Ferris, D.H., Hammarling, S.J., Martin, D.W. and Warham, A.J.P. (1983). Numerical solution of equations describing electromagnetic propagation in dielectric waveguides. NPL Report 16/83, National Physical Laboratory, Teddington, Middlesex.

Fix, G.J., Gulati, S. and Wakoff, G.I. (1973). On the use of singular functions with finite element approximations. *J. Comp. Phys.* **13**, pp. 209–228.

Fletcher, R. (1970). A new approach to variable metric algorithms. *Comput. J.* **13**, pp. 317–322.

Fletcher, R. (1971). A modified Marquardt subroutine for nonlinear least squares. Harwell Report AERE-E6799, A.E.R.E. Harwell, Oxfordshire.

Fletcher, R. (1975). An ideal penalty function for constrained optimisation. *J. Inst. Math. Appl.* **15**, pp. 319–342.

Fletcher, R. (1980). *Practical Methods of Optimization, Vol 1. Unconstrained Optimization*. Wiley, Chichester.

Fletcher, R. (1981). *Practical Methods of Optimization, Vol 2. Constrained Optimization*. Wiley, Chichester.

Fletcher, R. and Freeman, T.L. (1977). A modified Newton method for minimization. *J. Optim. Theory Appl.* 23, pp. 357-372.

Fletcher, R. and Powell, M.J.D. (1963). A rapidly convergent descent method for minimization. *Comput. J.* 6, pp. 163-168.

Fletcher, R. and Reeves, C.M. (1964). Function minimization by conjugate gradients. *Comput. J.* 7, pp. 149-154.

Foley, T.A. (1983). BSPLASH: A three-stage surface interpolant to scattered data. Preprint, Dept. of Mathematics, University of Nevada, Las Vegas.

Forsythe, G.E. and Moler, C.B. (1967). *Computer Solution of Linear Algebraic Systems*. Prentice-Hall, Englewood Cliffs, N.J.

Fox, L. (1977). Finite-difference methods for elliptic boundary-value problems. In *The State of the Art in Numerical Analysis* (ed. D.A.H. Jacobs) pp. 799-881. Academic Press, London.

Fox, L. (1980). Numerical methods for boundary-value problems. In *Computational Techniques for Ordinary Differential Equations* (eds. I. Gladwell and D.K. Sayers) pp. 175-216. Academic Press, London.

Franke, R. (1979). A critical comparison of some methods for interpolation of scattered data. Report TR-NPS-53-79-003, Naval Postgraduate School, Monterey, California.

Franke, R. (1982a). Scattered data interpolation: tests of some methods. *Math. Comp.* 38, pp. 181-200.

Franke, R. (1982b). Smooth interpolation of scattered data by local thin plate splines. *Comp. Math. Appl.* 8, pp. 273-281.

Franke, R. (1984). Private communication.

Franke, R. and Nielson, G. (1980). Smooth interpolation of large sets of scattered data. *Int. J. Numer. Methods Engng.* 15, pp. 1691-1704.

Friedman, A. (1964). *Partial Differential Equations of Parabolic Type*. Prentice-Hall, Englewood Cliffs, N.J.

Fritsch, F.N. (1982). PCHIP final specifications. Report UCID-30194, Lawrence Livermore Laboratory, Livermore, California.

Fritsch, F.N. and Butland, J. (1982). A method for constructing local monotone piecewise cubic interpolants. Report UCRL-87559, Lawrence Livermore Laboratory, Livermore, California.

Fritsch, F.N. and Carlson, R.E. (1980). Monotone piecewise cubic interpolation. *SIAM J. Numer. Anal.* 17, pp. 238-246.

Fritsch, F.N. and Carlson, R.E. (1983). BIMOND: Monotone bivariate interpolation code. Report UCID-30197, Lawrence Livermore Laboratory, Livermore, California.

Gantmacher, F.R. (1959). *Theory of Matrices, Vol. II*. Chelsea, New York.

Garbow, B.S. (1978). Algorithm 535: The QZ algorithm to solve the generalized eigenvalue problem for complex matrices. *ACM Trans. Math. Software* 4, pp. 404-410.

Garbow, B.S., Boyle, J.M., Dongarra, J.J. and Moler, C.B. (1977). *Matrix Eigensystem Routines - EISPACK Guide Extension. Lecture Notes in Computer Science* 51. Springer-Verlag, Berlin.

Gardner, G., Harvey, A.C. and Phillips, G.D.A. (1980). Algorithm AS 154. An algorithm for exact maximum likelihood estimation of autoregressive-moving average models by means of Kalman filtering. *Appl. Stat.* 29, pp. 311-322.

Gay, D.M. (1981). Computing optimal locally constrained steps. *SIAM J. Sci. Stat. Comp.* 2, pp. 186-197.

Gear, C.W. (1971a). Algorithm 407: DIFSUB for the solution of ordinary differential equations. *Comm. ACM* 14, pp. 185-190.

Gear, C.W. (1971b). *Numerical Initial Value Problems in Ordinary Differential Equations*. Prentice-Hall, Englewood Cliffs, N.J.

Gear, C.W. (1981). Numerical solution of ordinary differential equations: Is there anything left to do ? *SIAM Rev.* 23, pp. 10-24.

Gear, C.W. and Saad, Y. (1983). Iterative solution of linear equations in ODE codes. *SIAM J. Sci. Stat. Comp.* 4, pp. 583-601.

Gentleman, W.M. (1974a). Basic procedures for large, sparse or weighted linear least squares problems. *Appl. Stat.* 23, pp. 448-454.

Gentleman, W.M. (1974b). Regression problems and the QU decomposition. *Bull. IMA* 10, pp. 195-197.

Gentleman, W.M. (1976). Row elimination for solving sparse linear systems and least squares problems. In *Numerical Analysis. Dundee 1975* (ed. G.A. Watson) *Lecture Notes in Mathematics* 506, pp. 122-133. Springer-Verlag, Berlin.

Gentleman, W.M. and Sande, G. (1966). Fast Fourier transforms - for fun and profit. *AFIPS. Conf. Proc.* 28, pp. 563-579.

George, J.A. and Heath, M.T. (1980). Solution of sparse linear least squares problems using Givens rotations. *Lin. Alg. App.* 34, pp. 69-83.

George, J.A. and Liu, J.W. (1981). *Computer Solution of Large Sparse Positive Definite Systems*. Prentice-Hall, Englewood Cliffs, N.J.

George, J.A. and Ng, E. (1983). On row and column orderings for sparse least squares problems. *SIAM J. Numer. Anal.* 20, pp. 326-344.

George, A., Liu, J. and Ng, E. (1980). User guide for SPARSPAK. Report CS-78-30 (revised), Dept. of Computer Science, University of Waterloo, Canada.

Gibbs, N.E., Poole, W.G. and Stockmeyer, P.K. (1976a). An algorithm for reducing the bandwidth and profile of a sparse matrix. *SIAM J. Numer. Anal.* 13, pp. 235-251.

Gibbs, N.E., Poole, W.G. and Stockmeyer, P.K. (1976b). A comparison of several bandwidth reduction algorithms. *ACM Trans. Math. Software* 2, pp. 322-330.

Gill, P.E. and Murray, W. (1972). Quasi-Newton methods for unconstrained optimization. *J. Inst. Math. Appl.* 9, pp. 91-108.

Gill, P.E. and Murray, W. (1974). Newton-type methods for unconstrained and linearly constrained optimization. *Math. Progr.* 7, pp. 311-350.

Gill, P.E. and Murray, W. (1977). Modification of matrix factorizations after a rank-one change. In *The State of the Art in Numerical Analysis.* (ed. D.A.H. Jacobs) pp. 55-83. Academic Press, London.

Gill, P.E. and Murray, W. (1978). Algorithms for the solution of the nonlinear least-squares problem. *SIAM J. Numer. Anal.* 15, pp. 977-992.

Gill, P.E., Murray, W. and Wright, M.H. (1981). *Practical Optimization*. Academic Press, London.

Gill, P.E., Murray, W., Saunders, M.A. and Wright, M.H. (1983). User's guide for SOL/NPSOL: a Fortran package for nonlinear programming. Report SOL83-12, Dept. of Operations Research, Stanford University, California.

Gill, P.E., Murray, W., Saunders, M.A. and Wright, M.H. (1984). Recent developments in constrained optimisation. *NAG Newsletter* 1/84, pp. 8-20.

Gladwell, I. (1979). The development of the boundary-value codes in the ordinary differential equations chapter of the NAG Library. In *Codes for Boundary-Value Problems in Ordinary Differential Equations* (eds. B. Childs, M. Scott, J.W. Daniel, E. Denman and P. Nelson) *Lecture Notes in Computer Science* 76, pp. 122-143. Springer-Verlag, Berlin.

Gladwell, I. (1980). A survey of subroutines for solving boundary value problems in ordinary differential equations. In *Computational Techniques for Ordinary Differential Equations* (eds. I. Gladwell and D.K. Sayers) pp. 273–303. Academic Press, London.

Godolphin, E.J. (1977). A procedure for estimating seasonal moving average models based on large sample estimation of the correlogram. *J. Roy. Stat. Soc. B* 39, pp. 238–247.

Golub, G.H. (1965). Numerical methods for solving linear least squares problems. *Numer. Math.* 7, pp. 206–216.

Golub, G.H. and Kahan, W. (1965). Calculating the singular values and pseudo-inverse of a matrix. *SIAM J. Numer. Anal.* 2, pp. 202–224.

Golub, G.H. and Pereyra, V. (1973). The differentiation of pseudoinverses and nonlinear least squares problems whose variables separate. *SIAM J. Numer. Anal.* 10, pp. 413–432.

Golub, G.H. and Reinsch, C. (1970). Singular value decomposition and least squares solutions. *Numer. Math.* 14, pp. 403–420.

Golub, G.H. and Van Loan, C.F. (1980). An analysis of the total least squares problem. *SIAM J. Numer. Anal.* 17, pp. 883–893.

Golub, G.H. and Van Loan, C.F. (1983). *Matrix Computations*. North Oxford Academic, Oxford.

Golub, G.H. and Wilkinson, J.H. (1976). Ill-conditioned eigensystems and the computation of the Jordan canonical form. *SIAM Rev.* 18, pp. 578–619.

Golub, G.H., Klema, V.C. and Stewart, G.W. (1976). Rank degeneracy and least squares problems. Technical Report STAN-CS-76-559, Stanford University, California.

Golub, G.H., Luk, F.T. and Overton, M. (1981). A block Lanczos method for computing the singular values and corresponding singular vectors of a matrix. *ACM Trans. Math. Software* 7, pp. 149–169.

Golub, G.H., Nash, S. and Van Loan, C.F. (1979). A Hessenberg-Schur method for the problem AX+XB=C. *IEEE Trans. Auto. Cont.* AC-24, pp. 909–913.

Golub, G.H., Underwood, R. and Wilkinson, J.H. (1972). The Lanczos algorithm for the symmetric $Ax=\lambda Bx$ problem. Technical Report STAN-CS-72-270, Stanford University, California.

Goutis, C.E. and Ibrahim, M.K. (1983). New algorithm for spectrum estimation and an associated VLSI design. *IEE Proc.* 130, Pt F, pp. 251–255.

Gragg, W.B. (1965). On extrapolation methods for ordinary initial-value problems. *SIAM J. Numer. Anal.* 2, pp. 384–403.

Graham, I.G. (1980). *The Numerical Solution of Fredholm Integral Equations of the Second Kind*. Ph.D. Thesis, University of New South Wales.

Greenough, C. and Robinson, K. (1982). NAG Finite Element Library, Level 0 and Level 1. SERC Rutherford Appleton Laboratory, Chilton, Didcot, Oxfordshire.

Gregory, J.A. and Delbourgo, R. (1982). Piecewise rational quadratic interpolation to monotonic data. *IMA J. Numer. Anal.* 2, pp. 123–130.

Gustaffson, I. (1979). On modified incomplete Cholesky factorisation methods for the solution of problems with mixed boundary conditions and problems with discontinuous material properties. *Int. J. Numer. Methods Engng.* 14, pp. 1127–1140.

Haber, S. (1970). Numerical evaluation of multiple integrals. *SIAM Rev.* 12, pp. 481–526.

Hageman, L.A. and Young, D.M. (1981). *Applied Iterative Methods*. Academic Press, New York.

Hall, G. and Watt, J.M. (eds.) (1976). *Modern Numerical Methods for Ordinary Differential Equations*. Clarendon Press, Oxford.

Hall, G. and Williams, J. (eds.) (1982). Proceedings of a one-day colloquium on the numerical solution of ordinary differential equations. Numerical Analysis Report No. 84, Dept. of Mathematics, University of Manchester.

Hammarling, S.J. (1982). Numerical solution of the stable, non-negative definite Lyapunov equation. *IMA J. Numer. Anal.* 2, pp. 303-323.

Hammarling, S.J. (1983). How to live without covariance matrices: Numerical stability in multivariate statistical analysis. *NAG Newsletter* 1/83, pp. 6-31.

Hammarling, S.J. (1985). The singular value decomposition in multivariate statistics. *ACM Signum Newsletter* 20, No.3, pp. 2-25.

Hammarling, S.J., Long, E.M.R. and Martin, D.W. (1983). A generalized linear least squares algorithm for correlated observations, with special reference to degenerate data. NPL Report DITC 33/83, National Physical Laboratory, Teddington, Middlesex.

Hammarling, S.J., Kenward, P.D., Symm, G.J. and Wilkinson, J.H. (1981). Subroutine UKGE3R/D. NPL Linear Algebra Subroutine Library, National Physical Laboratory, Teddington, Middlesex.

Hannan, E.J. (1970). *Multiple Time Series*. Wiley, New York.

Hannan, E.J. and Kavalieris, L. (1984). Multivariate linear time series models. *Adv. Appl. Prob.* 16, pp. 492-561.

Hannan, E.J. and Rissanen, J. (1982). Recursive estimation of mixed autoregressive moving average order. *Biometrika* 69, pp. 81-94.

Harder, R.L. and Desmairais, R.N. (1972). Interpolation using surface splines. *J. Aircr.* 9, pp. 189-191.

Harrison, P.J. and Stevens, C.F. (1976). Bayesian forecasting. *J. Roy. Stat. Soc. B* 38, pp. 205-247.

Harvey, A.C. (1984). A unified view of statistical forecasting procedures. *J. Forecasting* 3, pp. 245-283.

Hävie, T. (1979). Generalized Neville type extrapolation schemes. *BIT* 19, pp. 204-213.

Hayes, J.G. (1974). Numerical methods for curve and surface fitting. *Bull. Inst. Math. Appl.* 10, pp. 144-152.

Hayes, J.G. (1978). Data-fitting algorithms available, in preparation, and in prospect, for the NAG library. In *Numerical Software - Needs and Availability* (ed. D.A.H. Jacobs) pp. 183-202. Academic Press, London.

Heath, M.T. (1982). Some extensions of an algorithm for sparse linear least squares problems. *SIAM J. Sci. Stat. Comp.* 3, pp. 223-237.

Heath, M.T., Laub, A.J., Paige, C.C. and Ward, R.C. (1985). Computing the singular value decomposition of a product of two matrices. Report ORNL-6118, Oak Ridge National Laboratory, Tennessee.

Heller, D. (1978). A survey of parallel algorithms in numerical linear algebra. *SIAM Rev.* 20, pp. 740-777.

Hemker, P.W. and Schippers, H. (1981). Multiple grid methods for the solution of Fredholm integral equations of the second kind. *Math. Comp.* 36, pp. 215-231.

Henrici, P. (1979). Fast Fourier methods in computational complex analysis. *SIAM Rev.* 21, pp. 481-527.

Hestenes, M.R. (1969). Multiplier and gradient methods. *J. Optim. Theory Appl.* 4, pp. 303-320.

Hiebert, K.L. (1981). An evaluation of mathematical software that solves nonlinear least squares problems. *ACM Trans. Math. Software* 7, pp. 1-16.

Hillmer, S.C. and Tiao, G.C. (1979). Likelihood function of stationary multiple autoregressive moving-average models. *J. Amer. Statist. Assoc.* 74, pp. 652-660.

Hindmarsh, A.C. (1980). LSODE and LSODI, two new initial value ordinary differential equation solvers. *ACM-SIGNUM Newsletter* 15, No. 4, pp. 10-11.

Hindmarsh, A.C. (1981). ODE solvers for use with the method of lines. In *Advances in Computer Methods for Partial Differential Equations* (eds. R. Vichnevetsky and R.S. Stepleman) pp. 312-316. IMACS, New Brunswick, N.J.

Hockney, R.W. and Jesshope, C.R. (1983). *Parallel Computers.* Adam Hilger, Bristol.

Holt, J.N. and Fletcher, R. (1979). An algorithm for constrained non-linear least squares. *J. Inst. Math. Appl.* 23, pp. 449-463.

Horn, M.K. (1983). Fourth- and fifth-order, scaled Runge-Kutta algorithms for treating dense output. *SIAM J. Numer. Anal.* 20, pp. 558-568.

Huebner, K.H. (1975). *Finite Element Method for Engineers.* Wiley, New York.

Hull, T.E., Enright, W.H. and Jackson, K.R. (1976). User's guide for DVERK - a subroutine for solving non-stiff ODE's. Report 100, Dept. of Computer Science, University of Toronto.

Hull, T.E., Enright, W.H., Fellen, B.M. and Sedgwick, A.E. (1972). Comparing numerical methods for ordinary differential equations. *SIAM J. Numer. Anal.* 9, pp. 603-637.

Hunter, D.B. (1972). The numerical evaluation of Cauchy Principal Values of integrals by Romberg integration. *Numer. Math.* 21, pp. 185-192.

Hutton, A.G. (1979). Progress in re-development of a finite element wall model. *Appl. Math. Model.* 3, pp. 322-326.

Ioakimidis, N.I. and Theocaris, P.S. (1980). A comparison between the direct and the classical numerical methods for the solution of Cauchy type singular integral equations. *SIAM J. Numer. Anal.* 17, pp. 115-118.

Irons, B.M. (1970). A frontal solution program for finite element analysis. *Int. J. Numer. Methods Engng.* 2, pp. 5-32.

Jackson, C.P. and Cliffe, K.A. (1981). Mixed interpolation in primitive variable finite element formulations for incompressible flow. *Int. J. Numer. Methods Engng.* 17, pp. 1659-1688.

Jacobs, D.A.H. (ed.) (1977). *The State of the Art in Numerical Analysis.* Academic Press, London.

Jacobs D.A.H. (1982). The exploitation of sparsity by iterative methods. In *Sparse Matrices and their Uses* (ed. I.S. Duff) pp. 191-222. Academic Press, London.

Jakeman, A.J. and Young, P.C. (1980). Refined instrumental variable methods of recursive time series analysis, Part II. *Internat. J. Control* 29, pp. 621-644.

Jenkins, G.M. and Watts, D.G. (1968). *Spectral Analysis and its Applications.* Holden-Day, San Francisco.

Jennings, A. (1966). A compact storage scheme for the solution of symmetric linear simultaneous equations. *Comput. J.* 9, pp. 281-285.

Jennings, A. and Malik, G.M. (1978). The solution of sparse linear equations by conjugate gradient methods. *Int. J. Numer. Methods Engng.* 12, pp. 141-158.

Joyce, D.C. (1971). Survey of extrapolation processes in numerical analysis. *SIAM Rev.* 13, pp. 435-490.

Kågström, B. (1977). Numerical computation of matrix functions. Report UMINF-58.77, Dept. of Information Processing, University of Umeå.

Kågström, B. (1983). On computing the Kronecker canonical form of regular $(A-\lambda B)$-pencils. In *Matrix Pencils* (eds. B. Kågström and A. Ruhe) *Lecture Notes in Mathematics* 973, pp. 30-57. Springer-Verlag, Berlin.

Kågström, B. and Ruhe, A. (1980). An algorithm for numerical computation of the Jordan normal form of a complex matrix. *ACM Trans. Math. Software* 6, pp. 398-419.

Kahaner, D.K. (1971). Comparison of numerical quadrature formulas II. In *Mathematical Software* (ed. J.R. Rice) pp. 229-259. Academic Press, New York.

Kaufman, L. (1977). Some thoughts on the QZ algorithm for solving the generalized eigenvalue problem. *ACM Trans. Math. Software* 5, pp. 442-450.

Kaufman, L. (1984). Banded eigenvalue solvers on vector machines. *ACM Trans. Math. Software* 10, pp. 73-85.

Kaufman, L. and Pereyra, V. (1978). A method for separable nonlinear least squares problems with separable nonlinear equality constraints. *SIAM J. Numer. Anal.* 15, pp. 12-20.

Kay, S.M. and Marple, S.L. (1981). Spectrum analysis - a modern perspective. *IEEE Proc* 69, pp. 1380-1419.

Keller, H.B. (1968). *Numerical Methods for Two-Point Boundary Value Problems*. Blaisdell, Waltham, Massachusetts.

Keller, H.B. (1974). Accurate difference methods for nonlinear two point boundary value problems. *SIAM J. Numer. Anal.* 11, pp. 305-320.

Kincaid, D. and Grimes, R. (1977). Numerical studies of several adaptive iterative algorithms. Report No. 126, Center for Numerical Analysis, University of Texas, Austin, Texas.

Kincaid, D.R., Respess, J.R., Young, D.M. and Grimes, R.G. (1982). ITPACK2C : A FORTRAN package for solving large sparse linear systems by adaptive accelerated iterative methods. *ACM Trans. Math. Software* 8, pp. 302-322.

Kleiner, B., Martin, R.D. and Thomson, D.J. (1979). Robust estimation of power spectra. *J. Roy. Stat. Soc. B* 41, pp. 313-351.

Kogbetliantz, E.G. (1955). Solution of linear equations by diagonalization of coefficients matrix. *Quart. J. Appl. Math.* 13, pp. 123-132.

Kourouklis, S. and Paige, C.C. (1981). A constrained least squares approach to the general Gauss-Markov model. *J. Amer. Stat. Assoc.* 76, pp. 620-625.

Krenk, S. (1978). Quadrature formulae of closed type for the solution of singular integral equations. *J. Inst. Math. Appl.* 22, pp. 99-107.

Kronrod, A.S. (1965). Nodes and weights of quadrature formulas. Consultants Bureau, New York. Translated from Russian.

Kublanovskya, V.N. (1970). On an approach to the solution of the generalized latent value problem for λ-matrices. *SIAM J. Numer. Anal.* 7, pp. 532-537.

Kumar, S. (1983). *Extrapolation to the Limit*. M.Sc. Thesis, University of the South Pacific, Suva, Fiji.

Lambert, J. (1973). *Computational Methods in Ordinary Differential Equations*. Wiley, London.

Lancaster, P. (1966). *Lambda-matrices and Vibrating Systems*. Pergamon Press, Oxford.

Lancaster, P. (1977). A review of numerical methods for eigenvalue problems non-linear in the parameter. In *Numerik und Anwendungen von Eigenwertaufgaben and Verzweigungsproblemen* (eds. E. Bohl, L. Collatz and K.P. Hadeler) pp. 43-67. Birkhauser Verlag, Basel and Stuttgart.

Laub, A.J. (1979). A Schur method for solving algebraic Riccati equations. *IEEE Trans. Auto. Cont.* AC-24, pp. 913-921.

Laurie, D.P. (1982). Algorithm AS 175. Cramer-Wold factorisation. *Appl. Stat.* 31, pp. 86-93.

Lemaréchal, C. (1982). Nondifferentiable optimization. In *Nonlinear Optimisation 1981* (ed. M.J.D. Powell) pp. 85-90. Academic Press, London.

Lentini, M. and Pereyra, V. (1977). An adaptive finite difference solver for nonlinear two-point boundary problems with mild boundary layers. *SIAM J. Numer. Anal.* 14, pp. 91-111.

Levenberg, K. (1944). A method for the solution of certain non-linear problems in least squares. *Quart. Appl. Math.* 2, pp. 164–168.
Levinson, N. (1947). The Wiener (root mean square) error criterion in filter design and prediction. *J. Math. Phys.* 25, pp. 261–278.
Levy, S. and Fullagar, P.K. (1981). Reconstruction of a sparse spike train from a portion of its spectrum and application to high-resolution deconvolution. *Geophys.* 46, pp. 1235–1243.
Lewis, J. (1977). Algorithms for sparse matrix eigenvalue problems. Technical Report STAN-CS-77-595, Dept. of Computer Science, Stanford University, California.
Lewis, J.G. (1982a). Algorithm 582: The Gibbs-Poole-Stockmeyer and Gibbs-King algorithms for reordering sparse matrices. *ACM Trans. Math. Software* 8, pp. 190–194.
Lewis J.G. (1982b). Implementation of the Gibbs-Poole-Stockmeyer and Gibbs-King algorithms. *ACM Trans. Math. Software* 8, pp. 180–189.
Ljung, G.M. and Box, G.E.P. (1979). The likelihood function of stationary autoregressive-moving average models. *Biometrika* 66, pp. 265–270.
Lyness, J.N. (1971). The calculation of Fourier coefficients by the Mobius inversion of the Poisson summation formula – Part III. Functions having algebraic singularities. *Math. Comp.* 25, pp. 483–494.
Lyness, J.N. (1976a). An error functional expansion for n-dimensional quadrature with an integrand function singular at a point. *Math. Comp.* 30, pp. 1–23.
Lyness, J.N. (1976b). Applications of extrapolation techniques to multidimensional quadrature of some integrand functions with a singularity. *J. Comp. Phys.* 20, pp. 346–364.
Lyness, J.N. (1976c). Quid, Quo, Quadrature? In *The State of the Art in Numerical Analysis.* (ed. D.A.H. Jacobs) pp. 535–562. Academic Press, London.
Lyness, J.N. (1978). Quadrature over a simplex. *SIAM J. Numer. Anal.* 15, pp. 122–133 and pp. 870–887.
Lyness, J.N. (1983). QUG2 – Integration over a triangle. Technical Memorandum ANL/MCS-TM-13, Argonne National Laboratory, Argonne, Illinois.
Lyness, J.N. (1985). The Euler Maclaurin expansion for the Cauchy Principal Value integral. *Numer. Math.* 46, pp. 611–622.
Lyness, J.N. and Delves, L.M. (1967). On numerical contour integration round a closed contour. *Math. Comp.* 21, pp. 561–577.
Lyness, J.N. and Genz, A. (1980). On Simplex trapezoidal rule families. *SIAM J. Numer. Anal.* 17, pp. 126–147.
Lyness, J.N. and Kaganove, J. (1976). Comments on the nature of automatic quadrature routines. *ACM Trans. Math. Software* 2, pp. 65–81.
Lyness, J.N. and McHugh, B.J.J. (1970). On the remainder term in the N-dimensional Euler Maclaurin expansion. *Numer. Math.* 15, pp. 333–344.
Lyness, J.N. and Monegato, G. (1980). Quadrature error functional expansion for the Simplex when the integrand function has singularities at vertices. *Math. Comp.* 34, pp. 213–225.
Lyness, J.N. and Ninham, E.W. (1967). Numerical quadrature and asymptotic expansions. *Math. Comp.* 21, pp. 162–178.
Machura, M. and Sweet, R.A. (1980). A survey of software for partial differential equations. *ACM Trans. Math. Software* 6, pp. 461–488.
Madsen, K. and Powell, M.J.D. (1975). A FORTRAN subroutine that calculates the minimax solution of linear equations subject to bounds on the variables. Technical Report #R-7954, A.E.R.E. Harwell, Oxfordshire.
Manteuffel, T.A. (1979). Shifted incomplete Cholesky factorisation. In *Sparse Matrix Proceedings 1978* (eds. I.S. Duff and G.W. Stewart) pp. 41–61. SIAM Press, Philadelphia.

Markham, G. (1983). The application of multi-grid techniques to preconditionings for conjugate-gradient type methods. Report TPRD/L/AP 127/MA83. CERL, CEGB, Leatherhead.

Markowitz, H.M. (1957). The elimination form of the inverse and its application to linear programming. *Manag. Sci.* 3, pp. 225-269.

Marple, S.L. and Nuttall, A.H. (1983). Experimental comparison of three multichannel linear prediction spectral estimators. *IEE Proc.* 130, Pt F, pp. 218-229.

Marquardt, D.W. (1963). An algorithm for least squares estimation of non-linear parameters. *SIAM J. Appl. Math.* 11, pp. 431-441.

McKeown, J.J. (1975). Specialised versus general purpose algorithms for minimising functions that are sums of squared terms. *Math. Progr.* 9, pp. 57-68.

Meijerink, J.A. and van der Vorst, H.A. (1977). An iterative solution method for linear systems of which the coefficient matrix is an M-matrix. *Math. Comp.* 31, pp. 148-162.

Meijerink, J.A. and van der Vorst, H.A. (1979). Incomplete decompositions as preconditioning for the conjugate gradient algorithm. In *Conjugate Gradient Methods and Similar Techniques* (ed. I.S. Duff). Harwell Report AERE-R9636. HMSO, London.

Mélard, G. (1984). A fast algorithm for the exact likelihood of autoregressive moving average models. *Appl. Stat.* 33, pp. 104-114.

Milanazzo, F., Barrodale, I. and Zala, C.A. (1984). An upper bound for the condition number of convolution matrices. Report #DCS-43-IR, Dept. of Computer Science, University of Victoria, B.C.

Miller, K. (1981). Moving finite elements. *SIAM J. Numer. Anal.* 18, pp. 1019-1057.

Miller, R.K. (1971). *Nonlinear Volterra Integral Equations.* Benjamin, California.

Mitchell, A.R. and Griffiths, D.F. (1980). *The Finite Difference Method in Partial Differential Equations.* Wiley, Chichester.

Mitchell, A.R. and Wait, R. (1977). *The Finite Element Method in Partial Differential Equations.* Wiley, London.

Modi, J.J. and Parkinson, D. (1982). Study of Jacobi methods for eigenvalues and singular value decomposition on DAP. *Comput. Phys. Comm.* 26, pp. 317-320.

Moler, C.B. and Stewart, G.W. (1973). An algorithm for generalized matrix eigenvalue problems. *SIAM J. Numer. Anal.* 10, pp. 241-256.

Monegato, G. and Lyness, J.N. (1979). On the numerical evaluation of a particular singular two-dimensional integral. *Math. Comp.* 33, pp. 993-1002.

Moré, J.J. (1978). The Levenberg-Marquardt algorithm : Implementation and theory. In *Numerical Analysis. Proceedings, Biennial Conference. Dundee 1977* (ed. G.A. Watson) *Lecture Notes in Mathematics* 630, pp. 105-116. Springer-Verlag, Berlin.

Munksgaard, N. (1979). Solving sparse sets of linear equations by preconditioned conjugate gradients. Harwell Report CSS-67, A.E.R.E. Harwell, Oxfordshire.

Navot, I. (1961). An extension of the Euler-Maclaurin summation formula to functions with a branch singularity. *J. Math. and Phys.* 40, pp. 271-276.

Nazareth, L. (1980). Some recent approaches to solving large residual nonlinear least squares problems. *SIAM Rev.* 22, pp. 1-11.

Newbold, P. (1974). The exact likelihood function for a mixed autoregressive-moving average process. *Biometrika* 61, pp. 423-426.

Nicholls, D.F. and Hall, A.D. (1979). The exact likelihood function of multivariate autoregressive-moving average models. *Biometrika* 66, pp. 259-264.

Nikolai, P.J. (1979). Algorithm 538: Eigenvectors and eigenvalues of real generalized symmetric matrices by simultaneous iteration. *ACM Trans. Math. Software* 5, pp. 118-125.

Nuttall, A.H. (1983). Analysis of a generalised framework for spectral estimation, Part 2. *IEE Proc.* 130, Pt F, pp. 242-245.

O'Brien, D.M. and Holt, J.N. (1981). The extension of generalised cross-validation to a multi-parameter class of estimators. *J. Aust. Math. Soc. B* 22, pp. 501-514.

Oden, J.T. (1972). *Finite Elements of Non-Linear Continua*. McGraw-Hill, New York.

Oren, S.S. (1974). On the selection of parameters in self scaling variable metric algorithms. *Math. Progr.* 7, pp. 351-367.

Osborn, D.R. (1977). Exact and approximate maximum likelihood estimators for vector moving average processes. *J. Roy. Stat. Soc. B* 39, pp. 114-118.

Osborne, M.R. (1976). Nonlinear least squares - the Levenberg algorithm revisited. *J. Aust. Math. Soc. B* 19, pp. 343-357.

Overton, M.L. (1982). Algorithms for nonlinear ℓ_1 and ℓ_∞ fitting. In *Nonlinear Optimisation 1981* (ed. M.J.D. Powell) pp. 91-102. Academic Press, London.

Paige, C.C. (1972). Computational variants of the Lanczos method for the eigenproblem. *J. Inst. Math. Appl.* 10, pp. 373-381.

Paige, C.C. (1974). Bidiagonalization of matrices and solution of linear equations. *SIAM J. Numer. Anal.* 11, pp. 197-209.

Paige, C.C. (1976). Error analysis of the Lanczos algorithm for tridiagonalizing a symmetric matrix. *J. Inst. Math. Appl.* 18, pp. 341-349.

Paige, C.C. (1978). Numerically stable computations for general univariate linear models. *Commun. Statist. - Simula. Computa.* B7(5), pp. 437-453.

Paige, C.C. (1979a). Fast numerically stable computations for generalised linear least squares problems. *SIAM J. Numer. Anal.* 16, pp. 165-171.

Paige, C.C. (1979b). Computer solution and perturbation analysis of generalized linear least squares problems. *Math. Comp.* 33, pp. 171-183.

Paige, C.C. (1985a). The general linear model and the generalized singular value decomposition. *Lin. Alg. App.* 70, pp. 269-284.

Paige, C.C. (1985b). Computing the generalized singular value decomposition. *SIAM J. Sci. Stat. Comp.* To appear.

Paige, C.C. and Saunders, M.A. (1975). Solution of sparse indefinite systems of linear equations. *SIAM J. Numer. Anal.* 12, pp. 617-629.

Paige, C.C. and Saunders, M.A. (1981). Towards a generalized singular value decomposition. *SIAM J. Numer. Anal.* 18, pp. 398-405.

Paige, C.C. and Saunders, M.A. (1982a). LSQR: An algorithm for sparse linear equations and sparse least squares. *ACM Trans. Math. Software* 8, pp. 43-71.

Paige, C.C. and Saunders, M.A. (1982b). Algorithm 583. LSQR: Sparse linear equations and least squares problems. *ACM Trans. Math. Software* 8, pp. 195-209.

Paige, C.C. and Van Loan, C.F. (1981). A Schur decomposition for Hamiltonian matrices. *Lin. Alg. App.* 41, pp. 11-32.

Parlett, B.N. (1980). *The Symmetric Eigenvalue Problem*. Prentice-Hall, Englewood Cliffs, N.J.

Parlett, B.N. and Reid, J.K. (1981). Tracking the progress of the Lanczos algorithm for large symmetric eigenproblems. *IMA J. Numer. Anal.* 1, pp. 135-155.

Parlett, B.N. and Scott, D.S. (1979). The Lanczos algorithm with selective orthogonalization. *Math. Comp.* 33, pp. 217-238.

Parzen, E.A. (1967). On empirical multiple time series analysis. In *Proceedings of the Fifth Berkeley Symposium on Mathematical Statistics and Probability* (eds. L. LeCam and J. Neyman) pp. 305-340. Univ. of California Press, Berkeley, California.

Parzen, E.A. (1968). Statistical spectral analysis. Technical Report 11, Dept. of Statistics, Stanford University, California.

Parzen E.A. (1969). Multiple time series modelling. In *Multivariate Analysis* (ed. P.R. Krishnaiah) pp. 389-409. Academic Press, New York.

Patterson, T.N.L. (1968). The optimum addition of points to quadrature formulae. *Math. Comp.* 22, pp. 847-856. Note also Errata, *Math. Comp.* 23, p. 892.

Pereyra, V. (1979). PASVA3: An adaptive finite difference FORTRAN program for first order nonlinear, ordinary boundary problems. In *Codes for Boundary-Value Problems in Ordinary Differential Equations* (eds. B. Childs, M. Scott, J.W. Daniel, E. Denman and P. Nelson) *Lecture Notes in Computer Science* 76, pp. 67-88. Springer-Verlag, Berlin.

Peters, G. and Wilkinson, J.H. (1970). $Ax=\lambda Bx$ and the generalized eigenproblem. *SIAM J. Numer. Anal.* 7, pp. 479-492.

Peters, G. and Wilkinson, J.H. (1979). Inverse iteration, ill-conditioned equations and Newton's method. *SIAM Rev.* 21, pp. 339-360.

Petzold, L.R. (1980). Automatic selection of methods for solving stiff and nonstiff systems of ordinary differential equations. Technical Report SAND80-8230, Sandia National Laboratories, Albuquerque, New Mexico.

Piccolo, D. and Tunnicliffe Wilson, G. (1984). A unified approach to ARMA model identification and preliminary estimation. *J. Time Ser. Anal.* 5, pp. 183-204.

Piessens, R., de Doncker-Kapenga, E., Überhuber, C.W. and Kahaner, D.K. (1983). *QUADPACK - A Subroutine Package for Automatic Integration.* Springer-Verlag, Berlin.

Polak, E. (1971). *Computational Methods in Optimization: A Unified Approach*. Academic Press, New York.

Pouzet, P. (1962). *Étude, en vue de leur traitement numérique, d'équations intégrales et intégro-differentielles du type de Volterra.* Doctoral Thesis, University of Strasbourg.

Powell, M.J.D. (1969). A method for nonlinear constraints in minimisation problems. In *Optimization* (ed. R. Fletcher) pp. 283-298. Academic Press, London.

Powell, M.J.D. (1970). A hybrid method for nonlinear equations. In *Numerical Methods for Nonlinear Algebraic Equations* (ed. P. Rabinowitz) pp. 87-114. Gordon and Breach, London.

Powell, M.J.D. (1982). VMCWD: A Fortran subroutine for constrained optimization. Report DAMTP 1982/NA4, University of Cambridge.

Priestley, M.B. (1980). State-dependent models : A general approach to non-linear time series analysis. *J. Time Ser. Anal.* 1, pp. 47-71.

Priestley, M.B. (1981). *Spectral Analysis and Time Series.* Academic Press, London.

Proskurowski, W. (1983). A package for the Helmholtz equation in nonrectangular planar regions. *ACM Trans. Math. Software* 9, pp. 117-124.

Pryce, J.D. and Hargrave, B.A. (1977). The scaled Prüfer method for one-parameter and multi-parameter eigenvalue problems in o.d.e's. *IMA Numer. Anal. Newsletter* 1, No.3, pp. 13-15.

Raith, K., Schnepf, E. and Schonauer, W. (1982). A new automatic mesh selection strategy for the solution of boundary-value problems. *Notes on Numer. Fluid Mech.* 5. Vieweg. pp. 261-270.

Reinsch, C.H. (1967). Smoothing by spline functions. *Numer. Math.* 10, pp. 177–183.

Rice, J.R. (1978). ELLPACK 77: Users' guide. Report CSD-TR-289, Dept. of Computer Science, Purdue University, Lafayette, Indiana.

Rice, J.R. (1983). *Numerical Methods, Software and Analysis.* McGraw-Hill, New York.

Riddell, I. (1981). *The Numerical Solution of Integral Equations.* Ph.D. Thesis, Dept. of Statistics and Computational Mathematics, University of Liverpool.

Riddell, I. and Delves, L.M. (1980). The comparison of routines for solving integral equations. *Comput. J.* 23, pp. 274–285.

Roberts, F.D.K. (1976). An algorithm for minimal degree Chebyshev approximation on a discrete set. *Int. J. Numer. Methods Engng.* 10, pp. 619–635.

Roberts, F.D.K. and Barrodale, I. (1978). Solution of the constrained L_∞ linear approximation problem. Report #DM-132-IR, Dept. of Mathematics, University of Victoria, B.C.

Roberts, F.D.K. and Barrodale, I. (1980). An algorithm for discrete Chebyshev linear approximation with linear constraints. *Int. J. Numer. Methods Engng.* 15, pp. 797–807.

Robertson, H.H. (1976). Numerical integration of systems of stiff ODE's with special structure. *J. Inst. Math. Appl.* 18, pp. 249–263.

Rockafellar, R.T. (1974). Augmented Lagrange multiplier functions and duality in nonconvex programming. *SIAM J. Control Optim.* 12, pp. 268–285.

Romberg, W. (1955). Vereinfachte numerische integration. *Kgl. Norske Vid. Selsk. Forh.* 38, pp. 30–36.

Ruhe, A. (1973). Algorithms for the non-linear eigenvalue problem. *SIAM J. Numer. Anal.* 4, pp. 674–689.

Ruhe, A. and Wedin, P.A. (1980). Algorithms for separable non linear least squares problems. *SIAM Rev.* 22, pp. 318–337.

Rutishauser, H. (1970). Simultaneous iteration method for symmetric matrices. *Numer. Math.* 16, pp. 205–223.

Sabin, M.A. (1980). Contouring - a review of methods for scattered data. In *Mathematical Methods in Computer Graphics and Design* (ed. K.W. Brodlie) pp. 63–86. Academic Press, London.

Sage, A.P. and Melsa, J.L. (1971). *Estimation Theory, with Applications to Communications and Control.* McGraw-Hill, New York.

Scargle, J.D. (1977). Absolute value optimization to estimate phase properties of stochastic time series. *IEEE Trans. Inf. Theory* IT-23, pp. 140–143.

Schippers, H. (1980). The automatic solution of Fredholm equations of the second kind. Report NW 99/80, Mathematisch Centrum, Amsterdam.

Schumaker, L.L. (1976). Fitting surfaces to scattered data. In *Approximation Theory II* (eds. G.G. Lorentz, C.K. Chui and L.L. Schumaker) pp. 203–268. Academic Press, New York.

Scott, D.S. (1980). Solving sparse quadratic λ-matrix problems. Technical Report ORNL/CSD-48, Union Carbide Corporation, Nuclear Division, Oak Ridge, Tennessee.

Scott, D.S. (1981). Solving sparse symmetric generalized eigenvalue problems without factorization. *SIAM J. Numer. Anal.* 18, pp. 102–110.

Scott, D.S. (1982). The LASO Package. Technical Report, University of Texas, Austin, Texas.

Scott, D.S. and Ward, R.C. (1982). Solving symmetric-definite, quadratic λ-matrix problems without factorization. Technical Report, Union Carbide Corporation, Nuclear Division, Oak Ridge, Tennessee.

Shampine, L.F. (1983). Efficient extrapolation methods for ODEs. *IMA J. Numer. Anal.* 3, pp. 383-395.

Shampine, L.F. (1984a). Interpolation for Runge-Kutta methods. Technical Report SAND83-2560, Sandia National Laboratories, Albuquerque, New Mexico.

Shampine, L.F. (1984b). Some practical Runge-Kutta formulae. Technical Report SAND84-0812, Sandia National Laboratories, Albuquerque, New Mexico.

Shampine, L.F. and Gear, C.W. (1979). A user's view of solving stiff ordinary differential equations. *SIAM Rev.* 27, pp. 1-17.

Shampine, L.F. and Watts, H.A. (1980). DEPAC - design of a user oriented package of ODE solvers. Technical Report SAND79-2374, Sandia National Laboratories, Albuquerque, New Mexico.

Shepard, D. (1968). A two dimensional interpolation function for irregularly spaced data. In *Proc. 23rd Nat. Conf. ACM.* pp. 517-523. Brandon/Systems Press Inc., Princeton, N.J.

Sincovec, R.F. and Madsen, N.K. (1975). Software for nonlinear partial differential equations. *ACM Trans. Math. Software* 1, pp. 232-260.

Sloan, I.H. (1979). The numerical solution of Fredholm equations of the second kind by polynomial interpolation. Technical Note BN905, University of Maryland.

Smith, B.T., Boyle, J.M., Dongarra, J.J., Garbow, B.S., Ikebe, Y., Klema, V.C. and Moler, C.B. (1976). *Matrix Eigensystem Routines - EISPACK Guide. Lecture Notes in Computer Science* 6. Springer-Verlag, Berlin. 2nd edition.

Smith, I.M. (1982). *Programming the Finite Element Method.* Wiley, London.

Steiger, W.L. (1980). Linear programming via L_1 curve-fitting beats simplex. *Abstr. AMS* 1, #80T-C26, pp. 385-386.

Stewart, G.W. (1972). On the sensitivity of the eigenvalue problem $Ax=\lambda Bx$. *SIAM J. Numer. Anal.* 9, pp. 669-686.

Stewart, G.W. (1973). Error and perturbation bounds associated with certain eigenvalue problems. *SIAM Rev.* 15, pp. 727-764.

Stewart, G.W. (1976a). Algorithm 506. HQR3 and EXCHNG: Fortran subroutines for calculating and ordering the eigenvalues of a real upper Hessenberg matrix. *ACM Trans. Math. Software* 2, pp. 275-280.

Stewart, G.W. (1976b). Simultaneous iteration for computing invariant subspaces of non-Hermitian matrices. *Numer. Math.* 25, pp. 123-136.

Stewart, G.W. (1976c). A bibliographical tour of the large sparse generalized eigenvalue problem. In *Sparse Matrix Computations* (eds. J. R. Bunch and D.J. Rose) pp. 113-130. Academic Press, London.

Stewart, G.W. (1979). Assessing the effects of variable error in linear regression. Technical Report 818, University of Maryland.

Stewart, G.W. (1983). A method for computing the generalized singular value decomposition. In *Matrix Pencils* (eds. B. Kågström and A. Ruhe) *Lecture Notes in Mathematics* 973, pp. 207-220. Springer-Verlag, Berlin.

Stewart, G.W. (1984). Rank degeneracy. *SIAM J. Sci. Stat. Comp.* 5, pp. 403-413.

Stewart, W.J. and Jennings, A. (1981). A simultaneous iteration algorithm for real matrices. *ACM Trans. Math. Software* 7, pp. 184-198.

Strang, G. and Fix, G.J. (1973). *An Analysis of the Finite Element Method.* Prentice-Hall, Englewood Cliffs, N.J.

Stüben, K. (1982). MG01: A multigrid program. IMA Report No. 82.02.02. GMD, St. Augustin, Bonn, W. Germany.

Stüben, K. and Trottenberg, U. (1982). Multigrid methods: Fundamental algorithms, model problem analysis and applications. In *Multigrid Methods* (eds. W. Hackbusch and U. Trottenberg) *Lecture Notes in Mathematics* 960, pp. 1-176. Springer-Verlag, Berlin.

Swartztrauber, P.N. (1977). The methods of cyclic reduction, Fourier analysis and the FACR algorithm for the discrete solution of Poisson's equation on a rectangle. *SIAM Rev.* 19, pp. 490–501.

Symm, G.T. (1974). Problems in two-dimensional potential theory (Ch. 20). Potential problems in three dimensions (Ch. 24). In *Numerical Solution of Integral Equations* (eds. L.M. Delves and J. Walsh) pp. 267–274, 312–320. Clarendon Press, Oxford.

Szyld, D.B. (1983). *A Two-level Iterative Method for Large Sparse Generalized Eigenvalue Calculations*. Ph.D. Thesis, Institute for Economic Analysis, New York University.

Taylor, H.L., Banks, S.C. and McCoy, J.F. (1979). Deconvolution with the L_1 norm. *Geophys.* 44, pp. 39–52.

te Riele, H.J.J. (1982). Collocation methods for weakly-singular second kind Volterra integral equations with non-smooth solutions. *IMA J. Numer. Anal.* 2, pp. 437–450.

Thomas, K.S. (1975). On the approximate solution of operator equations. *Numer. Math.* 23, pp. 231–239.

Thomas, K.S. (1976). On the approximate solution of operator equations. Part II. Preprint, University of Oxford.

Thomasset, F. (1981). *Implementation of Finite Element Methods for Navier Stokes Equations*. Springer-Verlag, New York.

Tsalyuk, Z.B. (1979). Volterra integral equations. *J. Sov. Math.* 12, pp. 715–758.

Tunnicliffe Wilson, G. (1983). The estimation of time series models. Technical Report 2528, Mathematics Research Centre, University of Wisconsin-Madison, Madison, Wisconsin.

Turner, M.J., Clough, R.W., Martin, H.C. and Topp, L.J. (1956). Stiffness and deflection analysis of complex structures. *J. Aero. Sci.* 23, pp. 805–823.

van der Houwen, P.J. (1980). Convergence and stability analysis of Runge-Kutta type methods for Volterra integral equations of the second kind. *BIT* 20, pp. 375–377.

van der Vorst, H.A. (1981). Iterative solution of certain sparse linear systems with a non-symmetric matrix arising from a P.D.E. problem. *J. Comp. Phys.* 44, pp. 1–19.

Van Dooren, P. (1979). The computation of Knonecker's canonical form of a singular pencil. *Lin. Alg. App.* 27, pp. 103–140.

Van Dooren, P. (1981a). A generalized eigenvalue approach for solving Riccati equations. *SIAM J. Sci. Stat. Comp.* 2, pp. 121–135.

Van Dooren, P. (1981b). The generalized eigenstructure problem in linear system theory. *IEEE Trans. Auto. Cont.* AC-26, pp. 111–129.

Van Dooren, P. (1982). Algorithm 590. DSUBSP and EXCHQZ: FORTRAN subroutines for computing deflating subspaces with specified spectrum. *ACM Trans. Math. Software* 8, pp. 376–382.

Van Dooren, P. (1983). Reducing subspaces – definitions, properties and algorithms. In *Matrix Pencils* (eds. B. Kågström and A. Ruhe) *Lecture Notes in Mathematics* 973, pp. 58–73. Springer-Verlag, Berlin.

Van Huffel, S., Vandewalle, J. and Staar, J. (1984). The total linear least squares problem: Formulation, algorithm and applications. Presented at ISCAS '84. Katholieke Universiteit Leuven, ESAT Laboratory, 3030 Heverlee, Belgium.

Van Loan, C.F. (1976). Generalizing the singular value decomposition. *SIAM J. Numer. Anal.* 13, pp. 76–83.

Van Loan, C.F. (1982a). A symplectic method for approximating all the eigenvalues of a Hamiltonian matrix. Numerical Analysis Report No. 71, Dept. of Mathematics, University of Manchester.

Van Loan, C.F. (1982b). Using the Hessenberg decomposition in control theory. In *Algorithms and Theory in Filtering and Control*. (ed. D.C. Sorenson and R.J. Wets) *Mathematical Programming Study* **18**, pp. 102-111. North Holland, Amsterdam.

Van Loan, C.F. (1983). A generalized SVD analysis of some weighted methods for equality constrained least squares. In *Matrix Pencils* (eds. B. Kågström and A. Ruhe) *Lecture Notes in Mathematics* 973, pp. 245-262. Springer-Verlag, Berlin.

Van Loan, C.F. (1984). Computing the CS and generalized singular value decompositions. Technical Report CS-604, Cornell University.

Varah, J.M. (1979). On the separation of two matrices. *SIAM J. Numer. Anal.* **16**, pp. 216-222.

Vu, T.V. (1983). Numerical methods for smooth solutions of ordinary differential equations. Report UIUCDCS-R-83-1130, Dept. of Computer Science, University of Illinois.

Wahba, G. (1977). The approximate solution of linear operator equations when the data are noisy. *SIAM J. Numer. Anal.* **14**, pp. 651-667.

Wahba, G. (1980). Numerical and statistical methods for mildly, moderately and severely ill posed problems with noisy data. Technical Report No. 595, University of Wisconsin-Madison, Madison, Wisconsin.

Wahba, G. and Craven, P. (1979). Smoothing noisy data with spline functions; estimating the correct degree of smoothing by the method of generalised cross validation. *Numer. Math.* 31, pp. 377-403.

Wait, R. and Martindale, I. (1984). Finite elements on the DAP. In *The Mathematics of Finite Elements and Its Applications V* (ed. J.R. Whiteman) pp. 113-122. Academic Press, London.

Walker, A.M. (1961). Large-sample estimation of parameters for moving average models. *Biometrika* **48**, pp. 343-357.

Walsh, J. (1977). Boundary-value problems in ordinary differential equations. In *The State of the Art in Numerical Analysis* (ed. D.A.H. Jacobs) pp. 501-533. Academic Press, London.

Ward, R.C. (1975). The combination shift QZ algorithm. *SIAM J. Numer. Anal.* **14**, pp. 600-614.

Ward, R.C. (1981). Balancing the generalized eigenvalue problem. *SIAM J. Sci. Stat. Comp.* 2, pp. 141-152.

Watkins, D.S. (1982). Understanding the QR algorithm. *SIAM Rev.* **24**, pp. 427-440.

Werner, B. (1981). Complementary variational principles and non-conforming Trefftz elements. *Numer. Behand. Diff.* 3, pp. 180-192.

Wesseling, P. (1982). A robust and efficient multigrid method. In *Multigrid Methods* (eds. W. Hackbusch and U. Trottenberg) *Lecture Notes in Mathematics* **960**, pp. 614-630. Springer-Verlag, Berlin.

Whittle, P. (1983). *Prediction and Regulation by Linear Least-Squares Methods*. Blackwell, Oxford. 2nd edition (revised).

Wiener, N. (1949). *Extrapolation, Interpolation and Smoothing of Stationary Time Series*. Wiley, New York.

Wilkinson, J.H. (1965). *The Algebraic Eigenvalue Problem*. Clarendon Press, Oxford.

Wilkinson, J.H. (1977). Some recent advances in numerical linear algebra. In *The State of the Art in Numerical Analysis* (ed. D.A.H. Jacobs) pp. 109-135. Academic Press, London.

Wilkinson, J.H. (1978a). Singular-value decomposition - Basic aspects. In *Numerical Software - Needs and Availability* (ed. D.A.H. Jacobs) pp. 109 - 135. Academic Press, London.

Wilkinson, J.H. (1978b). Linear differential equations and Kronecker's canonical form. In *Recent Advances in Numerical Analysis* (eds. C. de Boor and G.H. Golub) pp. 231-265. Academic Press, New York.

Wilkinson, J.H. (1979). Kronecker's canonical form and the QZ algorithm. *Lin. Alg. App.* 28, pp. 285–303.

Wittrick, W.H. and Williams, F.W. (1971). A general algorithm for computing natural frequencies of elastic structures. *Quart. J. Mech. and App. Math.* 24, pp. 264–284.

Wolkenfelt, P.H.M. (1981). *The Numerical Analysis of Reducible Quadrature Methods for Volterra Integral and Integro-differential Equations.* Thesis, University of Amsterdam. Mathematisch Centrum, Amsterdam.

Wolkenfelt, P.H.M., van der Houwen, P.J. and Baker, C.T.H. (1981). Analysis of numerical methods for second kind Volterra equations by imbedding techniques. *J. Integr. Equa.* 3, pp. 61–82.

Wright, S.J. (1984). Ph.D. Thesis, University of Queensland, Australia.

Wright, S.J. and Holt, J.N. (1985a). Algorithms for nonlinear least squares with linear inequality constraints. *SIAM J. Sci. Stat. Comp.* 6, pp. 1033–1045.

Wright, S.J. and Holt, J.N. (1985b). An inexact Levenberg-Marquardt method for large sparse nonlinear least squares. *J. Aust. Math. Soc. B* 26, pp. 387–403.

Young, D.M. and Kincaid, D.R. (1980). The ITPACK package for large sparse linear systems. Report CNA-160. University of Texas, Austin, Texas.

Young, P.C. and Jakeman, A.J. (1979). Refined instrumental variable methods of recursive time series analysis, Part I. *Internat. J. Control* 29, pp. 1–30.

Zienkiewicz, O.C. (1977). *The Finite Element Method.* McGraw-Hill, London. 3rd edition.

Zlamal, M. (1968). On the finite element method. *Numer. Math.* 12, pp. 394–409.

Zlatev, Z. and Thompsen, P.G. (1982). Sparse matrices – Efficient decomposition and applications. In *Sparse Matrices and their Uses* (ed. I.S. Duff) pp. 367–375. Academic Press, London.

INDEX